Vergil's *Georgics* and the Traditions of Ancient Epic

Vergil's *Georgics* and the Traditions of Ancient Epic

The Art of Allusion in Literary History

Joseph Farrell

New York Oxford
Oxford University Press
1991

Oxford University Press

Oxford New York Toronto
Dehli Bombay Calcutta Madras Karachi
Petaling Jaya Singapore Hong Kong Tokyo
Nairobi Dar es Salaam Cape Town
Melbourne Auckland

and associated companies in
Berlin Ibadan

Published by Oxford University Press, Inc.,
200 Madison Avenue, New York, New York 10016

Oxford is a registered trademark of Oxford University Press

Library of Congress Cataloging-in-Publication Data
Farrell, Joseph, 1955–
Vergil's Georgics and the traditions of ancient epic : the art of allusion in literary history / Joseph Farrell.
p. cm. Includes bibliographical references.
ISBN 0-19-506706-1
1. Virgil. Georgica. 2. Virgil—Knowledge—Literature. 3. Epic poetry, Classical—History and criticism. 4. Allusions. I. Title.
PA6804.G4F37 1991 871′.01—dc20 90-36166

1 3 5 7 9 8 6 4 2

Printed in the United States of America
on acid-free paper

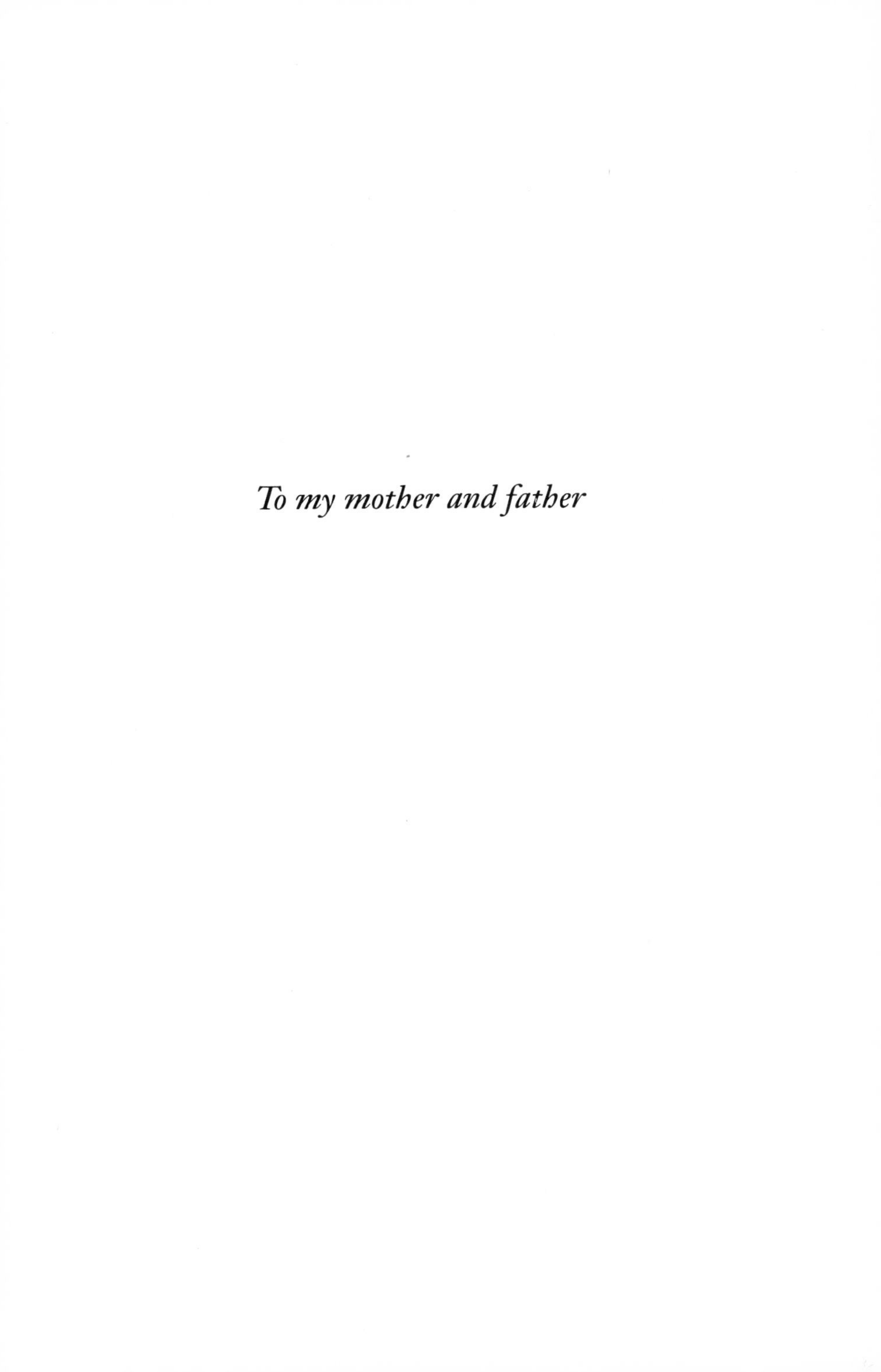

To my mother and father

The text of Cervantes and that of Menard are verbally identical, but the second is almost infinitely richer. (More ambiguous, his detractors will say; but ambiguity is a richness.) It is a revelation to compare the *Don Quixote* of Menard with that of Cervantes. The latter, for instance, wrote (*Don Quixote*, Part One, Chapter Nine):

> ...truth, whose mother is history, who is the rival of time, depository of deeds, witness of the past, example and lesson to the present, and warning to the future.

Written in the seventeenth century, written by the "ingenious layman" Cervantes, this enumeration is a mere rhetorical eulogy of history. Menard, on the other hand, writes:

> ...truth, whose mother is history, who is the rival of time, depository of deeds, witness of the past, example and lesson to the present, and warning to the future.

History, *mother* of truth; the idea is astounding. Menard, a contemporary of William James, does not define history as an investigation of reality, but as its origin. Historical truth, for him, is not what took place; it is what we think took place. The final clauses—*example and lesson to the present, and warning to the future*—are shamelessly pragmatic.

Equally vivid is the contrast in styles. The archaic style of Menard—in the last analysis, a foreigner—suffers from a certain affectation. Not so that of his precursor, who handles easily the ordinary Spanish of his time.

Jorge Luis Borges, "Pierre Menard, Author of Don Quixote"

* * *

di meliora piis, erroremque hostibus illum!

Georgics 3.513

PREFACE

OMNIA IAM VULGATA. When I first became interested in the *Georgics*, it was perfectly possible to justify a new volume on the poem by remarking on the paucity of such studies, especially in English. That was barely over a decade ago; and now I find myself adding to what has become in that short time a bibliographical avalanche. Today a new book requires a more compelling justification than in that *prisca aetas.*

Examining the *Georgics* scholarship of recent years, one notices certain pronounced trends. Although we have had a few valuable shorter contributions—articles and monographs—on specific problems, the chief critical dialogue on this formerly neglected poem has taken place in book-length studies offering global interpretations. While it is perhaps natural that such works should have the most impact, it is curious that the dialogue should have been carried on at this level in the conspicuous absence of basic aids to scholarship—a decent commentary, for instance, or a sufficient number of studies devoted to more limited aspects of the poem. Indeed, it is remarkable how much the best students of the *Georgics*—I think particularly of Putnam and Ross, whose interpretations complement and reinforce one another in so many ways—managed to achieve under these circumstances, and it will be clear in what follows how much I owe to each of them. But now we have at last one incisive scholarly commentary, by Richard Thomas, and another, by the late Sir Roger Mynors, shortly on the way. It seems unlikely that such tools as these will fail to stimulate and facilitate further research; and a good deal, which as a practical matter cannot be handled in works of general interpretation, remains to be done. In such a context, a book on the intertextuality of the *Georgics* makes sense. It is clear from the more general studies and especially from Thomas' recent commentary how much allusion, imitation, and reference contribute to the poem's texture and meaning; but a full-scale interpretive study of this subject has been lacking, while useful shorter studies of its particular aspects remain surprisingly few. I hope that this book will help to meet both these needs.

There is a further, much broader justification for a book such as this, a sense in which it is not "just another book on the *Georgics.*" The history of Latin literature as the best scholars conceive of it—again I think especially of Ross—emphasizes the Neoteric rediscovery of Callimachus as negligible. There are of course problems with such an extreme position: in particular, the *Aeneid* can hardly be explained simply as the work of a Callimachean epigone. My hypothesis in this book is that we can read the pattern of allusion in the *Georgics* as a Vergilian essay in literary history, an essay that renders much more intelligible the course that he followed in his career—from Theocritus in the *Eclogues* to Hesiod, Aratus, Lucretius, and Homer in the *Georgics*, and finally to Apollonius and a somewhat different Homer in the *Aeneid.* I see this Vergilian essay not as a record of, but as a meditation by the poet on his own intellectual dev ment. In my analysis, I have tried to be as clear as possible about s difficult issues, and have accepted, *argumenti causa*, the posi- scholars as a starting point. By this process I have run the ng the dichotomy between the "Neoteric" and "Roman" ublican and Augustan poetry. Some such distinction id exist in Vergil's mind, and it is the great merit l scholarship to have tried to define it as pre- ss, Vergil himself probably did not see the dichotomous as modern scholars have epresents a significant confluence of ut one that was not altogether s show. Thus I am aware that a starting point for my own ematic view of a subtler, t seemed to me, how- eve nuing an established literary ignoring it in an attempt t scratch. The result has co I hope that readers will agr

Several friend ne categories are not mutually exclusi by their thoughtful criticism, and have bu encouragement. What faults remain can surely b my own inability fully to appreciate their advice. My i gil was fostered by Joseph P. Foley, and my choice of a schola confirmed by Nathan Dane II. Both have been dead some years w; but while writing the book, I have thought of them often, and I hope that they would have found something of themselves in it. My interest in this particular topic goes back to graduate school, and its initial result was my dissertation, an

Erratum

The top of p. viii should read as follows: "the single most important factor in determining the nature of Augustan poetry. In comparison to this influence, all others are judged as almost" [and then continue with "negligible" as it now appears]. At the bottom of p. viii, eliminate the last two lines [which are repeated at the top of p. ix].

something of themselves in it. My interest in this particular topic goes back to graduate school, and its initial result was my dissertation, an effort to describe the methods and to discern the motives that inform Vergil's frequent references to Lucretius in the *Georgics*. Although the present work is almost entirely new,* it is a natural outgrowth of the earlier one, and it seems right that I should thank those who helped me to get started on what became a rather long journey: Agnes Michels, who directed the dissertation, and Kenneth Reckford, who served as second reader. Each was a tremendous source of wise counsel and has continued to provide welcome encouragement. More recently, Elaine Fantham and James E. G. Zetzel refereed this book for Oxford University Press and suggested numerous material improvements to the draft that I had submitted—doing so, I might add, in record time, despite numerous other scholarly and administrative responsibilities. Thanks as well to Rachel Toor, who accepted and edited the book, and to her staff for their professional and expeditious handling of the project. Parts of the book were read in various drafts to audiences in Bryn Mawr, New York, Philadelphia, Washington D.C., and at Wesleyan University, and I am grateful to all who offered their comments and criticism: I would especially mention Julia Gaisser, Richard Hamilton, and above all Richard Thomas, who generously read a draft of the whole and offered a wealth of good and timely advice. My work on the book itself began shortly after I joined the Department of Classical Studies at the University of Pennsylvania, and the experience of writing it will be forever linked in memory to specific courses, conversations, and running *disputationes*. My special thanks to students and colleagues, past and present, who provided the professional circumstances and intellectual atmosphere that made this book possible. *Primus inter pares* would be Nico Knauer, whose own work on Vergilian intertextuality has been so valuable to me as to so many others, and whose personal encouragement during these past few years has been inestimable. For help, counsel, and support of various kinds, my best thanks to Peter Bing, Rosaria Munson, Jim O'Donnell, Martin Ostwald, Bob Palmer, Mark Possanza, Matthew Santirocco, and Wesley Smith. No one has contributed more to the book than Ralph Rosen, who has read every word of every draft, discussed at length individual points ranging from basic methodology to typographic design, consigned many a substandard argument *emendaturis ignibus*, and

* I have adapted from the dissertation some arguments pertaining to Lucretius, and have so noted in the appropriate places.

substantially improved most of what remains. To him and to Ellen Rosen as well my best thanks for the support of a rare and wonderful friendship. Finally, to my long-suffering family, who must have looked many times to the sky and wondered, *quem das finem, rex magne, laborum?* Ann de Forest bettered the book in many places as only a good writer can, believed in it through good times and bad, endured countless episodes of authorial mania and depression, and cheerfully provided the sort of soothing sustenance that transcends mere gratitude. Our daughter Flannery has enriched the experience of writing the book and hastened its completion in ways that I would never have dreamed possible, and which I now hope that she too will someday understand. Last, thanks to our extended families, both Ann's and mine, for all their support and encouragement; particularly to my parents, who have looked on for over half my life now with who knows what mixture of emotions as I become more and more inextricably involved with the unfathomable pleasures of Vergil. It is to them that I dedicate this, the most tangible result of that long and continuing fascination, *χάλκεα χρυσείων*, in partial return for so much more.

PHILADELPHIA
AUGUST 1990

CONTENTS

PART II
VERGIL'S PROGRAM OF ALLUSION IN THE *GEORGICS*

PART III
THE *GEORGICS* IN LITERARY HISTORY

ABBREVIATIONS

The following standard abbreviations have been used:

L&S	Lewis and Short, *A Latin Dictionary* (Oxford 1879)
LSJ	Liddell–Scott–Jones, *A Greek–English Lexicon*[9] (Oxford 1940)
OLD	Glare et al., *Oxford Latin Dictionary* (Oxford 1968–1982)
RE	Pauly–Wissowa–Kroll, *Real-Encyclopädie der classischen Altertumswissenschaft* (Stuttgart 1893–)
Σ	scholia

Titles and abbreviations of ancient works are modeled on those in LSJ, L&S, and *OLD*, with some modifications. Modern secondary literature, other than the few exceptions noted above, is cited by the author's last name and the date of publication,* followed in the case of specific references by volume number (if necessary), pagination, and, in the case of commentaries and other reference works, line number (e.g. Ross 1986; Putnam 1979.37; Schanz–Hosius 1927–1935.1.314; Conington–Nettleship 1898.336 *ad G.* 3.553). Complete publication information can be found in the bibliography.

* There are a few exceptions in the case of conference proceedings published in a year different from that in which the papers were delivered, later editions, reprints, and facsimiles (e.g. Scaliger 1561, Sellar 1897, Hardie 1985).

PART I

Vergil's Allusive Artistry

CHAPTER 1

Introduction: On Vergilian Intertextuality

The Problem

IN *GEORGICS* 1, when Vergil gives a lengthy account of the signs by which farmers may predict the weather, he borrows abundantly from an earlier poem, the *Phaenomena* of Aratus. When in the second book he reflects upon his own activity in composing the *Georgics*, he calls the poem an "Ascraean song," alluding to the works of Hesiod. Both of the remaining books conclude with elaborate imitations, Book 3 of a passage from Lucretius' *De Rerum Natura*, Book 4 of several episodes drawn from Homer's *Iliad* and *Odyssey*. At other points Vergil imitates, borrows from, and alludes or refers to a wealth of other sources, both at length and in passing. Few passages of any extent fail to disclose a noticeable debt to some particular model. From beginning to end of the four books of *Georgics*, in Vergil's words we catch the echo of another's voice.

While it may be an exaggeration to cite the *Georgics* as the most allusive poem of antiquity, such a claim would not be essentially misleading. All three of Vergil's uncontested works are imitative, allusive, referential in the extreme. We know this in part, of course, because the long and hazardous road by which classical literature has reached us failed to destroy either Vergil or most of his favorite sources and models. While we can only guess or surmise the extent of the debt owed by the lyric poetry of Horace to that of Alcaeus, presumably his chief model in *Odes* 1–3, we can map the *Aeneid* in detail against the archaic and Hellenistic epics that were its chief models. Nor is this all. Even where the tradition has begrudged us, for instance, the complete texts of Ennius and Naevius, we have the means to assess with some confidence their general influence on Vergil. The reason is that even in antiquity imitation was seen as an unusually important element of Vergilian poetics. Ancient scholars devoted a large part of their time and energy to demonstrating this belief by collating texts, collecting parallel passages, and inventing anecdotes to illustrate Vergil's compulsively imitative working methods. Thus to some extent the *Georgics* appears to us outstandingly imitative simply because it is a Vergilian poem. Even in the context of Vergil's *oeuvre*, though, the *Georgics* is exceptional. The *Eclogues*, despite their mysteries, have in at least one respect always been quite clear: they are a deliberate effort to recreate in contemporary Roman terms the *Idylls* of Theocritus. In the same way, the *Aeneid* resurrects and unites both the *Odyssey* and the *Iliad* into a single line of action. The *Georgics* is different. Here there is no main model, no possibility of explaining the entire poem even superficially as an updated revision of Hesiod or of Lucretius: no one source eclipses all the others, as Homer does Apollonius in the *Aeneid* and as Theocritus does everyone in the *Eclogues*. Uniquely among Vergil's works, the *Georgics* ranges broadly over the traditions of ancient poetry, using allusion to draw together various strands of these traditions into a seamlessly unified and convincing whole. If not the most allusive poem that antiquity produced, it must be one of the most creatively and ambitiously imitative that literature has yet seen.

The aim of this book is to address these issues in an attempt to define more clearly the place of the *Georgics* in literary history. In order to do so, we shall draw on a number of critical approaches, old and new, often employed in unusual combinations. This is a point that calls for further elaboration.

Methods

Critical assessment of Vergil's allusiveness has had an extremely long history, beginning shortly after the poet's death and the posthumous publication of the *Aeneid*.[1] But the passage of time has produced no real consensus about how the reader should view this essential quality of Vergilian poetry. Nevertheless, certain useful generalizations can be made. From antiquity until about the middle of the nineteenth century,

1. In its main outlines the history of Vergil criticism is the history of *Aeneid* criticism. While it is not always possible to understand critical issues in the *Eclogues* and the *Georgics* in terms proper to the *Aeneid*, it is an essential part of my thesis that Vergil's allusive technique remains basically the same throughout his career. I hope to demonstrate this point with respect to the *Eclogues* on another occasion; its truth with respect to the *Georgics* will become clear in the course of my argument.

the identification and discussion of Vergilian allusions was conducted within the parameters of classical theory. This theory recognizes two aspects of allusion, *μίμησις* or *imitatio* and *ζήλωσις* or *aemulatio*.[2] Imitation is in a sense the more basic aspect, the process by which one takes the first steps towards Parnassus, selecting a model and attempting to copy its best features. It is by the second step, emulation, that the poet proves himself. He does so by attempting to surpass his model in some way.[3] This might involve exercising his critical faculty to identify the model's imperfections and to purge them from his imitation, or relying on his *inventio* to endow his own version with a quality that his model had overlooked. A good example is provided by Horace in his attempt to rival Catullus' mastery of the Sapphic meter. Catullus' memorable *Carmen* 11 begins:

> Furi et Aureli, comites Catulli,
> sive in extremos penetrabit Indos,
> litus ut longe resonante Eoa
> **tunditur unda**

Horace makes use of this effect in a similar context:

> Septimi, Gadis aditure mecum et
> Cantabrum indoctum iuga ferre nostra et
> barbaras Syrtis, ubi Maura semper
> **aestuat unda**
>
> *Carmen* 2.6.1–4

The allusion is obvious, both in its imitative and its emulative aspects. Horace copies Catullus' formal treatment of waves sounding upon the shore, and we may even say that he follows his model by choosing an impressively onomatopoeic verb to connote the waves' sound. But he also asserts his independence in a contest of *inventio*: where Catullus' waves pound the beach, Horace's seethe against it. By the terms of classical theory, the reader will judge whether Horace has not only challenged, but surpassed his model. (In my own view, the cost of Horace's independence, the loss of Catullus' exquisite assonance ***tunditur unda***, was too high.)

2. On these and similar terms in ancient usage see Reiff 1959.

3. Russell (1979.10), however, quite properly cautions against too rigid a distinction between these terms, which on the one hand are complementary, but on the other are often nearly synonymous.

It is clear that the poets themselves played this game, which was the cornerstone of their training. In ancient criticism, however, the idea that each poem must be measured against its models is taken to extremes. Books were written to castigate Vergil for his thefts, and stories told to illustrate the poet's confidence in his own artistic superiority.[4] The critical habit of umpiring the contest between poet and model gained new life in the Renaissance, when the rigid canons of ancient and medieval Latin culture were challenged by the rediscovery of Greek. Throughout the commentaries and literary criticism of the fifteenth, sixteenth, and seventeenth centuries we see scholars engaged in the twin tasks of identifying Vergil's Greek models and judging the poet's success in surpassing them. At first Vergil held his own. By the end of the eighteenth century, however, the tide had begun to turn. Led by the enthusiastic celebration of Homer's "original genius," nineteenth-century scholars and laymen, especially in Germany, had come to denigrate virtually all of Latin literature, and particularly Vergil, as derivative and unimaginative. This anachronistic judgment, which was also misguided given what we now know about Homeric originality, represents an extreme reaction from the position of earlier critics, who, if they were apt to focus too narrowly on the twin concepts of *imitatio* and *aemulatio*, at least played within the established rules. Instead, the romantic criticism of the nineteenth century refused to recognize that all classical Latin poetry was written by poets whose most basic literary assumptions included these concepts; or, if they did recognize this fact, they nevertheless arbitrarily faulted the poets for sacrificing their genius to such an artificial scheme.

Despite a general critical distaste for the derivative nature of Latin poetry, nineteenth-century *Quellenforschung* ransacked the earlier commentaries for parallel passages in the relentless pursuit of all possible sources or models from which that poetry derived. The result was a quantity of dissertations and *Programmschriften* listing passages in which Roman poets adapted their (primarily Greek) models. At the center of this effort was Vergil. Anyone interested will have no trouble in quickly locating a series of monographs on Vergil's borrowings from Homer,

4. Vergil is portrayed as defending his Homeric *furta* with the observation that *facilius esse Herculi clavem quam Homero versum subripere* (Donatus *Vita Vergilii* 46 [Hardie 1966.18]; cf. Macrobius *Sat.* 5.3.16). Conversely, he is made to apologize for borrowing from Ennius by telling someone, while reading that author, *aurum in stercore quaero* (Cassiodorus *Inst.* 1.1.8; cf. John of Salisbury *Policraticus* 281.4. The story is also alluded to in an interpolated passage of the Donatus *Vita:* see Brummer 1912.31).

Hesiod, Apollonius, Aratus, Lucretius, Ennius, and others. The shortcomings of such studies are well known. Their very real value consists in the fact that they do collect a good deal of the material relevant to the study of Vergilian allusion, at least where the *Eclogues* and *Georgics* are concerned. Indeed, the student of these poems in particular has been well served by the labors of Paul Jahn, who over a number of years carried out a thorough investigation of Vergil's sources in the *Eclogues* and *Georgics*.[5] His results were largely incorporated into his revision of the Ladewig–Schaper–Deuticke commentary.[6] His critical assumptions are straightforward, and his analysis sheds little light on the poetry of the *Georgics*.[7] But Jahn's merit is to have summed up the positivistic aspect of research into the sources of the *Eclogues* and *Georgics* since antiquity, and his collection of material remains basic to all those interested in the subject.

Traditional *Quellenforschung*, then, in spite of its critical limitations, has left us with an ample collection of material on which to base the study of Vergilian allusion. What is needed is a suitable means of dealing with this mass of data. The twentieth century has seen a variety of approaches, each with its own strengths and weaknesses. A judicious combination of the best elements found in the available methods will, in my opinion, provide the best means of addressing the questions that I have posed.

Systematic Allusion

Jahn's limited view of the *Georgics* as a mere compilation tends to support the popular nineteenth-century notion that Vergil was a poet of limited imagination and originality. Other scholars, however, while using a method similar to Jahn's have produced a much more complex and interesting picture of Vergilian allusion. One of the most frequently cited books on Vergil to appear in the past fifty years is G. N. Knauer's *Die Aeneis und Homer*. This book is used primarily for its valuable lists of parallels between the *Aeneid* and the Homeric poems. No one, however, really seems to have understood or come to terms with the book's

5. Jahn 1897–1899 covers the *Eclogues;* Jahn 1903a deals with *G.* 1.1–350, 1903b with all of Book 2, 1905a with *G.* 3.49–170, 1904 with *G.* 4.1–280, and 1905b with *G.* 4.281–558.

6. 9th edition = Jahn 1915.

7. He assumes that Vergil composed each section of the poem with a different model (*Muster*) in mind, and that various other, less important sources (*Quellen*) contributed secondary details throughout these passages.

most important implications.[8] In my view, Knauer's main contribution has been to disclose fully three facets of Vergil's Homeric imitation in the *Aeneid*, namely, its extent, its systematic nature, and its analytical basis. Formerly one supposed that in writing the poem Vergil simply availed himself of the standard conventional elements of epic in general, looking for inspiration and guidance to Homer as the "founder" of the genre. This approach remains familiar to students of later European epic: because Milton wished to compose an epic for his time in the manner of Homer and Vergil, *Paradise Lost* is fully equipped with the standard elements of the genre—scenes of celestial deliberation, invocations of the Muse, and so forth—some of them virtual translations of the poet's ancient models. This critical approach is thus valid as far as it goes; but that is not very far. With the *Aeneid*, thanks to Knauer, it is now necessary to think in terms of a very close relationship between individual lines or passages and one or more Homeric models, not just in the case of a few scenes, but from one end of the poem to the other. In a matter of such scope, different scholars may reasonably disagree on specific points of interpretation; but it may be regarded as proven, I think, that Vergil offered the *Aeneid* to the world as an imitation of both Homeric epics *in toto*; that individual points of contact between the *Aeneid* and one or both of Homer's epics, far from being isolated occurrences or decorative generic references, are integral parts of a tightly woven fabric of intertextuality; and that Vergil created this new fabric only after carefully and painstakingly unravelling the tapestries of both Homeric epics, strand by strand, becoming in the process intimately familiar with the peculiarities of each as well as with their similarities to one another.

One might be tempted to say that the *Aeneid* is, in the magnitude and quality of its allusiveness, unique in literature. Other highly allusive works, such as Joyce's *Ulysses*, approach the *Aeneid* in sophistication and extent of allusion; but in fact the close metrical relationship between the *Aeneid* and the Homeric poems makes possible a range of allusiveness unavailable either to the prose novelist or to the poet (e.g. Dante) who uses a different metrical scheme. At the same time, one realizes that such a tour de force can hardly have been altogether unprecedented, that the mastery of allusive technique so evident in the *Aeneid* argues for prior experience on Vergil's part—particularly as each of Vergil's previous major works had an explicit "model." It is this aspect of Knauer's

8. This point has recently been made by Cairns 1989.ix.

work that I shall adopt as the foundation of the method employed in this study: the concept of an extensive, systematic program of allusion based on an analytical reading of major sources. In the chapters that follow, I shall argue that we find in the *Georgics* an approach similar both in kind and in scope, to the system of Homeric allusion that Knauer has uncovered in the *Aeneid*.

At the same time, I must acknowledge the limitations that I have found in Knauer's approach. In a subsequent study of Homeric influence on the *Georgics*, Knauer applied the same basic method that he had first followed in studying the *Aeneid*.[9] Finding allusion to Homer largely confined to the poem's final episode, the "Aristaeus," he concluded that it was only as Vergil finished the poem that he realized the possibilities contained in this kind of allusion, and that the realization derived in some way from the fact that Homer was involved. There is no question of what this ambitious Homeric allusion means or why Vergil made it the conclusion of the *Georgics*. For Knauer, it is the very contact with Homer that produced in Vergil the ability to create a masterpiece like the *Aeneid*. The evidence, in spite of the imagination that he brings to bear on the analysis of specific parallels, is basically quantitative. Like Jahn, but even more so, Knauer combed the work of his great predecessors, the major Vergil commentators since antiquity, especially those of the Renaissance, to determine where and in what way Vergil followed his master.

The critical judgment that Knauer brings to bear on specific passages is something Jahn lacks; in this, Knauer is more the follower of Richard Heinze. But here too there are limitations. Knauer tends to treat Vergil's imitation of Homer as if it were a self-evident imperative: "If, *without requiring the reasons*, we assume that Vergil really wanted from the very beginning to incorporate both Greek epics into his poem...."[10] It is not quite fair to say that Knauer's reasoning is circular, but it is obvious that his working hypothesis colors his analysis and then becomes an important element of his conclusion.

The contrast that Knauer sees between the *Georgics*, in which he believes Vergil only gradually became aware of the allusive methods that Homer called for, and the *Aeneid*, which exploits these methods to the full, is instructive. The more completely Vergil imbibed the inspiration of the *Iliad* and the *Odyssey*, the more "Homeric" his poetry became;

9. Knauer 1981.890–918.
10. Knauer 1964b.64–65; emphasis mine.

consequently, the more capable he himself became of consummate poetic expression. But while we may agree that the meeting of two great poetic minds can be the decisive factor in the genesis of great literature, we must also admit that Knauer does not explain how the magnitude of Vergil's achievement is related to his astonishing Homeric *agon*, except in a purely technical sense. Indeed, Knauer's critics have complained that his approach reduces Vergilian allusiveness to a matter of technique and nothing more, that his *Aeneid* comes to resemble those centos of later antiquity that were created out of Vergil's own text.

For the most part, these criticisms strike me as misplaced: essentially, they attack Knauer for not having done something that he never intended to do. A more serious criticism, in my view, is that Knauer exaggerates the similarities between Vergil and Homer, often explaining differences between the *Aeneid* and its models by referring to the exigencies of Vergil's allusive program: in order to include this or that element of the *Iliad* or the *Odyssey*, he was forced to represent it in this or that unusual way. The implicit goal of such reasoning is to make even pointed differences seem like similarities. One comes away with the impression that Knauer wants his technical analysis of Vergil's imitative program to determine critical reaction to the *Aeneid* as a poem of ideas, and to outline a reading of the *Aeneid* that differs little from a fairly straightforward reading of the Homeric poems. This of course is not possible. It is permissible to observe that, on one level, Aeneas must kill Turnus because Achilles slays Hector, and that, on another level, these acts of vengeance are demanded by the deaths of Pallas and Patroclus respectively. It does not follow, however, that the actions of Aeneas and Achilles at this point are in every way morally equivalent, or, more important, that both deaths call for similar reactions on the part of the reader. The interpreter who addresses these questions will want all the information he can get concerning literary antecedents of the death of Turnus. But while this information will inform his reaction to the scene, it must not control it. To have it otherwise is simply to close one's eyes to the ironic element made possible by allusion.

This is what many have missed in Knauer's book. Inevitably, therefore, alongside his more positivist approach there has grown up a much suppler critical method. In fact, it is not quite proper to speak of this reaction to the critical techniques of traditional philology as a single method; in the more recent work on allusion in classical literature, there has been a notable and healthy diversity of theory and practice, some of it essentially compatible with work like Knauer's, and some deliberately distinguished from it. Despite this diversity, however, the most important work of the past quarter century—and the tendency has accelerated in the past few years—aligns itself with an approach first promulgated by Giorgio Pasquali.

The Art of Allusion

In his brief but seminal article on what he called the "art of allusion" (*arte allusiva*),[11] Pasquali showed that allusion is an essential component of all Latin poetry, that, far from indicating a failure of imagination, allusive artistry is actually an important affective element in the poet's repertoire, and that the reader who does not allow it to play upon his imagination misses an important part of the poetry's effect. First of all, Pasquali distinguishes allusion from reminiscences, which may be unconscious, or from imitations, which the poet may hope will go unnoticed: allusions, he argues, do not produce the desired effect except upon a reader who clearly recalls the texts to which they refer. He illustrates this view of allusion with examples from Horace and Vergil in particular—the *Georgics* itself is mentioned a number of times—as well as from Italian literature of the *Ottocento*. Pasquali focuses on passages in which allusion contributes a very specific effect, passages in which something is left unsaid, something that the reader will not perceive unless he or she recognizes the allusion as such and correctly interprets its relevance in context. As an example, he cites a pair of lines from the *Aeneid* concerning the punishment of a corrupt and venal politician in the pit of Tartarus:

> **vendidit hic** auro patriam dominum**que** poten**tem**
> impos**uit, fixit leges pretio atque refixit**
>
> *Aeneid* 6.621–622

This passage, as we know from Macrobius (*Sat.* 6.1.39), alludes to Varius' *De Morte*:

> **vendidit hic** Latium populis agros**que** Quiri**tum**
> erip**uit, fixit leges pretio atque refixit**
>
> Varius *De Morte* fr. 1
> (Morel–Büchner 1982.130)

11. Pasquali 1942.

In Vergil, the referent of *hic* is not clear; the exemplum is evidently offered not specifically, but as a type of damnable behavior. Pasquali agrees with most scholars that Varius, on the other hand, has someone very specific in mind, namely, M. Antonius.[12] He therefore argues that Vergil, too, is referring to Antonius, declining to name openly Augustus' most hated foe[13] but, in the mind of the reader who knows Varius' poem, unmistakably indicting him in particular. Because the passage of Varius has come down to us without context, we cannot be sure that Pasquali is right. Nevertheless, his point is well taken: we have noticed such passages throughout Augustan poetry. Their point, I might add, is not just to convey information, but also, and perhaps even more commonly, to contribute rather subtler poetic effects. If we consider Ovid's lines about Narcissus, for example,

> **multi illum** *iuvenes*, **multae** cupi**ere puellae,**
> sed—fuit in tenera tam dura superbia forma—
> **nulli illum** *iuvenes*, **nullae** tetig**ere puellae.**
>
> *Metamorphoses* 3.353–355

it will be a dull reader indeed who fails to see in them an allusion to Catullus' hexameter wedding hymn:

> ut flos in saeptis secretus nascitur hortis,
> ignotus pecori, nullo convulsus aratro,
> quem mulcent aurae, firmat sol, educat imber:
> **multi illum** *pueri*, **multae** optav**ere puellae;**
> idem cum tenui carptus defloruit ungui,
> **nulli illum** *pueri*, **nullae** optav**ere puellae.**
>
> *Carmen* 62.39–44

It will be an equally dull reader who needs this allusion, or any other means, to inform him of Narcissus' fate. The point of the allusion is not to impart information. Rather, recognizing these lines as an allusion to Catullus' flower simile, and knowing the metamorphosis that awaits Ovid's hero, the reader experiences the pleasure of ironic recognition, pleasantly aware that he may count himself an initiate of the literary mysteries that Ovid celebrates.

12. Servius *ad loc.* (Thilo 1881–1884.2.87–88) argues that Vergil has C. Scribonius Curio (*tribunus plebis* 50: see *RE* s.v. "Scribonius [11]") in mind; on Antonius see Norden 1957.291–292 and Austin 1977.198.

13. In conformity with the sentence of *damnatio memoriae*?

Allusion and Literary History

Pasquali's elegant paper is more suggestive than definitive; consequently, its influence has been felt not just widely but in various ways. Crucial to Vergilian criticism and to recent literary history of the Augustan period in general has been the study of allusion in Hellenistic and Neoteric poetry. The most vocal and influential champion of this approach is Giuseppe Giangrande, who cites Pasquali as his guide. Together with his students, however, Giangrande has taken the study of learned poetic allusion much farther than Pasquali's essay seems to call for, and in a quite different direction as well. The primary, perhaps the only interest of the *Giangrandisti* is not in affective allusion, but in lexical matters, i.e. in how the Hellenistic poets used Homer as a source of poetic language. Following Pfeiffer, they envision a world of scholar-poets for whom poetry was simply a direct, practical, and entertaining medium of scholarly debate on points of mythology, geography, history, astronomy, all branches of learning—and, in particular, grammar.[14] Thus Giangrande has written on Hellenistic allusions that point to textual variants in the Homeric poems,[15] on the creation of new forms and extended meanings based on Homeric usage,[16] and so forth. By the same token, the commentaries on Hellenistic poetry produced by his students routinely focus on the use of *hapax legomena*, formulaic language, and other notable features of Homeric style.[17]

Besides concentrating on lexical data, the most characteristic feature of Giangrande's approach is his emphasis on the allusive technique that he calls *oppositio in imitando*.[18] In Giangrande's view, Hellenistic poetry was written for connoisseurs of an aesthetic that valued close familiarity with the details of classic texts and expected modern poets to imitate these details in unexpected ways. Thus, to cite some of Giangrande's own examples, "If Homer says ἐς ἠέλιον καταδύντα (formula), Apollonius will say (*Arg*. 1.725) ἐς ἠέλιον ἀνιόντα; if Homer has the verb ἐκκυλίνδω, Callimachus will use εἰσκυλίνδω (*Hymn*. 4.33)," and so on.[19] This activity became a frequent vehicle for learned debate: scholar-poets, by imitating, say, specific textual variants uncovered in their Homeric research, would

14. Giangrande 1967.85 uses the "Pasqualian term" *arte allusiva*, while citing Herter 1929 as "the best methodological introduction to this literary feature."

15. Giangrande 1970.

16. Giangrande 1967.

17. See, for instance, Williams 1978 and Bulloch 1985.

18. E.g. Giangrande 1967.85, citing Kuiper 1896.114.

19. Giangrande 1967.85.

express their opinions about the original readings, sometimes referring with pointed contrast to the readings favored by others.[20]

Giangrande and his school focus rather narrowly on one specific aspect of learned allusion, and, as a rule, do not treat of allusion as would a literary critic. This approach in its pure form can and has been applied directly to Vergil; indeed, we shall learn of one or two examples in the following pages. In general, however, it is not in itself a technique of the first importance for understanding Vergil's allusive style. More important is the work of other scholars who use rigorous philological methods approximating those of Giangrande, but who are willing to take the further step of reading allusion as literary-historical evidence, and who have thus had a major impact on the study of Latin poetry. The original manifesto of this group was Wendell Clausen's paper "Callimachus and Latin Poetry."[21]

Clausen's article exemplifies the aesthetic that it celebrates. Brief and suggestive, like Pasquali's paper, it contains no new information or labored argument; rather it succinctly and gracefully states an important idea. Traditional literary history had seen Augustan literature as the product of a politically and aesthetically conservative movement inspired by the masterpieces of archaic and classical Greek literature in preference to the more modernist aesthetic of the Hellenistic world that was adopted by the Roman Neoteric poets of the mid-first century B.C. Clausen, on the other hand, stresses the element of continuity between Neoteric and Augustan poetry, and suggests that the aesthetic similarity between these successive generations of Roman poets stems from their devotion to an individual teacher, himself a recognized Hellenistic poet. It was Parthenius of Nicaea, Clausen asserts, who, after coming to Rome as a prisoner of war sometime between 73 and 65 B.C. introduced the Neoteric poets—Cinna, Calvus, and Catullus—and those of the next generation—Gallus and Vergil—to Callimachus; and it was this encounter that both produced the so-called "Catullan revolution" and gave Augustan poetry its essential character. It is an idea that Clausen has long held, and time has produced many opportunities for restatement and development.[22]

Where allusion is concerned, Clausen sees more than Giangrande. He accepts the notion that allusive artistry is a vital part of Hellenistic

20. See, in general, Giangrande 1973.

21. Clausen 1964.

22. See, for example, Clausen 1982.178–187.

Greek and Latin poetry; but, for Latin poetry at least, he finds in it a significance that goes well beyond the scholarly formation of a highly artificial *Dichtersprache*. For Clausen, allusion in Latin poetry is laden with artistic ideology. For Catullus or Vergil to quote Callimachus or Theocritus, or to allude to Hesiod or Aratus with the learning, wit, and delicacy that they had learned from these poets, was to declare independence from an archaizing native tradition that aimed mainly at the glorification of some patron's military exploits in verse modeled on the most obvious passages in Homer. Not that these poets considered allusion to Homer or to of the other archaic poets reactionary per se. What counted was the manner of the allusion—learned, subtle, recherché. Thus Vergil's and Horace's imitation of archaic models such as Hesiod, Homer, Sappho, and Alcaeus cannot be taken as evidence of a traditionalist reaction to Neotericism on the part of the leading Augustan poets; rather the manner in which these allusions are fashioned demonstrates, in Clausen's view, the essentially Callimachean and Neoteric basis of Augustan poetics.

The development of certain implications inherent in Clausen's ideas has been carried out largely by his followers. The most ambitious exponent of the literary-historical notions inherent in Clausen's approach is his student D. O. Ross, Jr. In his dissertation,[23] which was later published in revised form,[24] Ross defined the essential characteristics of Neoteric poetry more rigorously than had been done before by using the stylometric method of Bertil Axelson[25] to identify two distinct poetic styles used by Catullus, that of the polymetrics (*Carmina* 1–60) and the longer poems (*Carmina* 61–68), and that of the epigrams (*Carmina* 69–116). The former he identifies as a Neoteric innovation based on Alexandrian ideals, the latter as deriving from an older, well established Latin tradition. Ross's findings can be and have been challenged on some points;[26] but by bringing out the contrast between these styles, Ross establishes the essential characteristics of Neoteric poetry on a sound philological basis. In a subsequent and closely related investigation, Ross investigated Neoteric language and thematic motifs in the works of the Augustan poets, particularly in Vergil's *Eclogues* and Propertius' *Monobiblos*, to illustrate the impact that the Neoteric tradition had upon

23. Ross 1966.
24. Ross 1969.
25. Axelson 1945.
26. Some cogent individual objections are made by Benediktson 1977, but the main lines of Ross's argument remains intact.

the next generation of Roman poets.[27] This work on the Latin *Dichtersprache* parallels the activity of Giangrande and his school by revealing the philological basis of poetic artistry in Neoteric and Augustan Rome. But Ross, following Clausen, expands his approach to include ideological allusion; and he has done more than any other scholar to show how allusion can become the instrument of literary history.

In his most recent book, a study of the *Georgics*, Ross largely turns his attention to issues other than those that had concerned him in the past.[28] But the application of his literary-historical views to the *Georgics* has been carried out chiefly by his student, Richard Thomas. Thomas' commentary on the poem contains a wealth of sophisticated discussion in which Vergilian allusion discloses a consistent attitude towards the literary past.[29] The attitude expressed is essentially identical to that elaborated by Clausen and Ross, i.e. Callimachean and Neoteric in the extreme.[30] As a preliminary to the commentary, Thomas had already published a paper in which he uses the *Georgics* as a basis for establishing a typology of allusion—or, as he prefers to call it, reference—for Latin poetry in general.[31] The paper proceeds from relatively simple to quite complex forms of allusion, and in the process manages to indicate some of the thematic richness that allusion can create.[32] In general, though, Thomas' interest is in allusion as an illustration by the poet of his mastery over the literary past. In discussing what he calls "conflation" or "multiple reference," he observes that in one particular passage (*G.* 1.231–246)

> Virgil has overlaid a primary "translation" of technical Hellenistic sources with reference to the Homeric original as well as to previous Roman translations of that technical material. Archaic, Hellenistic, and Roman are...conflated into a single version, not just as an exercise in cleverness or erudition (and this is perhaps what, apart from its greater complexity, distingüishes Virgilian from Alexandrian reference), but rather as a demonstration of the eclectic and comprehensive

27. Ross 1975.
28. Ross 1987.
29. Thomas 1988.
30. See also Thomas 1985 for his observations on the Callimacheanism of the *Aeneid*, and cf. Clausen 1987.14.
31. Thomas 1986.
32. The discussion of "self-reference" (pp. 182–186) is especially suggestive.

> nature, and perhaps of the superiority, of the new version: the tradition has become incorporated into a new version.[33]

Thomas is making a fine distinction, and one that is difficult to illustrate; but, in my experience of Vergil, it is absolutely true. Vergil does match and surpass the Alexandrians in the complexity of his allusion, but never does learned display seem to be the point, as it so often does in Callimachus, Apollonius, and the rest. Where they are concerned with using their formidable learning to control the classic texts of the past, and thereby to carve themselves a niche in literary history, Vergil uses allusion to create his own vision of that history, to make the past anew, to call into being the tradition to which he wishes to be heir.

In this study, I shall adopt the basic working assumption of these three scholars, that allusion in Vergil (and the other poets of his age) frequently serves ideological purposes, and therefore offers useful literary-historical evidence. One closely related assumption that these scholars share, however, I cannot accept without modification. All three of them argue that the Callimachean and Neoteric outlook was the decisive factor in the development of Latin poetry in the first century, and that it remained the dominant source of inspiration until the end of the Augustan period. The allusive style that they find in Latin poetry from Catullus to Ovid, and many other features as well, they consistently trace to the poets' adherence to Callimachean principles. My view differs from theirs in two respects. First, many features of Augustan poetry, particularly the decisions of Vergil and Ovid to attempt epic poetry on a grand scale, seem to me fundamentally incompatible with the premise of an exclusively or even predominantly Callimachean ideology. Second, I believe that in the allusive program of the *Georgics* it is possible to discover a literary-historical outlook that derives from a rich diversity of sources, one that greatly helps to explain the character of subsequent Latin poetry. What I hope to show, then, is that their approach can be further expanded so as to enrich and advance our understanding of Augustan poetry.

Allusion and Poetic Memory

Despite the fact that they claim methodological descent from Pasquali, both Giangrande and his followers and Clausen et al. tend to ignore the

33. Thomas 1986.197–198.

affective quality of allusion. This is especially true of Giangrande's school, for whom allusion remains more or less exclusively a vehicle for scholarly debate. Clausen, Ross, and Thomas do consider the aesthetics of allusion; but their primary interest is in an aesthetic response, difficult to define, that is closely tied to the excitement of philological discovery and thus accessible only to those who revel in such abstract joys. This attitude is clearly exemplified by Thomas' paper on allusion in Catullus' "Peleus and Thetis" (*Carmen* 64).[34] Thomas concludes this paper by asking a very basic question: "Why does Catullus begin a poem on the wedding of Peleus and Thetis with an account of the Argo's voyage?"[35] His answer: "It does not seem preposterous to suggest that Catullus found the story of the Argo, in the appropriate range of its previous treatments, an irresistible starting point for a poem with whose central themes it traditionally had little in common."[36] The restricted focus of Thomas' paper was addressed by J. E. G. Zetzel in a subsequent study of the same poem.[37] Zetzel quotes Thomas' paper to the effect that "a great deal of the intent of the New Poetry is to modify, conflate, and incorporate prior treatments. Through this method the poet rejects, corrects or pays homage to his antecedents and—the ultimate purpose—presents his own and superior version."[38] "In other words," Zetzel retorts,

> the purpose of literary allusions in Catullus is, quite simply, to make literary allusions. The goal of the learned poet is no more than to demonstrate his learning.
>
> No one would deny that the *poeta doctus* was interested in displaying his erudition, or that at least a part of the pleasure of writing and reading such poetry was to feel the warm glow of superiority to less learned poets and readers. But a poetry that existed primarily for the purpose of displaying learning would be remarkably sterile; and while it may be an apt characterization of, for example, Lycophron or Nicander, it seems scarcely adequate to Catullus 64 or to Callimachus himself. While some poets were, to an extraordinary degree, self-conscious in their deliberate manipulation of the details of language and meter, this

34. Thomas 1982a.
35. Thomas 1982a.163
36. Thomas 1982a.164.
37. Zetzel 1983a.
38. Thomas 1982a.163.

> technical mastery was not an end in itself, for either the Alexandrians or their Roman imitators.[39]

Zetzel's paper goes on to consider allusion to Ennius in Catullus. Where Thomas sees "polemical" reference to Ennius as the anti-type of the New Poet, Zetzel sees a creative engagement with Rome's literary past, one that informs his own poetry, particularly *Carmen* 64, at the level of theme as well as that of diction, style, and general erudition. Well acquainted with the more philologically oriented approaches employed by Giangrande and Clausen, Zetzel nevertheless sees the need to consider allusion as a poetic device, and not simply as a vehicle of scholarly controversy.

Other scholars, too, have taken an approach similar to Zetzel's and have applied it to Vergil in particular.[40] Undoubtedly the most ambitious attempt to define the poetic value of allusion in Latin poetry, especially that of Vergil, is the work of G. B. Conte and his school. Conte proceeds from a carefully developed theory that regards imitation as, on the one hand, analogous to the class of rhetorical figures known as tropes and, on the other, as an overt form of cultural memory encoded in texts. His analyses show a fine feel for the affective element in allusion. While Conte himself has made the fullest theoretical defense of the critical methods employed by the members of this school, Alessandro Barchiesi has also directly confronted the issues that concern us here. In his illuminating and suggestive study of Homeric imitation in the *Aeneid*, Barchiesi explicitly cites the theoretical work of Conte for his general approach.[41] In particular, he raises the question of how to proceed in light of Knauer's research. Citing Knauer's "monumental work" as "the conclusion, as well as the definitive assessment, of a learned activity that extends from the ancient investigations into Vergilian plagiarism down to the great commentaries of the modern era," Barchiesi goes on to state his belief that an opportunity remains for anyone who wishes to study the functions that the Homeric poems assume as models for the *Aeneid*.[42] He recognizes that the massive amount of philological evidence that Knauer has adduced "seems to have definitively established

39. Zetzel 1983a.253.

40. Lyne 1987 deserves special mention in this regard: see his stimulating, if not uniformly convincing, chapter on "Allusion" (pp. 100–144).

41. Barchiesi 1984.9 n.1.

42. Barchiesi 1984.9.

the constitutive character of Homeric imitation in the *Aeneid*," but he nevertheless insists with justice on the need to examine this aspect of the poem from other perspectives, asserting that "the price would be too high if we therefore had to accept *en bloc* Knauer's method and critical ideology."[43]

The main value of Conte's and Barchiesi's work, in my opinion, lies in the careful, sensitive analysis of individual passages. This point is illustrated by Conte's discussion of the relationship among three passages of Homer, Catullus, and Vergil:[44]

> *ἄνδρα μοι ἔννεπε, Μοῦσα, πολύτροπον, ὃς μάλα* **πολλὰ**
> *πλάγχθη, ἐπεὶ Τροίης ἱερὸν πτολίεθρον ἔπερσε.*
> **πολλῶν** *δ᾽ ἀνθρώπων ἴδεν ἄστεα καὶ νόον ἔγνω,*
> **πολλὰ** *δ᾽ ὅ γ᾽ ἐν πόντῳ πάθεν ἄλγεα ὅν κατὰ θυμόν.*
>
> *Odyssey* 1.1–4

> ***Multas*** *per* gente*s* *et* ***multa*** *per aequora vectu*s
>
> *Carmen* 101.1

> ***Quas*** ego te terra*s* *et* ***quanta*** *per aequora vectu*m
> accipio, **quantis** iactatum, nate, periclis!
>
> *Aeneid* 6.692–693

Conte writes affectingly of the "hidden reserve of literary energy whose full potential is released" only when these texts are brought together.[45] Homer, of course, establishes the pattern of Odysseus, the much-traveled, long-suffering hero. Catullus then casts himself in such a role, and does so in a context—a funeral epigram for his brother—that allows the reader to draw further connections between Catullus' experience and other Homeric themes, such as the general tragedy of the Trojan war, Odysseus' "*Nekyia*," etc.[46] Finally, Vergil's Anchises casts his son as both Odysseus and Catullus—at once a storm-tossed survivor and a pious kinsman come to honor his dead.

Conte's analysis is, in my opinion, a convincing, sensitive reading of the relationship among these passages. His work and that of his followers should serve as a useful protreptic to those who favor more traditionally

43. Barchiesi 1984.16 n. 9.
44. Conte 1986.32–39.
45. Conte 1986.33.
46. Conte 1986.32.

philological approaches. On the other hand, I cannot accept their work as a model without the following qualifications:

1. Both Conte and Barchiesi evidently regard the general theory of allusion that Conte has worked out as the essential basis of their practical criticism. I, however, find a certain gap between Conte's theoretical framework, which in any event leaves certain questions unanswered, and his practice. Conte writes that

> Pasquali's approach reveals a privileging of the moment of *intentionality* in the "poetic memory." His method creates a substantial opposition between inert material and intentional elements. These latter, precisely because of their intentionality, enter the realm of creativity.... Within an approach of this kind, the position of the author becomes predominant and the notion of the centrality of the text as a unified, complete, and interlocking system is inevitably weakened. In the philological tradition this imbalance in favor of the author is decidedly unfruitful.... If one concentrates on the text rather than on the author, on the relation between texts (intertextuality) rather than on imitation, then one will be less likely to fall into the common philological trap of seeing all textual resemblances as produced by the intentionality of a literary subject whose only desire is to emulate. The philologist who seeks at all costs to read intention into imitation will inevitably fall into a psychological reconstruction of motive, whether it is homage, admiring compliment, parody, or the attempt to improve on the original. If poetic memory is reduced to the impulse to emulate, the production of the text will be devoted to the relationship between two subjectivities, and the literary process will center more on the personal will of two opposing authors than on the structural reality of the text.[47]

The problem of authorial intention can never be far from the thoughts of those who study allusion. In a general way, the concept of the "intentional fallacy" as conceived by W. K. Wimsatt and Monroe C. Beardsley, who coined the term, seems unarguable. Their main point, in the celebrated essay of the same title, is that an author's intention remains inaccessible, and that it makes no sense to fault a poet for not achieving goals that the critic has imputed to him.[48] This point, however novel it may once have appeared, with the added perspective of some forty-five years has come

47. Conte 1986.26–27.
48. Wimsatt–Beardsley 1946.

to seem self-evident. In that intervening time, however, the concept has extended itself to the point that many regard the issue of authorial intention as beyond the scope of literary criticism. They believe that the author's opinion about the meaning of his work is merely one among many possible competing readings, and thus should not be held as privileged. Even the author's role in composing his work is now viewed by many as incidental, paling in significance beside the claims of other factors, such as the immanent influence of form, genre, metrical tradition, etc. Conte's theory of poetic memory clearly owes a lot to these ideas. And yet, like others who have studied allusion, he seems to ask himself certain questions: Is this an allusion, and not merely a chance parallelism or reminiscence? Did the poet intend the reader to notice it as such? If so, precisely how did he expect the reader to take it? One realizes, of course, that in many cases these questions can receive no definitive answer, and not just for lack of evidence: often the very indeterminacy seems intended. But at other times, a reader simply wonders if he is really in tune with his author.

Conte is no exception to this rule. Consider the following statement:

> Before the allusion can have the desired effect on the reader, it must first exert that effect on the poet. The more easily the original can be recognized—the more "quotable" (because memorable) it is—the more intense and immediate its effect will be. The reader's collaboration is indispensable to the poet if the active phase of allusion is to take effect. Thus allusion will occur as a literary act if a sympathetic vibration can be set up between the poet's and the reader's memories when these are directed to a source already stored in both.[49]

What sort of a process does Conte envision? One in which allusion has a *desired* effect. Desired by whom? Evidently by the poet, whose experience of poetic memory the reader, if he reads correctly, shares. Thus in the discussion of Homer, Catullus, and Vergil cited above, Conte ascribes to Catullus a specific purpose in alluding to Homer ("to make Odysseus' mythical journey well up through his words") while at the same time denying him another ("He certainly has no intention of competing with Homer").[50] Similarly, Conte speaks of Vergil's "motive" in alluding to Catullus as "not emulation but a *desire* to pay tribute to the

49. Conte 1986.35.
50. Conte 1986.36.

methods of a poetic he values and *wishes* to be identified with."[51] Barchiesi, too, invokes the notion of authorial intention.[52] This procedure is at odds with the theory to which both scholars adhere; but it is, as I have said, perfectly understandable. Literary theorists have not yet worked out this complex issue; but the student of allusion, even if he cannot adequately define its place in criticism, cannot simply wish the notion of authorial intention away.[53] I raise the issue only because I recognize the practical value of the work done by Conte and his school and wish to adopt some of their key aims without accepting the theoretical underpinnings of their approach. My own position, which is at odds with Conte's theory if not his practice, is that the student of allusion is on some level concerned with a poet's intentions.[54]

2. Conte elsewhere states that "the reader must be poetically knowledgeable. At some point in its development...poetry seems to start joking and playing with its readers, offering them participation 'in aenigmate' (in mystery). But readers can attain to that source of pleasure only if they have acquired sufficient poetic 'doctrine.'"[55] I believe that Conte touches here on one of the essential pleasures of reading allusive poetry, the initiate's sense of privileged belonging. But he leaves utterly open the question of how a reader becomes qualified. He seems to me to be operating, like Pasquali, with a cultural ideal in mind, someone who has read widely and remembers well, who is quick to recognize the similarities between texts. But how much can we expect of this reader? In the second passage quoted above, Conte invokes the notion of memorability: the more memorable an allusion is, the more forceful and immediate its effect will be. He supports this statement with observations on the titular aspect of first lines in ancient poetry; and in the example that he adduces, Catullus and Vergil both imitate just such highly memorable passages. But a very important characteristic of "learned" allusion, as

51. Conte 1986.37; the emphasis is mine.

52. See, e.g., Barchiesi 1984.16.

53. See the useful remarks of Hirsch 1967.11–12 *et passim*. For a clear-headed statement about the place of authorial intention in criticism of the visual arts, with obvious parallels to literary criticism, see Summers 1989.400–401.

54. At the same time, I am unwilling to equate intention with other aspects of the author's psyche, in the manner of Harold Bloom. Such an approach can of course yield fascinating and valuable results, as in the case of Greene 1982; but the state of mind in which the author confronts the literary past will not be the focus of this study.

55. Conte 1986.57

Giangrande, Clausen, and others have shown, is precisely the tendency to avoid such obvious models, or at least to seek out less obvious ones in addition to the better known. There are, indeed, allusions in Vergil, not to mention the Alexandrians, that go to great lengths to conceal themselves from the reader who is acquainted only with the most memorable passages of the literary past, allusions that are accessible only to readers who control a considerable scholarly apparatus, but which, once identified, are unmistakably there. It is, therefore, misleading to limit the discussion of allusion in these poets only to those passages that conform to a concept of literary decorum demonstrably alien to the practice of the ancient poets. We must, in other words, become readers who are sensitive to and appreciative of, not only allusions to the opening lines of the *Odyssey*, but to Apollonius' variations on Homeric *hapax legomena* as well.

3. This brings me to my final point. Conte and his followers, not unlike Giangrande, Clausen, and most of the others whose work I have discussed, focus on allusions as individual points of contact between two texts. Thomas, who has done so much to plumb the complexity of Vergilian allusions as literary-historical artifacts, is an exception, but only a partial one, because he too concentrates on individual passages within Vergil's text. I believe that a more synthetic approach is called for. In the first place, the very nature of the evidence cries out for synthesis. On the one hand, there are those passages that interest Conte in which Vergil recalls some famous literary antecedent. On the other are the many much less obvious, some almost surreptitious ones involving obscure aspects not only of Homer, but of poets like Callimachus and Aratus, which interest Giangrande, Clausen, and the rest. Should not a general theory of allusion address both classes of evidence? It is, I believe, a shortcoming of both these approaches that they do not take a truly synoptic view of Vergilian allusion in its systematic aspect. Vergil alludes both to famous and to obscure passages; a few authors in particular repeatedly contribute both obvious and not so obvious material to his allusions. Is it not necessary to consider whether all of this material conforms to some general plan? I have already discussed Knauer's study of the *Aeneid*, citing this as its great contribution to Vergilian studies. In this study I will try to analyze the *Georgics* on a similar basis, *mutatis mutandis* and without losing sight of the important contributions made by the other scholars whose work I have discussed.

Conclusion

I am fully aware that these observations do not amount to a rigorous theoretical statement about method. The work of Thomas and Conte is particularly important in this regard. These scholars have challenged all students of Latin poetry by their efforts to establish a theoretical foundation for the study of allusion. This is a laudable aim, but, perhaps, a utopian one, not unlike what Rudolph Pfeiffer called for years ago: a complete typological study of Hellenistic allusion.[56] Whether such a study would be feasible is very doubtful. My own inability to subscribe wholeheartedly to either Thomas' or Conte's theory is due mainly to a belief that no single theory can be sufficient to account for the fascinating multiplicity of allusion that one finds in Hellenistic Greek and Latin poetry. Accordingly, this study is presented as a seriatim encounter with a number of particular cases, one which aims at a coherent reading of the *Georgics* alone. Interpretations are supported by analogy and general conclusions are drawn, but the actual basis of the work is empirical, and no attempt has been made to construct a universal or transferable theoretical approach.

The foregoing survey of criticism should also make clear the necessity of going beyond technical questions to explore the affective nature of allusion. The focus of this study, as its subtitle suggests, will be on the literary-historical nature of Vergilian allusion as it seeks to read in the allusive program of the *Georgics* a Vergilian essay in literary history; but it will not ignore, particularly as it draws to its conclusion, more purely aesthetic concerns. I am aware that I have left much to be said particularly under this last heading; but can anyone write on Vergil and not feel that way?

There are many points at which we might begin, but one seems especially appropriate. Having finished with these preliminaries, let us turn to examine a very special allusion—perhaps the most famous in the *Georgics*—the passage in which Vergil all but names the poet traditionally regarded as the model of his poem.

56. Pfeiffer 1955.72.

CHAPTER 2

Ascraeum Carmen

The "Model" of the *Georgics*

VERGIL TELLS HIS READERS more than is generally perceived in his ringing and eminently quotable boast,

> Ascraeumque cano Romana per oppida carmen.
>
> *Georgics* 2.176

In this single line the blend of disparate elements that is the *Georgics* comes clearly into focus. The contrast between the archaic and the contemporary, between the rustic and the urbane, between the fine arts of Greece and the practical arts of Rome, is neatly drawn in the opposing terms *Ascraeum carmen* and *Romana oppida.* Typically of Vergil, this symmetry is enlivened by inconcinnity. The epithet *Ascraeum* is more specific than *Romana;* for while Rome can boast many famous sons, Ascra claims only Hesiod. Hence the belief that Vergil here declares the *Georgics* to be a "Hesiodic" song. On this much, I believe, there is fairly general agreement. Moreover, these familiar views are, to a certain extent, patently unquestionable. Even if readers today, as some in antiquity evidently did, wish to take the word *Ascraeum* in a slightly different sense, it is obvious that to deny the fundamental relationship between Hesiod and Vergil, between *Works and Days* and *Georgics,* would be utterly absurd. That is why it is so astonishing to realize that the phrase *Ascraeum carmen,* far from being a declaration of Vergil's literary goals, actually signals the end of Hesiodic influence on his poem.

This statement, as will become clear from my defense of it, carries with it some important implications about the nature and plan of the *Georgics,* about Vergil's poetic technique, and about his literary goals and ideas in general. To trace these implications to the end will be a long and involved task, one that will ultimately entail rethinking a number of the most common assumptions about the literary history of Rome in the first century B.C. The trail, however, is fairly clear. It begins with Vergil's decision to choose as a "model" for the second of his three major works that extremely ambivalent figure, Hesiod of Ascra.

Two Hesiods

Servius has little doubt about Vergil's literary models:

> Vergilius in operibus suis diversos secutus est poetas: Homerum in Aeneide, quem licet longo intervallo secutus est tamen; Theocritum in Bucolicis, a quo non longe abest; Hesiodum in his libris, quem penitus reliquit. Hic autem Hesiodus fuit de Ascra insula.[1] Qui scripsit ad fratrem suum Persen librum quem apellavit Ἔργα καὶ Ἡμέρας, id est *Opera et Dies*. Hic autem liber continet, quemadmodum agri et quibus temporibus sint colendi. Cuius titulum transferre noluit, sicut "Bucolicorum" transtulit, sicut "Aeneidem" apellavit imitatione "Odyssiae"; cum tamen per periphrasin primo exprimit versu dicens: "indicabo *quo opere* et *quibus temporibus* ager colendus sit." Ingenti autem egit arti ut potentiam nobis indicaret ingenii coartando lata et angustiora dilatando; nam cum Homeri et Theocriti in brevitatem scripta collegerit, unum Hesiodi librum divisit in quattuor. Quod ratione non caret: nam omnis terra, ut Varro docet, quadrifariam dividitur.[2] Aut enim arvus est ager, id est sationalis, aut consitus, id est aptus arboribus, aut pascuus, qui herbis tantum et animalibus vacat, aut floreus, in quo sunt horti apibus congruentes et floribus....
>
> Servius *praefatio in G.* (Thilo 1887.128–129)

This notice, despite one or two factual errors, is quite consistent with the main stream of modern opinion, a situation that I find amazing. Servius

1. This remarkable error calls into question how well Servius (and his source?) really knew *Works and Days*. Hesiod's comments on the landlocked town of Ascra occur in the context of the "*Nautilia*" (*WD* 630–662), where his distrust of seafaring is made clear. On the larger significance of this episode see Rosen 1990.

2. Servius is mistaken here. At *DRR* 1.2.12ff. the characters of Varro's dialogue discuss what departments of rural industry truly belong to agriculture, actually rejecting such activities as raising livestock. At length Scrofa does make a four-part division of the subject (1.5.3), but his categories are very different from those of Servius, embracing (1) knowledge of the land itself, (2) the necessary equipment, (3) the operations to be performed, and (4) the proper schedule for performing them.

himself seems to feel the weakness of his argument that *Works and Days* was *the* model of the *Georgics:* it is certainly clear that he felt the point required detailed illustration. This passage is far more carefully argued and less discursive than the prefaces to Servius' *Eclogues* and *Aeneid* commentaries, where Vergil's *aemulatio* of Theocritus and of Homer is taken practically for granted. Here instead the relationship between *Georgics* and *Works and Days* is defended point by point and at some length. Ironically, by arguing so carefully in support of this relationship, Servius succeeds mainly in creating the impression that he regarded Hesiod as a somehow inadequate or unworthy model for Vergil. Hesiod is, after all, the only one of Vergil's three models whom Servius states that Vergil actually surpassed. Next, almost apologizing for the fact that the title *Georgics* is so dissimilar to that of *Works and Days,* Servius points to the paraphrase of ἔργα καὶ ἡμέραι in *Georgics* 1.1 (*quid* ***faciat*** *laetas segetes,* ***quo sidere*** *terram*). By so doing, he actually suppresses the fact that the poem's title is borrowed directly from a different poem, the *Georgica* of Nicander Colophonensis.[3] Furthermore, Servius alleges that Vergil's technique in imitating *Works and Days* contrasts sharply with the one that he employed in the *Eclogues* and was to follow again in the *Aeneid.* Whereas both the previous and the subsequent works are considerably shorter than their Greek exemplars,[4] in the *Georgics* Hesiod's "single book" is expanded into four. Finally, he ascribes Vergil's four-part division of the *Georgics* not to its primary model, *Works and Days,* but to a passage of Varro Reatinus concerning the four chief departments of agriculture. Thus Servius' statement, which subsequent generations of Vergilians have generally regarded as self-evident, upon closer inspection stands revealed as a piece of special pleading.

Obviously the plea has been successful. Hesiod continues to hold his place in the modern canon of Vergilian models.[5] Thus scholars have

3. Quintilian (*IO* 10.1.56) asserts simply that Vergil "followed" Nicander's poem. Macrobius (*Sat.* 5.22.10) cites *Theriaca* 359–371 as the source of *Georgics* 3.425, indicating that Nicander had not been utterly forgotten by the Roman literati of the fourth century.

4. In the case of the *Aeneid* this point is obvious; as for the *Eclogues,* Servius (*praefatio in Ecl.* [Thilo 1887.20–21]) observes that only seven poems conform to the generic norms established by ten of Theocritus' *Idylls,* while *Eclogues* 4, 6, and 10 are felt to be departures. The Theocritean canon is not specified, nor does Servius openly state that Vergil's decision to collect ten eclogues into a single book is based on a similar collection of the ten Theocritean poems that he designates as *meras rusticas.*

5. The remarks of Wilkinson (1969.36) still represent the *communis opinio:* "Having succeeded in becoming the Roman Theocritus, Virgil aspired to be the Roman Hesiod, partly because he was interested in rustic life, partly because Lucretius had shown how

been ready to make easy assumptions about the meaning of the phrase *Ascraeum carmen.* On this point, too, Servius leads the way:

> ASCRAEUM Hesiodicum; nam Hesiodus de civitate Ascra fuit.
>
> Servius *ad G.* 2.176 (Thilo 1887.237)

Similar remarks are apposed to other Vergilian periphrases.[6] Ancient opinion appears almost monolithic on this point. Yet an anonymous voice in the Servian appendix appears to represent an alternative point of view:

> ASCRAEUMQUE et rel. Hesiodum dicit de civitate Graeciae Ascra. ubi natus est Hesiodus, *ibi*[7] sunt natae Musae; unde sumpsit Hesiodus poema, inde ego sumpsi, sed ille Graecis cecinit, ego Romanis.[8]

This notice seems to take *Ascraeum* in a more ambitious sense than Servius does. There is still, of course, a direct reference to Hesiod, but with more emphasis on the sources of Hesiod's inspiration—on his relationship with the Muses—than on his tangible literary output. This view of Hesiod is a useful supplement to Servius' approach, and may even be more congruent with current critical efforts to understand the *Georgics.* A few scholars after all have shared Servius' dissatisfaction with Hesiod as a Vergilian model.[9] But it has taken a long time for this dissatisfaction to become a critically potent force. It was uniformly felt until well into this century that *Works and Days* was without any serious

---

great a didactic poem could be as poetry, and partly because Hesiod, so much admired by the Alexandrians of the third century, interested the neo-Callimachean poets of the Neoteric movement and had not yet been appropriated." For a more accurate view of the matter see Thomas 1988.1.4–9, especially p. 6.

6. Cf. Servius *ad Ecl.* 4.1 (Thilo 1887.44); 6.1 (pp. 64–65); 8.10, 21 (pp. 93, 95); 10.50 (p. 125).

7. *Ibi* is my conjecture for the MS reading *ubi.*

8. *Brevis exp. in G.* 2.176 (Hagen 1902.298); cf. Σ *Bern. ad loc.* (Hagen 1867.231).

9. Most recently Griffin 1986a.12: "Virgil goes far beyond Hesiod in giving his poem a pervasive moral and reflective colouring.... The influence of Lucretius, earnest and passionate, is really more important here than that of Hesiod." In another publication of the same year, however, Griffin voiced the traditional view, summing up the central problem of the *Georgics* with the question, "Could the homespun rustic verse of Hesiod be transformed into a Latin poem that would satisfy the aesthetic demands of Virgil and his audience?" and noting that "The first book of the *Georgics* has some close echoes of Hesiod, to establish the colouring of the whole" (Griffin 1986b.623).

doubt Vergil's single, obvious, and openly acknowledged model for the *Georgics*—that his goal in writing the poem was to become the "Roman Hesiod." There was no alternative. Only recently have some scholars advanced a revisionist interpretation of Hesiod's symbolic significance in Vergil's poem.

This view concerns the Alexandrian interest in Hesiod as an important literary ancestor of and useful model for contemporary poets in the Hellenistic period. The fact that such an interest existed is undeniable. It is important, however, to be clear about just what kind of model Hesiod was. Scholars have tended to view the ascendancy of Hesiod in Alexandria as involving a corresponding devaluation of Homer. This is of course untrue. If Hellenistic poets regarded Homer as in a sense inimitable, they certainly did not refrain from trying. What they avoided was writing poetry of the scale and scope of the *Iliad* and *Odyssey;* but Homeric characters, actions, and language became the basic elements of most Hellenistic poetry. Of Hesiod this is not true. Neither *Theogony* nor *Works and Days* was never used, as the Homeric poems were, as an inexhaustible mine—or, to use the common ancient metaphor, ocean[10] —of poetic raw material. Instead, they offered formal models for epic on a small scale, thematic models for inquiries into the origins of things, and generic models for handling both sublimely grand (*Theogony*) and utterly quotidian (*Works and Days*) themes while avoiding Homer's "kings and battles." The most important Alexandrian proponent of "Hesiodic" principles was Callimachus of Cyrene.[11] The ideals that Callimachus upheld and took Hesiod to exemplify were later adopted by a group of influential Roman poets who flourished in the middle of the first century B.C.[12] Callimachus' position with regard to Hesiod and the reception of this ideology at Rome has become the focal point for much of the criticism of Latin poetry produced in (roughly) the last thirty years. Vergil, of course, has received a great deal of this critical attention, and his collection of *Eclogues* in particular is no doubt much better understood today than in the recent past. This understanding derives in

10. See Williams 1978.85–89, 98–99.

11. See in general Reinsch-Werner 1976.

12. The essential evidence is collected and discussed by Wimmel 1960; Clausen 1964 recognizes allusions to Callimachus in Ennius' *Annales* prologue (frr. 2–10 Skutsch 1985.70–71) and Catulus (fr. 1 Morel–Büchner 1982.55–56), but admits of no continuous or significant Callimachean influence on Latin poetry before the arrival of Parthenius of Nicaea (probably after 65 rather than in 73) as the prisoner of Cinna. On this question see pp. 296–299.

large part from a willingness to apply the critical categories developed from the study of Callimachus and of Alexandrian poetry in general. The same is true of the *Aeneid*, where however the situation is obviously more complex. Nevertheless, with regard to the *Georgics* in particular, there is now great support for the view that the phrase *Ascraeum carmen* refers, no less and perhaps more than to the poet of *Works and Days*, to Hesiod the ideal poet as conceived by Callimachus along with his Alexandrian and Neoteric followers.[13]

We have then two distinct interpretations of the phrase *Ascraeum carmen*, each of them imputing to Vergil a different view of his great predecessor, each with its appeal. It is too easy from our perspective to see in the literary history of antiquity a kind of determinism, a guiding hand that makes such an unlikely endeavor as Vergil's imitation of Hesiod seem a natural occurrence. In fact, it was anything but. It was, so far as we know and in all probability, completely unprecedented in Latin literature, which had down to Vergil's time been dominated by Homer, tragedy, New Comedy, and the Hellenistic poets. For a poet of Vergil's day to resuscitate the unusual form and often baffling sentiments that comprise *Works and Days*, and even to catch something of the archaic poet's tone of voice, was no small achievement.[14] We cannot deny that the proud phrase *Ascraeum carmen* refers, above all, to this fact. On the other hand, no reader of the *Georgics* can seriously maintain that the whole poem derives essentially from *Works and Days* alone. Even the most "Hesiodic" sections of the *Georgics* adhere strictly to the principles of stylistic refinement and allusive subtlety that we identify as the hallmarks of Alexandrianism. Recognizing this fact, we can hardly ignore the likelihood that the phrase *Ascraeum carmen* connotes the Hesiod of Callimachus' *Aetia* as well as the Hesiod of *Works and Days*.

The problem as I see it is that these differing views of Hesiod as the "model" of the *Georgics* continue to be presented as alternative rather than as compatible or at least related interpretations of Vergil's meaning. Let me then state proleptically that I believe these two views on the the nature of Vergil's Hesiodic imitation to be absolutely congruent: the phrase *Ascraeum carmen* summarizes Vergil's achievement in having seriously and convincingly imitated the authentic, rustic, and archaic Hesiod

13. Ross (1975.136), for example, takes *G.* 2.174–176 to indicate "the Hesiodic precedent, not simply of subject matter, but in the tradition of the poet's role and descent as it was conceived and reformed by Callimachus and Gallus and Virgil himself...."

14. On this point see pp. 314–317.

in a work that on the whole owes much more to those sophisticated, highly urbane Hellenistic and Roman poets who pointed to him, as much in the spirit of play as of polemic, as their literary and spiritual ancestor. Further, the position of the phrase has point: deployed near the beginning of *Georgics* 2, it signals the end of this ambivalent imitative program, which by the time the phrase occurs has already given way to a new one.

The proof of this position emerges from an analysis of the phrase *Ascraeum carmen* in its literary-historical context. The specific object of this investigation will be a particular conceit by which certain Greek and Latin poets refer to the works and to what I may call the literary-historical personae both of themselves and, especially, of their contemporaries and predecessors.

Antonomasia and Literary-Historical Personae

The phrase *Ascraeum carmen* is unique in the *Georgics* because it alludes to one of Vergil's literary models practically by name and indicates the poet rather than the poetry.[15] But similar references to specific poets and to particular genres and styles of poetry are fairly common in both Greek and Latin verse. In the *Eclogues*, for example, Vergil uses the phrases *Sicelides musae* (4.1), *Syracosio versu* (6.1), *pastoris Siculi* (10.50–51), *Ascraeo seni* (6.70), *Sophocleo coturno* (8.10), and *Chalcidico versu* (10.50) to specify the poetry or poetic personae of Theocritus, Hesiod, Sophocles, and Euphorion. Horace uses, among others, the expressions *Dircaeus cygnus* (*Carm.* 4.2.25), *Aeolium carmen* (3.30.13) and *Parios iambos* (*Epist.* 1.19.23). In Propertius we find *Aoniam lyram* (1.2.28), *Aeolio plectro* (2.3.19), *Philitaea aqua* (3.17.40), *Cyrenaeis aquas* (4.6.4), and many others. These references are not at all difficult to understand, and so have attracted little attention. But if we take the trouble to define them as a formal class and to inspect their history, we may find them more interesting and informative than they seem at first glance.

The phrases are formally consistent within a small range of variation. All consist of a noun modified by a proper adjective. The noun employed may mean "poetry," either literally (*Aeolium* ***carmen*** Horace *Carm.* 3.30.13, *Syracosio* ***versu*** Vergil *Ecl.* 6.1) or figuratively (*Cyrenaeas*

15. The reference to Lucretius at *G.* 2.490–492 (*felix qui potuit, etc.*) is so much less explicit that some scholars (e.g. Leach 1981) have doubted that it refers to Lucretius at all.

aquas Propertius *Carm.* 4.6.4). The noun itself may denote either a specific type of verse (e.g. *Parios* ***iambos*** Horace *Epist.* 1.19.23) or else poetry in general (*Maeonii* ***carminis*** Horace *Carm.* 1.4.2); but in either case the phrase gains specificity from the accompanying epithet, either a poet's *ethnikon* or an adjectival form of his name. Thus ***Parios*** *iambos* points to the invectives of Archilochus, ***Aeolium*** *carmen* to the melos of Alcaeus, ***Syracosio*** *versu* to the pastorals of Theocritus; ***Pindarici*** *fontis* (Horace *Epist.* 1.3.10) is the source of Pindar's inspiration, ***Pindarico*** *ore* (Propertius *Carm.* 3.17.40) his style. In all these cases reference is made to the work of an important poet from the past. Other cases refer not to the poetry, but to the poet himself—again either literally (*Ascraeo* ***seni,*** *Ecl.* 6.70) or figuratively (*Dircaeus* ***cygnus*** Horace *Carm.* 4.2.25).

The conceit is also found in Greek, in both imperial poets and their predecessors. Here we find the same basic range of variations as in Latin along with one or two others and some differences in preferred usage.[16] For instance, passages in which the noun that stands metaphorically for "poetry" or "poet" is replaced by the more direct ἀοιδός are much more common. Sometimes the noun becomes a simple ἀνήρ or ἄνθρωπος, or else is eliminated altogether. In either case, the poet's identity is again made clear by the accompanying epithet, which in this variation is, for obvious reasons, invariably the poet's *ethnikon* (as in the examples given above). In its most extreme form, our conceit is no different from the normal Greek toponymic idiom, and gains literary-historical significance from context alone.[17]

16. The type involving an adjectival form of the poet's name, however, which is so common in Latin, is practically unknown in Greek; in fact, I cannot cite a single example.

17. Thus when Theocritus speaks of Simonides (pace Bacchylides) as ἀοιδὸς ὁ Κήϊος (*Id.* 16.44), we may be sure that he uses the phrase to indicate the poet's honored place in literary history. On the other hand, Callimachus' Κήϊον ἄνδρα (*Aetia* 3, fr. 64.9 [Pfeiffer 1949.67]), also in reference to Simonides, is in a different class. It occurs as part of an imaginary sepulchral epigram, where reference to the dead man's home was conventional. On the popularity of such epigrams in the Hellenistic period, see Bing 1988.39–40, 57–65, 67–70.

This Greek idiom is practically unavailable to Roman poets. It appears only twice, and both examples illustrate its obviously exotic flavor in Latin verse and its lack of literary-historical reference. Ennius uses the phrase *Graius homo* in referring to King Pyrrhus of Epirus (*Ann.* fr. 165 [Skutsch 1985.85]), and Lucretius applies the same phrase to Epicurus (*DRN* 1.66). Pyrrhus, of course, was the first great foreign general that the Romans ever faced. When Ennius calls him *Graio patre Graius homo rex*, not only the repeated adjective but the foreign idiom underlines this fact. Lucretius seems to quote the phrase from Ennius

The phenomenon itself, then, is clear enough. The question is, for what purpose is this conceit used? What can the parallels tell us about its meaning in *Georgics* 2? There are two main points. First, the history of the motif before Vergil supports the notion that his use of it marks him as a poet in the Alexandrian mold; and this in turn makes it seem the more likely that he uses it to designate the symbolic Hesiod of the Alexandrians in addition to the actual archaic poet as models for his own poetic endeavors. Second, the use of such phrases by Vergil in the *Eclogues*, measured against the practice of his contemporaries Horace and Propertius, shows that the phrase marks the end of Vergil's Hesiodic program, thus signaling an intention to tap new sources of inspiration. I shall address each of these points separately.

Greek Predecessors

The affinity of this conceit with Greek idiom, the fact that it normally designates a Greek poet, and its frequent use by many of the Greek poets, all make it clear that the conceit was imported to Rome at some point. Its obvious artificiality and "learned" character suggests a Hellenistic origin. The conceit certainly was popular in Alexandria and contemporary centers of culture, as we shall see; but its beginnings actually date to the threshold between the archaic and classical periods.

Both Pindar and Bacchylides use phrases that appear to have been models for the Hellenistic and Augustan examples quoted above.[18] These phrases have the essential noun + proper epithet form and are found in programmatic contexts. They do not, however, designate another poet as a generic or stylistic model, but refer instead to the poet's own characteristic style, to his poetic persona, or to the tradition within which he works. This is especially true of Pindar's self-conscious references to his Dorian style. In rousing himself to sing the praises of Hiero in the first Olympian ode, the poet bids himself "take from its peg the Dorian lyre":

---

in order to suggest a similarity between Epicurus and Pyrrhus: both men are represented as formidable Greek invaders of Italy. Thus, far from presenting Epicurus as a litterateur, the allusion supports the military imagery with which Lucretius cloaks the philosopher's intellectual achievements (on which see Buchheit 1971).

18. This is not surprising, since Callimachus and other literary theorists borrowed a good deal of their critical terminology from the archaic poets, and then passed it on to the Romans.

ἀλλὰ Δωρίαν ἀπὸ φόρμιγγα πασσάλου
λάμβαν'....

Olympian 1.17–18

Here Δωρίαν refers to the Dorian mode to which Pindar's ode was sung. In naming this mode, Pindar emphasizes the dignity of all epinician poetry and especially that of the song that he offers to Hiero.[19] In a second passage, the poet glories in his accomplishment of combining harp and flute accompaniment with the Doric meter in his third Olympian ode:[20]

Μοῖσα δ' οὕτω ποι παρέστα
μοι νεοσίγαλον εὑρόντι τρόπον
Δωρίῳ φωνὰν ἐναρμόξαι πεδίλῳ

ἀγλαόκομον· ἐπεὶ χαίταισι μὲν ζευχθέντες ἔπι στέφανοι
πράσσοντί με τοῦτο θεόδματον χρέος,
φόρμιγγά τε ποικιλόγαρυν καὶ βοὰν
αὐλῶν ἐπέων τε θέσιν....

Olympian 3.4–8

Both these phrases find formal and conceptual parallels in the Augustan period, especially in Horace. For example, Pindar's Δωρίῳ πεδίλῳ is close to *Lesbium pedem* (*Carmen* 4.6.35);[21] with Δωρίαν φόρμιγγα compare *Lesboum barbiton* (1.1.34), *Lesbio plectro* (1.26.11), along with Propertius' *Aoniam lyram* (1.2.28) and *Aeolio plectro* (2.3.19). The Horatian examples in particular point to a specific melic tradition, beginning with Sappho and Alcaeus, of which the poet feels himself to be a part. It is a tradition identified with specific forms—the Aeolic meters and especially the

19. As noted at Σ *in O.* 1.26g (Drachmann 1903–1927.3.26 = fr. 67 Snell–Maehler 1975.70): περὶ δὲ τῆς Δωριστὶ ἁρμονίας εἴρηται ἐν παιᾶσιν, ὅτι Δωρίον μέλος σεμνότατόν ἐστι. Cf. *P.* 8.20.

20. Cf. fr. 191 (Snell–Maehler 1975.130) = Σ *in P.* 2.128b (Drachmann 1903–1927.2.53), with Bergk's supplement (see Snell's comment *ad loc.*) ⟨αὐλὸς⟩ Αἰολεὺς ἔβαινε | Δωρίαν κέλευθον ὕμνων.

21. This phrase refers specifically to the *Carmen Saeculare*, written in Sapphics. The fact that this hymn was actually sung strengthens the parallel between Horace's phrase and that of Pindar.

Sapphic and Alcaic stanzas in which two-thirds of the *Odes* are written—and with specific themes, especially erotic and civic themes. This is very close to the Pindaric conception of Doric melos as the type most appropriate to the poetry of praise.

Even closer to Augustan practice is Bacchylides' boastful prediction that men will sing his praises as the "honey-voiced Cean nightingale" (μελιγλώσσου τις ὑμήνσει | χάριν Κηΐας ἀηδόνος 3.97–98). Here antonomasia refers not to the tradition within which the poet works, but becomes a kind of honorific title assumed by the poet himself. Both Pindar and Bacchylides elsewhere use birds of various kinds as metaphors of their poetic personae.[22] Only here, however, does this conceit take the form of a kenning. As in the case of Pindar's "Dorian lyre" and "meter," moreover, this title too is paralleled in the references of later poets not to themselves, but to their predecessors: thus Pindar is called Δωρίας **ἀηδόνος** (Auct. inc. *AP* 15.27.4; cf. Horace's *Dircaeus* **cygnus,** *Carmen* 4.2.25), Anacreon ὁ Τήϊος **κύκνος** (Antipater Sidonius *AP* 7.30), and so forth. Because these passages parallel Bacchylides' boast so directly, some sort of indebtedness seems likely. On the other hand, I would not suggest that all such later examples necessarily derive from specific models. It is more reasonable to suppose that they reflect a widespread conceit among the early Greek melic poets of using kennings to put their seal on a poem or to designate their poetry in general, their individual poetic style. One need not assume that all archaic poets to whom formally similar references are attested in later times explicitly claimed these or like titles for themselves, or that they referred to their individual styles periphrastically. Once the formula had been established, new ones might be coined to suit new purposes. Thus some or all of the Hellenistic examples quoted above may be nonce formations based on an understanding of the conceit in general.

The generic topoi of the epinician ode may therefore be the origin of this conceit. There is, however, another possibility. While Pindar and Bacchylides use antonomasia only for self-reference, some archaic poets used similar forms to designate their predecessors as well. The *locus classicus* is a passage believed by later Greeks to carry the authority of Homer himself, the *sphragis* of the Homeric Hymn to Delian Apollo:

22. E.g. *N.* 3.80–81, 5.19–21, 8.19. See Steiner 1986.105–110.

χαίρετε δ' ὑμεῖς πᾶσαι· ἐμεῖο δὲ καὶ μετόπισθεν
μνσαθ', ὁππότε κέν τις ἐπιχθονίων ἀνθρώπων
ἐνθάδ' ἀνείρηται ξεῖνος ταλαπείριος ἐλθών·
"ὦ κοῦραι, τίς δ' ὔμμιν ἀνὴρ ἥδιστος ἀοιδῶν
ἐνθάδε πωλεῖται, καὶ τέῳ τέρπεσθε μάλιστα;"
ὑμεῖς δ' εὖ μάλα πᾶσαι ὑποκρίνασθαι ἀφήμως·
"τυφλὸς ἀνήρ, οἰκεῖ δὲ Χίῳ ἔνι παιπαλοέσσῃ
τοῦ πᾶσαι μετόπισθεν ἀριστεύσουσι ἀοιδαί."

[Homer] *Hymn* 3.166–173

It is Thucydides (3.104.4–6) who explicitly identifies the blind man of Chios as Homer. This identification, which has been challenged but without good reason,[23] is supported by subsequent literary references to Homer as ὁ Χῖος and by another archaic parallel as well. I refer to a well-known fragment ascribed variously to Semonides of Amorgos and to Simonides of Ceos:

ἓν δὲ τὸ κάλλιστον Χῖος ἔειπεν ἀνήρ·
"οἵη περ φύλλων γενεή, τοίη δὲ καὶ ἀνδρῶν."[24]

Simonides fr. 8 dub. (West 1972.114)

The quotation is from *Iliad* 6.146, and so for the scholar puts the identity of ὁ Χῖος ἀνήρ beyond doubt. But for its original audience, the phrase Χῖος ἀνήρ guaranteed the quotation, not, as today, vice versa. The poet quotes Homer as a figure of great repute. "The man of Chios," the τυφλὸς ἀνήρ of the Hymn to Apollo, was known to all. In both poems, antonomasia underlines the authority of the poet to whom it refers. There was no need to use his proper name.

These examples are not plentiful and their dating is insecure.[25] It nevertheless appears that self-referential antonomasia, such as Bacchylides'

23. E.g. by Wilamowitz 1916.368, 453, 456 and Allen–Halliday–Sikes 1936 *ad loc.*

24. For discussion see West.1974.179–180.

25. It is again Thucydides who provides the *terminus ante quem* for the *Hymn to Apollo*, which may therefore be a fifth-century composition. Most would, however, date it earlier: the argument of Burkert (1979.58–62) is rather attractive, although he adopts the minority view that the Delian and Pythian hymns comprise a unified composition (on which see especially Miller 1986). Janko (1982.99–115) sees the two halves of the hymn as separate compositions, but is in general agreement with Burkert as to the date of the Delian hymn. If the elegiac quotation is by Semonides, then the phrase Χῖος ἀνήρ can be dated well back into the archaic period; Simonides, on the other hand, is roughly contemporary with Pindar and Bacchylides.

Κηΐας ἀηδόνος, is no older than surviving "third-person" examples, such as Simonides' Χῖος ἀνήρ. There is no telling for sure which of these phenomena is original, which derivative: beyond this point relative chronology will not go.

I think it likely, however, that the idea of Homer as ὁ Χῖος ἀνήρ was the original or prototype of all subsequent examples of the conceit.[26] Homer is, after all, *the* poet—the only Greek poet, in fact, whose authority and popularity were such that we might expect him to be accorded such an epithet as "the man of Chios."[27] It seems to me more likely that Semonides or Simonides was using a common expression to introduce his Homeric aphorism, and that Pindar and Bacchylides modeled their own epithets on an existing cliché, than that an original lyric convention came later to be applied more broadly to Homer than to anyone else. Furthermore, if the scholiast on Pindar who explains the term Ὁμηρίδαι deserves any credence at all, then we must suppose that figures such as Cynaethus of Chios, whom the scholiast names as the actual author of the Hymn to Apollo, exploited the tradition that Homer was a Chian by birth as a means of strengthening their own reputation as *Homeridae*, i.e. descendants of Homer.[28] In view of the ancient controversy over Homer's true birthplace, the designation of Homer as ὁ Χῖος ἀνήρ would have suited their purpose admirably. Unfortunately, the earlier history of the motif cannot be traced; there is simply no evidence. Yet it is not implausible that what surfaces in the *sphragis* of the Hymn to Delian Apollo is just the end of a fairly lengthy, now lost tradition among the Chian *Homeridae*, one that regularly referred to Homer as ὁ Χῖος ἀνήρ. By the time of our extant sources, their promotion of this idea had been so successful that disinterested poets, such as Simonides, could also refer to Homer simply as "the Chian." On these grounds, I am tempted to conclude, at least provisionally, that ὁ Χῖος ἀνήρ was the original instance of literary-historical antonomasia, that it was used from a fairly early date, and that it amounts to an honorific title intended to connote the authority of the figure to whom it refers.

26. I have profited greatly from discussing this point with my friend Peter Bing.

27. One might compare the common English appellation of Shakespeare as "the bard of Avon."

28. Σ *in N.* 2.1c, 3.29.9 (Drachmann 1903–1927.3.29): Ὁμηρίδας ἔλεγον τὸ μὲν ἀρχαῖον τοὺς ἀπὸ τοῦ Ὁμήρου γένους, οἳ καὶ τὴν ποίησιν αὐτοῦ ἐκ διαδοχῆς ᾖδον· μετὰ δὲ ταῦτα καὶ οἱ ῥαψῳδοὶ οὐκέτι τὸ γένος εἰς Ὅμηρον ἀνάγοντες. On this and other questions pertaining to the provenance of the hymn see Burkert 1979.

The use of similar periphrases, whether in the first or third person, by other poets must have had much the same purpose. Simonides refers to ὁ Χῖος ἀνήρ as to an unimpeachable authority. Pindar mentions his Δωρίαν φόρμιγγα using much the same manner as when he addresses his Muse.[29] Bacchylides looks forward to being celebrated by future generations as Κήϊα ἀηδών. In all cases the use of antonomasia seems intended to connote the authority both of the designee and of the speaker, whether they are the same person or not. This hypothesis supports the primacy of ὁ Χῖος ἀνήρ in relation to other periphrases. The authority of Homer was supreme. Once the idea of his unique authority became attached to the expression Χῖος ἀνήρ, however, it was easy for other poets to claim authority for themselves or to ascribe it to others by inventing similar expressions.

In addition to those mentioned above, a further Bacchylidean example is of interest—one that refers not to Homer, but to Hesiod:

> Βοιωτὸς ἀνὴρ τᾶδε φών[ησεν, γλυκειᾶν
> Ἡσίοδος πρόπολος
> Μουσᾶν, ὃν ⟨ἂν⟩ ἀθάνατοι τι[μῶσι, τούτῳ
> καὶ βροτῶν φήμαν ἕπ[εσθαι.
>
> 5.191–194

The phrase Βοιωτὸς ἀνήρ parallels Simonides' Χῖος ἀνήρ precisely, both in its form and in the fact that it introduces a quotation from the poet it designates;[30] but several contextual features distinguish Bacchylides' phrase. Unlike Simonides' terse reference to Homer, this passage glosses the riddling antonomasia with the poet's name and, so to speak, title: "Hesiod, servant of the <sweet>[31] Muses." This formulation is a direct appeal to Hesiod's own statement that an ἀοιδός is the Μουσάων θεράπων (*Th.* 100)[32] and to his famous account of his initiation as a poet by the Muses themselves, who "taught him [a] beautiful song" (καλὴν ἐδίδαξαν ἀοιδήν *Th.* 22). Thus Bacchylides, in adducing Hesiod as an

29. Compare, for instance, *O.* 1.17–18 with 3.4–8.

30. The exact quotation does not occur in any of Hesiod's extant poetry, and is considered doubtful by Merkelbach–West 1967.172. Campbell 1967.433 compares Theognis 169 ὃν δὲ θεοὶ τιμῶσιν, ὁ καὶ μωμεύμενος αἰνεῖ. The question of authenticity, however, does not affect my argument: it is clear that Bacchylides himself did regard this gnome as Hesiodic.

31. This word is restored to Bacchylides' text on the basis of the parallel text in Hesiod.

32. Other occurrences of the phrase are noted by West 1966.188 *ad loc.*

authority, appeals to the same source of legitimacy for that authority as Hesiod had named on his own behalf, i.e. his privileged relationship with the Muses.

Hesiod was conceptually useful to later generations of poets primarily for just this reason: he made this relationship, the basis of his authority, quite explicit within the limits presented to him by convention. While the relationship between Homer and his Muse is evidently of the same sort, heroic epos does not allow its poets to establish the relationship explicitly through self-referential accounts of a symbolic initiation.[33] Probably because heroic epos lacked this convention, Homer's authority tended to be regarded by later generations as inherent, while Hesiod's could be traced back to a specific ἀρχή. Bacchylides expresses his understanding of this difference when he uses the phrase *πρόπολος Μουσῶν*. What is more, he underscores the distinction by casting his allusion to Hesiod in the form of a riddle. The word order of this passage in effect asks, "can you guess whom I mean by Βοιωτὸς ἀνήρ?" and then provides the answer, "Ἡσίοδος." This effect must have been heightened, if I am correct, by the fact that Βοιωτὸς ἀνήρ would have been understood as a deliberate variation on the more common Χῖος ἀνήρ, which Simonides had used in an extremely similar context. Thus by virtue of the similarity between these phrases, Bacchylides introduces the figure of Hesiod into his poem as, so to speak, "not-Homer." At the same time, he likens Hesiod's authority to that of Homer in degree, even if in no other way, by referring to Hesiod as Βοιωτὸς ἀνήρ, a phrase modeled on the one used by the *Homeridae* of Chios to connote the authority of their eponymous "ancestor," Homer, *ὁ Χῖος ἀνήρ*. Viewed in this light, Bacchylides' phrase takes its place in the long tradition linking these two great poets. This tradition, which extends throughout classical literature and is perhaps best represented by the *Certamen Homeri et Hesiodi*, is in essence not polemical, but celebratory, since it acknowledges these two as the pre-eminent poets of the Greeks. As such, the tradition embraces both Herodotus' assertion (2.53.2) that it was Hesiod and Homer who gave the Greeks their gods, and Ovid's humorous boast that he has attained the stature of both Hesiod and Homer.[34]

33. The first such account in heroic epos is that of Ennius (*Ann.* frr. 1–11 [Skutsch 1985.70–71]). On the *Dichterweihe* motif in general, see Kambylis 1965.

34. Laetus amans donat viridi mea carmina palma
praelata Ascraeo Maeonioque seni.
Ars Amatoria 2.3–4

The use of antonomasia to signal the authority of one's poetic predecessors was thus established by the end of the archaic period and continued to appear in classical times. Cratinus, for example, uses the phrase "Thasian brine" to designate his most important model for invective poetry, Archilochus of Thasos.[35] The conceit truly came into its own, however, in the Hellenistic era, when it became almost *de rigueur* to designate any poet by using antonomasia either in place of or in addition to his own name. The most outstanding witness to the conceit's popularity is also one of the very earliest surviving passages of Hellenistic verse. Hermesianax of Colophon produced, along with other more scantily attested works, three books of elegiac verse entitled *Leontion*, named after and addressed to his mistress. Of this poem's third book Athenaeus (13.597b) preserves a curiously amusing passage[36] that contains a catalogue of poets and philosophers—the greatest literary and intellectual figures of the classical past—and demonstrates that each of these great men had succumbed to the shafts of Eros. The poet's goal, of course, is to defend by appeal to *exempla* his passion for Leontion: if the greatest and most serious poets of the past have felt passion for a woman, why should he not admit his feelings and even make them the subject of his verse? The passage thus announces a point of view and employs a rhetorical stratagem that would become increasingly familiar in later times. What is interesting for our purposes, though, is the way in which Hermesianax designates his predecessors. Throughout the catalogue he shows a kind of obsession with the national origin of almost every man he names. This information has nothing to do with the apparent purpose of the passage; instead, the succession of *ethnika* serves a largely ornamental function.[37] Hesiod appears as "come from his Boeotian home" (21), which is then twice specified as Ascra (23–24). Ironically, he is not treated to antonomasia, as is practically everyone else mentioned in the poem. Orpheus is described as "imbued with the Thracian cithara" (2), Anacreon as "the man of Teios" (50), Sophocles

35. The reference occurs, appropriately, in the *Archilochi* (fr. 6 Kassel–Austin 1983.116):

> *εἶδες τὴν Θασίαν ἅλμην, οἷ' ἄττα βαΰζει;*
> *ὡς εὖ καὶ ταχέως ἀπετείσατο καὶ παραχρῆμα.*
> *οὐ μέντοι παρὰ κωφὸν ὁ τυφλὸς ἔοικε λαλῆσαι.*

On the significance of this passage and on the literary relationship between Cratinus and Archilochus, see Rosen 1988.40–49, especially pp. 42–43 and n. 22 on the identity of "the blind man."

36. Hermesianax fr. 7 Powell 1925.98–105.

37. This point is well brought out by Krevans 1983.

as "the Attic honeybee" (57), Philoxenus as "the man of Cythera" (69), Aristippus as "the man of Cyrene" (95).

It is obvious that Hermesianax is enjoying the conceit as an opportunity for rhetorical invention and learned display rather than borrowing examples of it from his predecessors. The popularity of antonomasia among other Hellenistic poets supports this observation. It is of course impossible to find uniformity of purpose in what became such a freely available conceit. I would argue, however, that the idea of ὁ Χῖος (ἀνήρ) as the prototype of all similar periphrases remained alive in the minds of the most sophisticated Hellenistic poets. It may even be that the Alexandrian use of the figure looked to a tradition of using antonomasia to designate Hesiod in particular as a poet of authority comparable to that of Homer himself. A pair of examples will serve to illustrate these two points.

The idea of differentiating a poet from Homer by means of antonomasia is clearly expressed in a self-referential epigram composed as a *sphragis* to Theocritus' *Idylls*. The speaker of the poem states emphatically that "the man of Chios is someone else," and that he is Theocritus of Syracuse:

> Ἄλλος ὁ Χῖος, ἐγὼ δὲ Θεόκριτος ὃς τάδ' ἔγραψεν
> εἷς ἀπὸ τῶν πολλῶν εἰμὶ Συρακοσίων,
> υἱὸς Πραξαγόραο περικλειτᾶς τε Φιλίννας·
> μοῦσαν δ' ὀθνείαν οὔτιν' ἐφελκυσάμαν.
>
> [Theocritus] *AP* 9.434

There are many questions about this epigram, the chief problem being the existence of a certain Theocritus of Chios.[38] Some have felt that the author of this poem wishes merely to distinguish the pastoral poet from this rather obscure figure.[39] This intention may be present; but it is difficult to believe in the light of other examples that a Greek poet could have used the expression ὁ Χῖος without expecting his readers to think first of Homer. Thus the epigram has two functions: namely, to distinguish the likeness of Theocritus and the collection of *Idylls* that accompanied it as the work of Theocritus of Syracuse, and to make a point concerning that poet's literary principles. The declaration that Theocritus came from "the masses of Syracuse" seems at first to clash

38. Mueller 1848.2.86–87; *RE* 5A 202 s.v. "Theokritos (2)."
39. So Gow 1952.2.549–550; but see Wilamowitz 1906.125.

with certain well-known programmatic utterances of Callimachus, it is true—even allowing for the fact that this information is presented as (auto)biography, not ideology. The idea is not, however, inconsistent with the deliberate lack of grandiloquence and general avoidance of pretension that is so much a part of Theocritus' style. More to the point, the epigram presents Theocritus as a *ζηλωτής* of no one: he did not appropriate another's Muse in writing his pastorals (line 4). This statement amounts to a modernist manifesto. It presents Theocritus as a new, original poet. Taken together with the opening disclaimer, "I am not the man of Chios," such a statement means, "I am not, like most, a mere epigone of Homer." It is this bold declaration—*ἄλλος ὁ Χῖος*, "Homer is somebody else, a style that is alien to me"—that gives meaning to the poem's final line and focus to its ideology.

The overtly homerizing poet, in Alexandrian terms, betrays a failure of imagination, and incurs the charge of hubristically challenging the greatest poet of all time. The intelligent poet's alternative is to become, like Hesiod, a servant of his own Muse. Thus to follow Hesiod instead of Homer is, paradoxically, to go one's own way.[40] This attitude appears clearly in another epigram, Callimachus' well-known salute to Aratus' *Phaenomena:*

> *Ἡσιόδου τὸ τ' ἄεισμα καὶ ὁ τρόπος· οὐ τὸν ἀοιδῶν*
> *ἔσχατον, ἀλλ' ὀχνέω μὴ τὸ μελιχρότατον*
> *τῶν ἐπέων ὁ Σολεὺς ἀπεμάξατο· χαίρετε, λεπταί*
> *ῥήσιες, Ἀράτου σύμβολον ἀγρυπνίης.*
>
> *Epigram* 27 (Pfeiffer 1953.88)

This little poem contains all the essential ingredients of the modernist orthodoxy practiced in third-century Alexandria. First, the epigram opens pointedly with Hesiod's name. Next, in mentioning Hesiod as Aratus' model, Callimachus enigmatically contrasts Homer—*τὸν ἀοιδῶν* | *ἔσχατον*—with the honey-sweet epos of Hesiod, *τὸ μελιχρότατον τῶν ἐπέων*. Finally Callimachus uses the *ethnikon* not of Hesiod, but of his follower, *ὁ Σολεύς*, employing a form of riddle, as Bacchylides had done in speaking of *ὁ Βοιωτός*, withholding Aratus' name until the final line of the poem. I would paraphrase the epigram as follows: "His style is that of Hesiod; he has imitated not the ultimate poet (Homer), but rather, I daresay, the sweetest of epic verses, has the man of Soloi; hail,

40. Bing 1988.83–89.

graceful utterances, emblem of Aratus' sleepless nights." The reference to Homer is oblique in the extreme, but the polemical nature of the epigram is established, I think, by the scholia to *Phaenomena*, which indicate that a controversy existed over whether Aratus was a follower of Hesiod or of Homer.[41] Callimachus must have written this epigram in order to claim Aratus for the Hesiodic camp.[42] Furthermore, while his method of doing so may have already become an Alexandrian commonplace, it is clear that he follows the tradition of, so to speak, "usurping" antonomasia from its "rightful owner," the man of Chios, to confer honor and authority upon some other poet. Coupled with the emphatic statement that Aratus wields the style of Hesiod himself, Callimachus' designation of Aratus as "the man of Soloi" seems to express a view similar to that of the Theocritus epigram discussed above—namely, that he is not the man of Chios, i.e. a "not-Homeric" poet.

It is paradoxical if this way of using antonomasia, having begun with the idea of Homer as ὁ Χῖος ἀνήρ, later became a way of distancing poets such as Hesiod and Aratus from Homer and the Homeric style. Yet in the hands of a true *doctus poeta* such as Callimachus, I believe that this is precisely what happened. It would, however, be going too far to claim that all examples of this conceit in the Hellenistic period share this motive for designating either Hesiod or alleged followers of Hesiod as non-Homeric poets. Not every poet who refers, let us say, to Philitas of Cos as ὁ Κῷος will have literary polemics in mind; the figure was appealing to poets on other levels as well. The same goes for any poet who alludes in such a way to one of the ancient classics—even Hesiod himself. It is not excessive to say that the standard way of alluding to Hesiod in Hellenistic poetry is not by his proper name, but by a phrase such as 'Ασκραῖος ἀνήρ, ποιμήν, etc. In fact, 'Ασκραῖος periphrases become so common in this period that they rival ὁ Χῖος ἀνήρ as the paradigm of the conceit. Because such phrases are so commonplace, they must have occurred spontaneously to many poets who were unconcerned with literary polemic. But I suggest that the sudden popularity of ὁ 'Ασκραῖος in Hellenistic times is due to the active promotion of Hesiod as a conceptual model by Callimachus and his followers. It is also probable that the almost constant use of antonomasia to designate

41. The controversy is attested by the *Isagoga bis excerpta in Arati Phaenomena* (Maass 1898.324, 326).

42. Callimachus' task was made easier by Aratus' signal imitations of Hesiod; see pp. 163–164.

Hesiod turned the conceit into a kind of code. The phrase ὁ ᾿Ασκραῖος refers to Hesiod in his aspect as a model of what the poet who does not slavishly imitate Homer can achieve. Phrases such as Callimachus' ὁ Σολεύς, used with ὁ ᾿Ασκραῖος in mind, refer to poets like Aratus as accomplished practitioners of an up-to-date, "Hesiodic" style of poetry.

The Neoterics

The next question is, how was this code passed on to the Augustans, who used it with such zest? There is justification for assuming that antonomasia along Alexandrian lines was cultivated by the Neoteric poets, particularly by Cinna, Catullus, possibly Cornificius, and Gallus.[43] More than this, it is possible to make an informed conjecture about the contexts in which these poets used the motif and about the significance which they attached to it. Finally, there is even good evidence to suggest that the use of the figure was promoted as an integral part of the deliberate Alexandrianizing movement that the Neoterics represent.

The earliest example of literary antonomasia in Latin occurs in a telling context, Cinna's version of the Callimachean epigram on Aratus' *Phaenomena*:

> haec tibi *Arateis* multum invigilata *lucernis*
> carmina, quis ignis novimus aetherios,
> levis in aridulo malvae descripta libello
> Prusiaca vexi munera navicula.
>
> Cinna fr. 11 (Morel–Büchner 1982.116)

The epigram is clearly a version rather than a translation of Callimachus, but the author has taken pains to put the identity of his model beyond doubt. The first line in particular recalls Callimachus (*multum invigilata lucernis* ≈ ᾿Αράτου σύμβολον ἀγρυπνίης) especially (for present purposes) in the matter of antonomasia (*Arateis lucernis* ≈ ὁ Σολεύς). Cinna's decision to imitate so closely the Callimachean epigram, an important manifesto of Alexandrian literary principles, amounts to a major ideological statement on his part, a declaration of allegiance to the principles followed by Callimachus and Aratus, the principles of both ὁ Σολεύς and

43. Cinna and Catullus, possibly along with Calvus, are identified by Lyne 1978b as the "Callimachean faction" of the Neoteric movement, while Cinna, Gallus and, again possibly, Calvus are singled out as those most likely to have been influenced by Parthenius.

ὁ Ἀσκραῖος. The fact that Cinna's version, much more clearly than the Callimachean original, was intended to appear at the front of a book roll containing *Phaenomena*[44] is in itself significant. Callimachus' poem, as the previously discussed scholia to *Phaenomena* indicate, is to be understood as a polemical utterance issued as part of a current literary controversy. It was probably written when *Phaenomena* was new and was just having its first impact upon the literary world. Cinna, on the other hand, could point to Aratus as a classic—not, of course, on the same order as Homer or Ennius or any of the Roman school classics, but as one of the great poets of an unusually fertile and comparatively recent period in Greek literature. Certainly, Aratus had already been well received in Rome; Cicero's translation was to be (presumably) only the first of many. At the same time, the act of presenting a friend, quite probably another poet, with a copy of Aratus inscribed in this way amounts to a prescriptive or programmatic statement, an artistic manifesto of sorts. By using the phrase *Arateis lucernis*, moreover, Cinna not only acknowledges ὁ Σολεύς as a poet in the mold of ὁ Ἀσκραῖος, but he actually improves on the original. In the Callimachean epigram, Ἀράτου ἀγρυπνίης cleverly hints not only at the nights which Aratus spent working on his poem, but at the nights he supposedly spent researching his material by scrutinizing the constellations in the night sky as well. Cinna's version embraces both these meanings[45] and, I would argue, one other: *Arateis invigilata lucernis* | *carmina* can be rendered "poems labored over by night at the light of (1) Aratus' lamps, (2) Aratus' stars, (3) Aratus' example."[46] If Cinna's *liber* contained a Latin translation of Aratus, then even a fourth meaning could be teased out of his phrase, "poems labored over and inspired by the light of Aratus' genius." The epigram could thus have been affixed to the version of Varro Atacinus or of some other third party, or even to one by Cinna himself. The point would be that the translator had been guided by the same principles as Aratus in producing the *Aratea* which Cinna was presenting to his friend. Even if the Latin epigram accompanied an original Greek *Phaenomena*, however, my point is essentially the same. Cinna's poem testifies to a celebration among friends of the artistic ideals that they shared. The phrase *Arateis lucernis*—a tangible link with the ideology of

44. But why the plural *carmina?* Does Cinna refer to *Phaenomena* and *Diosemiae* as two distinct poems? On this issue see p. 163 n. 56.

45. Despite Gow–Page 1965.2.209.

46. Cf. Varro *DLL* 5.9 *quod…ad Aristophanis lucernam…lucubravi.*

the Alexandrians—stands as a prominent programmatic declaration of these ideals.

The figure was known to Catullus as well. In his poem to Cornificius (*Carmen* 38) he complains that his love life is making him depressed.[47] In these straits, he asks his friend for consolation. Cornificius, of course, is himself a poet, one of Catullus' stripe, to judge by the scraps of his work that survive, and from the very fact that Catullus honors him with this request. What sort of poem does Catullus want? Something that matches his lugubrious mood—something *maestius* ***lacrimis Simonideis*** (8). The basic meaning is clear: Cornificius must outdo the poet of lament *par excellence*. Of course, *aemulatio* of Simonides itself does not imply anything about Alexandrian literary principles. But phrasing the request in this way, i.e. by means of antonomasia, is perfectly in line with Catullus' well-known adherence to Callimachean literary principles.

The next appearance of the motif again involves two Roman poets and their Greek predecessors. In *Eclogue* 10, the character Gallus, shaken by his unsuccessful love affair with Lycoris, resolves to abandon erotic elegy in favor of Theocritean pastoral:

> ibo et *Chalcidico* quae sunt mihi condita *versu*
> carmina ***pastoris Siculi*** modulabor avena.
>
> *Eclogue* 10.50–51

Chalcidico versu refers of course to the *Amores*, erotic elegies inspired by Euphorion of Chalcis, and *pastoris Siculi* is Theocritus, the natural model for Gallus' new bucolic project.[48] By itself this passage admits of no very sweeping conclusions. It discloses rather more information, however, in combination with the following lines of *Eclogue* 6:

> "hos tibi dant calamos (en, accipe!) Musae,
> *Ascraeo* quos ante *seni*, quibus ille solebat
> cantando rigidas deducere montibus ornos.
> his tibi Grynei nemoris dicatur origo,
> ne quis sit lucus quo se plus iactet Apollo."
>
> *Eclogue* 6.69–73

47. I assume that *meos amores* (6) has the usual idiomatic meaning (= *meam puellam*).
48. As noted by Servius *ad loc.* (Thilo 1887.125).

The addressee of these lines is of course Gallus again, the speaker Linus, the *divino carmine pastor* (line 67) who presides over Gallus' *Dichterweihe*. Both these passages have provoked rich and varied commentary, but in the present connection only a few points need mentioning: (1) both passages concern Gallus and his poetry; (2) both are programmatic in that they purport to define in advance the nature of a literary project; (3) both use antonomasia to refer to specific literary models; (4) all the examples of antonomasia found in these lines designate poets who are in some way connected to the Callimachean school. David Ross has argued at length that the "poetic genealogy" outlined in the *Eclogue* 6 as well as the pastoral element announced in *Eclogue* 10 must have been present in Gallus' poetry when Vergil wrote these lines; further that Gallus probably made his literary ambitions and ideals quite explicit, using the full panoply of Callimachean motifs and terminology to meditate openly upon the course that his literary career would take.[49] If this is so, it may be necessary to add antonomasia to the repertory of Alexandrian literary motifs introduced to Latin poetry by Gallus and his colleagues. In particular, the imitation of Callimachus' Aratus epigram by Cinna, Catullus' poem to Cornificius, the use of the phrase *Ascraeus senex* in the context of a highly symbolic, demonstrably Callimachean *Dichterweihe* which may itself have been derived from Gallus, and the occurrence of similar periphrases in the mouth of "Gallus" in *Eclogue* 10—all the evidence points in this direction. It seems highly likely that Cinna, Catullus, possibly Cornificius, and Gallus, perhaps along with their Neoteric associates, were the first Roman poets to use antonomasia in the Alexandrian fashion as a conceit expressive of their learned modernism. It even seems possible that they used it, at least on occasion, as did Callimachus in his Aratus epigram, as a component of their literary manifesto.

The case of Gallus depends, of course, on how closely Vergil is reporting Gallus' own words in the *Eclogues*. But whether Gallus discussed his own poetry in terms of antonomasia, or it is Vergil himself who imposes the conceit on Gallus, our conclusion must be basically the same. In the *Eclogues* we find Vergil using antonomasia in essentially the same way as did Callimachus, Cinna, and the rest. It seems likely that his Neoteric predecessors formed the conduit by which the learned Alexandrian use of the motif reached him; and this observation brings us to our initial

49. Ross 1975.34–37.

conclusion: that in the *Georgics*, when Vergil uses the phrase *Ascraeum carmen*, he does indeed refer to his poem not only as an imitation of Hesiod in particular, but as a "Hesiodic" song in a symbolic sense as well.

Augustan Usage

Programmatic antonomasia goes on to thrive in the Augustan period, when its chief practitioner is Propertius. It is he of all the Augustans who is most given to making programmatic statements again and again at virtually any point in a given poem or cycle of poems. One of his favorite means of doing so is to employ antonomasia, with which his text is thickly strewn: one finds *Aoniam lyram* (*Carmen* 1.2.28), *Orpheae lyrae* (1.3.42), *Aeolio plectro* (2.3.19), *Haemonio equo* (2.10.2), *Graios choros* (3.1.4), *Mygdoniis cadis* (4.6.8), and many others. Obviously a phrase such as *Graios choros* in itself lacks any reference to a particular writer or even genre. Yet even within the broad spectrum of delimiting terms that Propertius uses in periphrases, a certain consistency can be found. When he advises his (probably fictitious) friend Lynceus

> desine et *Aeschyleo* componere verba *cothurno*,
> desine, et ad mollis membra resolve choros,
>
> *Carmen* 2.34.41–42

one needn't infer that "Lynceus" was a translator of Aeschylus, or even that he wrote tragedies. Indeed, this couplet is better understood with reference to lines 25–32:

> Lynceus ipse meus seros insanit amores!
> Solum te nostros laetor adire deos.
> Quid tua *Socraticis* tibi nunc sapientia *libris*
> proderit aut rerum dicere posse vias?
> Aut quis *Erecthei* tibi prosunt *carmina* lecta?
> Nil iuvat in magno vester amore senex.
> Tu satius memorem Musis imitere Philitan
> et non inflati somnia Callimachi.

Despite the specific tragic themes listed in the intervening lines (33–40), the phrase *Aeschyleo coturno* stands revealed as essentially a synonym of

Socraticis libris and *Erecthei carmina.* Its specificity sharpens the contrast between the austere classical style that it designates and the more delicate (N.B. *mollis...resolve* 42) style favored by Callimachus and Philitas. It is this style or approach to poetry—the one practiced by the poet himself—that is embodied in his love of such periphrases. We need not worry that many other examples, such as the aforementioned *Aeschyleo cothurno* and *Erecthei carmina,* denote a style presented as antithetical to that of Propertius and his Alexandrian models. The mere use of the conceit in this programmatic context signals its ideological character.

Particular forms and genres are less important to Propertius than attitude and style. Thus his early programmatic statements lack all specificity (*Aoniam lyram* 1.2.28 and *Orpheae lyrae* 1.3.42, for example). Through antonomasia he attains a pregnant ambiguity of reference while he avoids committing himself to particular formal models. Later references, however, including the passage discussed above, but especially those found in Books 3 and 4, identify much more openly the poet's preferred exemplars, namely, Philitas and, above all, Callimachus. For example, Book 3 opens with a solemn invocation of these two revered figures:

> Callimachi manes et *Coi* sacra *Philitae,*
> in vestrum, quaeso, me sinite ire nemus.
> Primus ego ingredior puro de fonte sacerdos
> ***Itala*** per *Graios* ***orgia*** ferre *choros.*
> Dicite, quo pariter carmen tenuastis in antro,
> quove pede ingressi, quamve bibistis aquam?
>
> *Carmen* 3.1.1–6

Cf. *Carmen* 4.6.3–4:

> ***Serta*** *Philiteis* certet ***Romana*** *corymbis*
> et *Cyrenaeas* ***urna*** ministret *aquas.*

Carmen 3.3 goes even farther by developing an initiation scene, clearly based on the prologue to the *Aetia,* in which Propertius' Callimachean ideology is made even more explicit. The poet, *molli recubans Heliconis in umbra* (1), dreams that he has tried to drink from the same stream as "thirsty father Ennius," i.e. to attempt a *carminis heroi...opus* (16). In the following lines (17–18) Apollo stops him, noting that

> "Non hic ulla tibi speranda est fama, Properti:
> mollia sunt parvis prata terenda rotis."

Calliope later appears to amplify this theme (41–42):

> "Contentus niveis semper vectabere cycnis,
> nec tibi fortis equi ducet ad arma sonus.
> Nil tibi sit rauco praeconia classica cornu
> flare, nec *Aonium* tingere Marte *nemus*.

The poem ends with a chastened Propertius saying (51–52):

> talia Calliope, lymphis a fonte petitis,
> ora *Philitea* nostra rigavit *aqua*.

The antonomasia *Aonium nemus* points to no particular genre or poet, but rather to an ideology that excludes the epic genre and heroic themes. Although one can assume that Propertius' own elegies provide an example of what he means by *Aonium nemus*, the phrase is not (like *Philitea aqua*) coextensive with the elegiac form or with amatory themes. The Aonian grove is part of the symbolic literary topography of Helicon as it was fixed by Callimachus in the *Aetia* with reference to Hesiod's initiation scene in *Theogony*. By implying a connection between himself and the Aonian grove, Propertius is simply representing himself as a "Hesiodic," i.e. "Callimachean," poet. The same basic motive lies behind his other periphrases as well. Thus for Propertius, at least, antonomasia of this type functions generally as a kind of badge that marks its wearer as a *poeta doctus*. This attitude is like the one that we have observed in the Neoterics and in Callimachus himself. It is markedly different, however, from the practice of Propertius' own contemporaries.

Horatian antonomasia shares with the Propertian variety two more or less regular features: first, an affinity for programmatic contexts; second, an orientation toward Callimachean poetics. By way of excusing himself from composing a *Vipsaniad*, Horace informs Agrippa:

> Scriberis Vario fortis et hostium
> victor ***Maeonii carminis*** *alite*....
>
> *Carmen* 1.6.1–2

Similarly, when he begins his second book of odes with the banishment of *Ceae neniae* (2.1.38), Horace expresses his preference for lighter themes than those of his first book, which he likens to the solemn compositions of the lamentation poet par excellence, Simonides of Ceos.[50] In both these cases Horace defines his present project by contrasting it with a rejected program, which he designates by using antonomasia. This procedure resembles that of Propertius when he dispenses advice to "Lynceus." At the same time, there are two notable differences between the two poets' ways of handing the motif. First, although Propertius is not incapable of being quite specific when using antonomasia, he prefers generality. Most of the epithets that he uses do not concern real poets or poems, but rather an idealized mythical poetic realm inhabited by figures such as Orpheus; nor does he shy away from such extremes of vagueness as *Graios choros*. Here vagueness is not an impediment to understanding, but may even be construed as an advantage. The apparent clarity of phrases such as *Aeschyleo coturno* is at any rate entirely specious. Here Propertius' concern is demonstrably not with Aeschylus himself nor with tragedy per se, but rather with the high-minded style and approach to literature that these ciphers represent. This attitude is even more apparent in Propertius' assertion that Apollo has forbidden him *gracilis contemnere Musas* and ordered him *Ascraeum habitare nemus* (*Carmen* 2.13.3–4). In claiming to inhabit the Ascraean grove, Propertius cannot mean that he is a follower of Hesiod as a didactic, agricultural, or theogonic poet: his reference is clearly to Hesiod as ὁ 'Ασκραῖος, the symbol of Callimachean poetics. Literary genre is simply not a consideration; in all such passages, it is ideology that really matters. Horace, on the other hand, always preserves generic specificity: *Maeonii carminis alite* means Homeric epos, precisely what Callimachus and the Neoterics are said to have prided themselves in not writing. This procedure is typical of Horace, who refers periphrastically to his own poetry with great precision as well; thus *Parios iambos* means epodes in the manner of Archilochus, *Aeolium carmen* lyrics modeled on those of Alcaeus, etc. These references are not to be located at the Callimachean and non-Callimachean poles of a single axis, as are those of Propertius, but instead within a matrix of discrete genres, each one retaining its individuality.

In addition to specificity, Horace differs from Propertius in what may be called perspective. Propertius frequently uses antonomasia to represent an unrealized project or one that is under way, but not yet

50. Cf. Catullus' *lacrimis Simonideis* discussed on p. 48.

complete. Horace, on the other hand, normally applies the figure only to a program that has been fulfilled or to an exemplar from the past. One of the major themes of the *Odes* is of course Horace's *aemulatio* of Sappho and, especially, Alcaeus. This theme is sounded programmatically, but with almost ostentatious diffidence, at the end of the first poem of the collection:

> me gelidum nemus
> nympharumque leves cum Satyris chori
> secernunt populo, si neque tibias
> Euterpe cohibet nec Polyhymnia
> *Lesboum* refugit tendere *barbiton.*
>
> *Carmen* 1.1.30–34

The theme recurs in poem 1.26, again in a plea to the Muses to lend the charm of the Lesbian poets to Horace's song:

> Nil sine te mei
> prosunt honores: hunc fidibus novis,
> hunc *Lesbio* sacrare *plectro*
> teque tuasque dicet sorores.
>
> *Carmen* 1.26.8–12

Throughout the *Odes* Horace uses antonomasia to distance himself slightly from his own ambition. Not, significantly, until the final poem of the cycle does he actually lay claim to the title that he has earned:

> dicar...
> . . . .
> ...ex humili potens
> princeps *Aeolium carmen* ad ***Italos***
> deduxisse ***modos.***
>
> *Carmen* 3.30.10–14

Later, in the fourth book, Horace speaks freely and almost boastfully of his achievement (e.g. *Carm.* 4.6.35 *Lesbium pedem*, with reference to the *Carmen Saeculare;* 4.3.12 ***Aeolio carmine*** *nobilem*), even going so far as to coin a new example to express his pride (4.3.23 ***Romanae*** *fidicen* ***lyrae***). In the *Epistles* he refers with similar confidence to the successful program of imitation that he followed in the *Epodes:*

Parios ego primus *iambos*
ostendi Latio, numeros animosque secutus
Archilochi, non res et agentia verba Lycamben.

Epistula 1.19.23–25

The same policy obtains in cases where Horace speaks of some other poet's accomplishments. Antonomasia normally relegates to the past the endeavor that it designates. This is self-evidently the case in those passages that concern Sappho and Alcaeus. It is also true of the Roman poets who successfully imitated particular Greek models. I have already mentioned Horace's designation of Varius as ***Maeonii carminis*** *alite* (*Carmen* 1.6.1–2), where the phrase *Maeonii carminis* stands not as a symbol of Varius' ambition or intention to rival Homer, but rather as an acknowledgement that he had already done so and had acquitted himself with distinction. Elsewhere Horace uses antonomasia in a slightly different way to connote the presumption of a certain Titius:

Pindarici fontis qui non expalluit haustus,
fastidire lacus et rivos ausus apertos.
Ut valet? Ut meminit nostri? **Fidibus**ne **Latinis**
Thebanos aptare ***modos*** studet auspice Musa,
aut tragica desaevit et ampullatur in arte?

Epistula 1.3.10–14

Significantly, when Horace discloses his own Pindaric ambitions, he studiously avoids antonomasia; whereas Propertius writing in the same vein deliberately seeks it. The two following passages clearly illustrate the poets' contrasting practices:

Pindarum quisquis studet aemulari,
Iule, ceratis ope Daedalea
nititur pennis vitreo daturus
nomina ponto.

Monte decurrens velut amnis, imbres
quem super notas aluere ripas,
fervet immensusque ruit profundo
Pindarus ore.

Horace *Carmen* 4.2.1–8

Haec ego non humili referam memoranda coturno,
qualis *Pindarico* spiritus *ore* sonat.
Propertius *Carmen* 3.17.39–40

Thus while Propertius appears to preserve the regular Neoteric attitude toward the use of literary antonomasia, Horace seems to follow a more circumscribed approach, one that treats antonomasia not merely as a way of advertising his Callimachean literary principles in general, nor as a suitable means of announcing his unrealized literary ambitions. Rather, he uses the figure to designate a specific Greek model in order to indicate a standard of comparison for a work that has already been completed. work. Particularly noteworthy is his habit of using the figure to look back on his own *past* achievements from the perspective of several years even as he is engaged on a new project, as he does in speaking of Archilochus' *Parios iambos* in the *Epistles*, or to signal the successful completion of an imitative program, as he does in referring to Alcaeus' *Aeolium carmen* in *Carmen* 3.30.

Although Vergil uses antonomasia much less frequently than do Horace and Propertius, there are enough examples to show that his practice is equally precise as that of Horace and much more so than that of Propertius. In referring to contemporaries, Vergil follows the same path as Horace does in calling Varius ***Maeonii carminis*** *alite* (*Carmen* 1.6.2). For example, he addresses *Eclogue* 8 to Asinius Pollio[51] with these words:

Tu mihi seu magni superas iam saxa Timavi,
sive oram Illyrici legis aequoris, en erit umquam
ille dies, mihi cum liceat tua dicere facta?
En erit ut liceat totum mihi ferre per orbem
sola *Sophocleo* tua carmina digna *coturno?*
A te principium, tibi desinam: accipe iussis
carmina coepta tuis, atque hanc sine tempora circum
inter victricis hederam tibi serpere lauros.
Eclogue 8.6–13

The phrase *Sophocleo coturno* in this passage refers unambiguously to Pollio's attainments in the field of tragedy, where the actual imitation or

51. Servius *ad Ecl.* 8.6 (Thilo 1887.93), whose chronology of the *Eclogues* was badly confused (Zetzel 1984), thought that the addressee was Augustus. The identification of Pollio is found in Servius Danielis *ad Ecl.* 8.10 and 12 (Thilo 1887.93–94) and Junius Philargyrius *ad Ecl.* 8.6, 10, and 12 (Hagen 1902.143–146), and was argued by Scaliger 1606 *ad annum MMXX*, whence it became the standard view (Van Sickle 1981.21). The

translation of Sophoclean exemplars may be inferred. Contrast Propertius' injunction to "Lynceus," in which the similar phrase *Aeschyleo coturno*, which was probably inspired by Vergil, is equivalent to the commonplace *Socraticis libris* and even to the highly unspecific *Erecthei carmina*. Again in *Eclogue* 10 Vergil has Gallus acknowledge the model of his *Amores* by means of antonomasia, as we have seen. In addition to their specificity, these passages share the retrospective quality also noticed in Horatian periphrases. In all cases antonomasia is used to mark the accomplishments, rather than the ambitions, of a poet, either Pollio, Gallus, or Vergil. In *Eclogue* 8, Vergil's words make no sense unless one is to understand that Pollio has already attained some measure of success as a tragedian, just as Horace's *recusatio* to Agrippa is absurd unless Varius can be assumed to have distinguished himself as a writer of heroic epos. Horace himself in another passage confirms both suppositions:

Pollio regum
facta canit pede ter percusso; forte epos acer
ut nemo Varius ducit.

Sermo 1.10.42–44

Vergil's *Sophocleo coturno*, like Horace's *Maeonii carminis*, exemplifies antonomasia used to designate a poet's past accomplishments. The contexts of both phrases show the Alexandrian and Neoteric ancestry of the figure. Each occurs in a *recusatio:* Horace contrasts his lyric poetry with Homeric epic as practiced by Varius, while Vergil implicitly differentiates his pastorals from the Sophoclean tragedies of Pollio. To this extent Vergil and Horace agree with Propertius in *Carmen* 2.34, where antonomasia denotes a kind of poetry that is introduced as a foil for the present or proposed kind. But Propertius, who (as I have said) is probably imitating Vergil directly here, speaks very loosely both of the foil and of its practitioner, "Lynceus." Vergil and Horace are quite clear in each case about the foils and their practitioners. In Horace's *Carmen* 1.6 the poet in question is actually named. In *Eclogue* 8, Vergil supplies enough detail about his addressee—including the fact that he was a distinguished tragedian—to identify him. The *recusatio* formula demands a foil for the poem that is actually being written. Vergil and Horace do not abuse the

---

attempt of Bowersock 1971 to redate the poem to 35 instead of 39 and to identify its addressee as Octavian has found some favor, but is in my view seriously weakened by Bosworth 1972. The case for Octavian has recently been reargued by Mankin 1988. I intend to state more fully my own views on the matter at a later date.

conventions of antonomasia by alluding to vaguely defined or imaginary poetry; they allude clearly and specifically to the actual work of a real poet.

The same pattern can be observed in Vergil's allusions to Theocritus, his own chief model in the *Eclogues*. When speaking of the *Eclogues*, Vergil twice uses antonomasia in programmatic contexts:

> *Sicelides musae*, paulo maiora canamus!
> Non omnis arbusta iuvant humilesque myricae;
> si canimus silvas, silvae sint consule dignae.
>
> *Eclogue* 4.1–3

> Prima *Syracosio* dignata est ludere *versu*
> nostra neque erubuit silvas habitare Thalea.
> Cum canerem reges et proelia....
>
> *Eclogue* 6.1–3

Both periphrases point clearly to the pastoral genre and to Theocritus, its founder. But despite this obvious relationship to Vergil's bucolic program, both passages actually signal temporary departures from that program. Most of the individual *Eclogues*—particularly poems 2, 3, 5, 7, 8, and 10—borrow structural elements and translate passages from the *Idylls* without openly acknowledging their source. In *Eclogue* 3, for instance, Vergil indicates his debt to Theocritus via an introductory quotation:

> <Menalcas> Dic mihi, Damoeta, cuium pecus? an Meliboei?
> <Damoetas> Non, vero Aegonis; nuper mihi tradidit Aegon.
>
> *Eclogue* 3.1–2

> ⟨Βάττος⟩ Εἰπέ μοι, ὦ Κορύδων, τίνος αἱ βόες; ἦ ῥα Φιλώνδα;
> ⟨Κορύδων⟩ Οὔκ, ἀλλ᾽ Αἴγωνος· βόσκειν δέ μοι αὐτὰς ἔδωκεν.
>
> *Idyll* 4.1–2

What follows in each case is an amoebean singing contest between the first two speakers judged by a third party. Having thus established the Theocritean background of his poem, Vergil leaves himself free to vary subsequent details while requiring the reader to assess such variations as deliberate departures from a specified model. The point, however, is that Vergil's allusion to Theocritus remains tacit and so can be inferred only by a knowledgeable reader who must then realize that the whole poem is to be read as an imitation of Theocritus. By way of contrast, Vergil opens *Eclogues* 4 and 6 with patent references to his Theocritean program. But in his note on *Syracosio* (*Eclogue* 6.1) a recent commentator shrewdly remarks that "As in 4.1, Vergil delights in recalling the Theocritean origin of the genre at the start of an un-Theocritean piece."[52] There are in fact, as the same commentator states elsewhere, "virtually no Theocritean echoes" in the poem.[53] Instead both these *Eclogues* draw on a variety of other poets for thematic material.[54] Moreover, this material may on the whole fairly be characterized as anything but bucolic. Despite enough generally pastoral accouterments to preserve a scenic continuity with the rest of the cycle, both poems greatly expand and threaten to transgress the boundaries of the genre as they had been set by Theocritus. Thus the periphrastic allusions to Theocritus that begin the poems, if read as conventional programmatic statements, are misleading. Both passages actually presuppose Vergil's pastoral achievement. As such, they are entirely in line with Horace's retrospective periphrases *Aeolium carmen*, *Parios iambos*, *Maeonii carminis*, etc. Furthermore, insofar as they introduce poems that explore areas outside the traditional limits of pastoral poetry, they parallel Vergil's own use of *Chalcidico versu* to describe Gallus' verse at a point when he is portrayed as contemplating a thematic expansion of his elegies to include bucolic material. In short Vergil, like Horace, uses the figure of antonomasia with both generic and specific allusive intent, in retrospective rather than anticipatory programmatic contexts, and with the additional irony that the imitative activity so designated is effectively abandoned in favor of a new project which assumes these prior attainments, but which also greatly expands and builds upon them.

Conclusion

It is in the light of this evidence that Vergil's use of *Ascraeum carmen* at *Georgics* 2.176 takes on its true appearance. Similar periphrases in the *Eclogues* strongly suggest that here too the phrase refers not to Vergil's plan for the *Georgics* as a whole, but rather to an imitative program that has already been completed. If the parallelism is exact, the use here of

52. Coleman 1977.173.
53. Coleman 1977.203.
54. For further discussion of these poems see pp. 289–314.

antonomasia relegates this accomplishment to the past while signaling Vergil's intent to tap new sources of inspiration. This interpretation fits exactly the contrast between Greek and Roman elements in this passage upon which I have remarked; for, in the highly Roman context of the "*Laudes Italiae,*" Vergil has already begun, with notable effect, to draw upon native Italian models. It is into this context that Hesiod, Vergil's abandoned Greek model, intrudes himself via the exotic antonomasia *Ascraeum carmen.*

At this point we should recall our first conclusion, that Vergil's use of *Ascraeum carmen* refers to his imitation not only of the authentic Hesiod, but of the symbolic, Alexandrian one as well. Here we must ask whether Vergil means to bid both Hesiods farewell, and in what sense. To answer these questions it will be necessary to take a broader view of Vergil's goals in the *Georgics* and, especially, to examine more fully his use of literary sources in general. This will be the business of the following chapter.

CHAPTER 3

Vergil's Allusive Style

Vergil's Models

VERGIL HAS ALWAYS enjoyed a reputation for wide reading in all sorts of literature, thanks in part to the testimony of ancient scholars, but mainly to the many verbal echoes of previous Greek and Latin literature that have been found in his works. The authors that he knew and used extend from the most celebrated literary classics to genres of technical writing virtually unknown to other poets. But despite of his impressive range, there can be no doubt that his use of the epic tradition is an especially important element of his style.

By the criteria familiar to Vergil, the epic tradition consists of poetry in the epic meter, the dactyllic hexameter of Homer, Hesiod, Theocritus, Aratus, Ennius, and Lucretius. Although some moderns are uncomfortable in referring to such a diverse group of poets under the single rubric of "epic," this is a common practice among poets and critics of the Hellenistic and Roman periods.[1] With them, we may distinguish the heroic, the didactic, the bucolic, and the satiric modes of epic,[2] and may even go so far as to refer casually to the heroic mode simply as "epic" par excellence.[3] But when we consider the ancient view, essentially confirmed by modern scholarship, of how each of these sub-genres developed, we find that a common term serves to express the fundamental

1. See Manilius 2.1–52 and Quintilian *IO* 1.46–58, with the discussions of Steinmetz 1964.459–463 and Koster 1970.135–137.

2. As does Horace *Sermo* 1.10.43–48, correlating differences in content with appropriate differences in style: see Fraenkel 1957.130–132; Van Sickle 1978.114–115. For "didactic" (or "didascalic") as a critical category, see, e.g., Servius *praefatio in G.* (Thilo 1887.129.9) and Diomedes (Keil 1857.1.482–483). On "bucolic" as a mode of epic, see Halperin 1983, especially Part IV, pp. 193–248.

3. This tendency can be observed in Aristotle (see Koster 1970.42–80), which helps to explain its ascendancy in the modern period.

relationship among them. The genres themselves, in ancient theory, either antedated Homer or were invented by him; at any rate, his poetry subsumes them all.[4] Later epic poets, including all of those mentioned above, developed their own styles and, in some cases, "discovered" new epic sub-genres by being the first to treat them as discrete poetic categories. Thus, while Homer remained for the Greeks their greatest teacher, Hesiod was distinguished as the inventor of didactic poetry.[5] Similarly Theocritus, working in the bucolic mode, modeled his *Idylls* on well-chosen passages of Homer.[6] When we consider the matter in this way, it becomes clear that the idea of epic is no mere procrustean category, but a useful means of referring to an organically unified, if richly diverse, poetic tradition.

Ancient critics stress Vergil's deliberate, self-conscious involvement with this tradition. Despite the occasional influence of tragedy and other genres, epic is the only meter in which the mature Vergil worked, and the earlier epic poets are his most important models.[7] In the case of the *Eclogues* and the *Aeneid*, modern scholarship has gone far beyond the ancient testimony to disclose how meticulously Vergil worked out a relationship between his own poetry and his chosen epic models, Theocritus and Homer. What has become more and more clear in recent years is that Vergil does not indulge in superficial allusion to these authors occasionally or at whim in order to lend his own work a merely decorative Theocritean or Homeric color. Rather, he uses allusion more or less constantly throughout these works in order to create a complex intertextual relationship between model and imitation. By exploring this relationship, the reader participates in a creative dialogue, through which

4. Aristotle, for example, cites Homer as the earliest and greatest poet of both tragedy (the *Iliad* and the *Odyssey*) and comedy (*Margites*) at *Poetics* 1448b–1449a. For Homer as the source of all literature, see Williams 1978.87–89, 98–99; Cairns 1989.150.

5. See Koster 1970.10, 100, 127–128

6. Di Benedetto 1956; Van Sickle 1975 and 1976; Halperin 1983. On the epic character of the *Eclogues* see Schmidt 1972.38–45, 282–298; Van Sickle 1978.101–116.

7. In this statement and what immediately follows, I am aware that some will see evidence of the "generic fallacy." I beg their temporary indulgence. Eventually my argument will make it clear how indispensable I find the concept of *Kreuzung der Gattungen*, first articulated by Kroll 1924.202–224 and subsequently elaborated by many others, for all Vergil's poetry, especially the *Georgics*. Because this study focuses on epic, however, I have dealt mainly with crossing of the epic sub-genres mentioned above, rather than with Vergil's use of the non-epic genres. On this topic see Cairns 1989.129–248. We can expect further illumination from Richard Thomas, who has also now turned his attention to the problem.

the essential themes and ideas of the original work take on new meaning as they are analyzed and reintegrated into a new poetic structure.

In the case of the *Georgics*, Hesiod is an important part of this process, comparable in some ways to Theocritus and Homer. Both the *Eclogues* and the *Aeneid* contain a wealth of, on the one hand, individual lines and, on the other, extended dramatic and narrative elements that closely and obviously imitate particular passages in Theocritus and Homer. It is by pervasive, integrated allusion on both levels, that of versecraft and that of dramatic or narrative structure, that Vergil creates a privileged relationship with these authors in particular. In the *Georgics*, especially in Book 1, integrated allusion to *Works and Days*, comprising parallels of both versecraft and expository structure, is extremely prominent. But, unlike Theocritus in the *Eclogues* and Homer in the *Aeneid*, Hesiod cannot be seen as *the* model of the *Georgics:* the contribution of other authors at all levels is simply too important. Nor is it sufficient to speak of these other authors as subsidiary models, like Callimachus, Lucretius, and Calvus in *Eclogue* 6, or like Apollonius of Rhodes in *Aeneid* 4.[8] However important these authors may be, their influence is felt only in specific portions of a larger work and pales in comparison to that of Theocritus and Homer, which is pervasive. But Hesiod does not pervade the *Georgics*, overshadowing the poem's other models, in the same way; in fact, his influence on the poem resembles that of a secondary model in the *Eclogues* or the *Aeneid*, since it is virtually confined to *Georgics* 1.

The evidence assembled over the years by *Quellenkritiker* and others indicates a number of different sources for various episodes of the *Georgics*. None of these sources exerts a pervasive influence on the poem, as do Theocritus and Homer in the *Eclogues* and the *Aeneid*, but instead, like Hesiod, exert an essentially local influence similar to that of the subsidiary sources mentioned above. The most extensive concentrated imitations in the *Georgics* are the concluding episodes of Books 1 (351–464, "Weather Signs") and 3 (478–566, "The Plague of Noricum"), which are closely modeled on passages of Aratus (*Ph.* 733–1154, the "*Diosemiae*") and Lucretius (*DRN* 6.1090–1286, "The Plague of Athens") respectively. The even more ambitious finale of Book 4 (315–558, "Aristaeus") incorporates elements of several Homeric episodes (*Iliad*

8. A case for more extensive Apollonian influence on the *Aeneid* has now been made by Nelis 1988.

1.345–427 and 18.22–137, "Achilles and Thetis"; *Odyssey* 4.351–572 "Menelaus and Proteus"). Briefer episodes throughout the poem allude to these and other authors, including Hesiod.

Of course, the relationships among these individual episodes and their models have been fairly intensively studied. No one, however, has seriously examined these passages as parts of an allusive whole. The question can be put very simply: why in a poem that pretends to imitate Hesiod's *Works and Days* does Vergil allocate so much space to extensive and elaborate imitations of these other models?

In order to answer this question, we must attempt to understand the basic forms, methods, and purposes of Vergilian allusion. What follows is offered more as a précis than as a full analysis of that topic, written with a view towards those aspects of Vergil's technique that have the most bearing on the *Georgics*. A complete account must be both more detailed and wider in its purview than is possible here. Nevertheless, allusion in the *Georgics* is quite similar to what we see in the *Eclogues* and the *Aeneid* as well, and I will not hesitate to draw parallels with those two poems in order to illustrate and clarify the intertextual processes at work in Vergil's transitional masterpiece.

Technique

The intertextuality of Vergil's poems is richly and dazzlingly faceted. In many cases, the specific relationship between two poems that results in particular allusions is all but indefinable and impossible to recover. I believe, however, that in all cases, even those enticingly evasive allusions that seem to offer the reader insight into the poet's (and perhaps his own) unconscious, rather than any deliberate design—even these cases are ultimately the result of Vergil's analytical reading. Many have noted that Vergil's imagination was so thoroughly imbued with the work of this or that author that lines and phrases, words and images were apt to occur to him unbidden and to find their way into his own poetry. Even where the reminiscence is totally unforced, however, such allusions are guided by some element of the poet's imagination, an element very close to the essential quality of his poetic genius.

I hope that a sense of this richness will emerge as my argument develops. We can best begin to make sense of Vergilian allusion, however, by considering some comparatively obvious examples according to mainly formal criteria. Assuming that reminiscences did, as a natural part of composing, find their way into Vergil's poetry, it is hard to believe that the poet did not recognize many of these as he revised, ousting any that did not suit his purposes and polishing those that contributed to his overall plan. It is this stage of composition that we will examine first, trying to discern what it can tell us about the poet's overall plan for the *Georgics*.

It is coarse and grossly simplistic to imagine the source poem as being digested into its elemental components, which are then reused more or less *ad libitum* in creating a new poem. But the assumption does provide a useful and instructive point of departure, and it will in the end, though in a more sophisticated form, be borne out. It has the further advantage of focusing our attention on Vergil's place in the epic tradition. The essential element of this tradition is the epic verse itself; and Vergil liked to stress his debt to the tradition by borrowing particular lines and half-lines, intact or nearly so, from his predecessors. A few of these borrowings amount to exact quotations. Their effect, especially where Greek authors are concerned, is arresting:

Ἄλκανδρόν θ' Ἅλιόν τε Νοήμονά τε Πρύτανίν τε

Iliad 5.678

Alcandrumque Haliumque Noëmonaque Prytanimque

Aeneid 9.767

Γλαύκῳ καὶ Νηρεῖ **καὶ εἰν**αλίῳ **Μελικέρτῃ**[9]

Parthenius *AP* 6.164 (fr. 647
Lloyd-Jones and Parsons 1983.309–310)

Glauco et Panopeae **et Inoo Melicertae**

Georgics 1.437

ἢ Ἄθω ἢ Ῥοδόπαν ἢ Καύκασον ἐσχατόωντα

Idyll 7.77

aut Athon aut Rhodopen aut alta **Cerau**nia telo

Georgics 1.332

9. On the probable reading of this line, and for provocative speculation on other elements in Vergil's adaptation of it, see Thomas 1988.1.140–141 *ad G.* 1.437.

et **arguta lacus circumvolitavit hirundo**

Varro Atacinus fr. 22.4
(Morel–Büchner 1982.127)

aut[10] **arguta lacus circumvolitavit hirundo**

Georgics 1.377

sed quaedam **simulacra modis pallentia miris**

De Rerum Natura 1.123

ingens, et **simulacra modis pallentia miris**

Georgics 1.477

pulveris **exhalat nebulam,** nubis**que vol**antis

De Rerum Natura 5.253

quae tenuem **exhalat nebulam** fumos**que vol**ucris

Georgics 2.217

Of course, in any allusion, imitation is normally accompanied by variation: while some elements of the original are taken over unchanged, others are typically altered in one or more respects. But usually these varied elements maintain a degree of resemblance with their original; and an awareness of Vergil's preoccupation with epic models makes it possible to notice resemblances that might otherwise be missed. For example, virtually everyone recognizes the following line as an allusion to Lucretius:

heu! **magnum alterius** frustra **specta**bis **ace**rv**um**

Georgics 1.158

e terra **magnum alterius spectare** lab**or**em

De Rerum Natura 2.2

Here Vergil practically quotes his source, but he does not do so exactly. Three of six words are repeated, one of the three in a different form, the other two in a different *sedes.* The change from *spectare* to *spectabis* is of course dictated by circumstances, but the new position of the phrase *magnum alterius* requires some comment. Consider the overall prosody of the two lines: Both begin with a monosyllable and are metrically identical after the hepthemimeres. Between these boundaries is found the

10. Or *et?* See Skutsch 1985.778 n. 4.

repeated phrase *magnum alterius*, either preceeded (in Lucretius) or followed (in Vergil) by a spondee. It seems clear that Vergil has altered his source here in two respects: he has replaced Lucretius' *terra* with *frustra* (note homoeoteleuton), and has exchanged its *sedes* with that of *magnum alterius.* Scholars often ascribe such "revisions" to purely stylistic considerations. In Lucretius' version there is no real break at the penthemimeres, which Vergil has gained by transposing the phrase in question. But the prosodic element in this allusion is in my view more subtle and more important than this. Lucretius' conventional contrast between the safety of land and the danger of the sea is discordant with Vergil's characterization of farming as warfare; therefore it disappears along with the phrase *e terra.* The monosyllabic replacement *heu!* allows *magnum alterius* to move forward and make room for *frustra*, which plays a semantic role in Vergil's line similar to that of Lucretius' *terra:* Lucretius states that it is pleasant to observe someone else's trouble on the seas, so long as one is *on land*, i.e. to reflect upon trials in which one has no part. Vergil states that the man who does not work will gaze upon someone else's heap of produce, but *in vain. Terra* and *frustra* are the essential qualifying words in the two lines. Thus, by "moving" this element from its Lucretian position to one of much greater emphasis—between the penthemimeres and the hepthemimeres—Vergil thrusts this qualification into greater prominence. Finally, in addition to the three repeated words, there is also a close relationship between Vergil's *acervum* and Lucretius' *laborem;* for not only (like *frustra* and *terra*) are the words metrically identical and (unlike *frustra* and *terra*) do they occupy the same *sedes*, but they are causally related as well: in Vergil, the other man's *acervum* is the product of his *labor.*

It goes without saying that one need not ascribe such a deliberate process of ratiocination to Vergil himself as he composed the line in question. More probably he was guided primarily by the sound of the two verses. In the case of *acervum* ≈ *laborem*, it is hard to imagine that he did not consciously ponder the semantic relationship, but even here the metrical shape that the two words share, along with the assonance of *a*, *r*, and *m*, must have helped to determine the exact form that his variation took. It is also true that few informed readers are likely to miss this allusion even without making the effort to understand the exact formal relationship between the two lines. But by making the effort to describe as fully as possible the formal principles of *imitatio* and *variatio* in fairly obvious cases such as this, the critic gains insight into the workings of

Vergilian allusion, and thus into its poetic effect. The subtle differences between two lines as similar as these provide readers and critics with a good opportunity to measure with some precision the distance between Vergil and his model, and this opportunity in turn becomes an important index to the poem's meaning, the poet's thought. This poet, perhaps beyond all others enthralled and empowered by a sense of poetic tradition, spoke in his own voice while using the words of others. To understand him, we must occasionally resist the Siren call of that voice and attend to the alien words. It is here that the work of interpretation may begin. And if I refer to this interpretive undertaking as "measuring with precision" something as rich and complex as Vergilian poetic discourse, I do not mean to imply that we are engaged in a mechanical process. Vergil's palette of ironies and ambiguities is unmatched. Too often have critics viewed these subtle shades as monochromatic "homage" or "polemics." In the case of our Lucretius allusion, we find a number of contradictory reactions that succeed mainly in obscuring the relationship between *Georgics* and *De Rerum Natura*.[11] The besetting sin of most existing criticism of the poem is to focus a blinkered gaze on a single allusion and to guess at its point, or else to assume that the motive behind all Vergilian allusions involving a given author must be more or less the same and represent a consistent, easily defined attitude—a will to become the "Roman Hesiod," for instance, or to reject explicitly the seductive influence, felt powerfully in youth, of Lucretius' philosophical and poetic masterpiece. But nothing in Vergil is that simple.

To interpret these allusions successfully, we must first understand the technique in all its variety. We must then try to understand the allusions themselves as a unified whole. This is, of course, extremely difficult, since the allusions span four books of *Georgics* and involve a group of apparently disparate authors, with each one of whom Vergil creates a complex set of relationships. Above all, the connections between allusions involving one or more authors must be carefully traced. The critic's challenge then becomes the (to all Vergilians) familiar one of understanding this bewilderingly refracted diversity as a coherent unity. When it is put in this way, we see that the task at hand is essentially similar to that of understanding the development of a Vergilian character,

11. Farrington (1963.91) speaks of Vergil's "scoffing reminder of the necessity of hard work" as a "parody of" and a "retort" to the Lucretian passage that it imitates. Griffin (1981.32) sees the passage as "gently humorous."

theme, or image; and it is this similarity that bears out the legitimacy of our method.

Vergil's allusive technique is basically identical at all levels: the same principles of analysis may be used to describe allusions consisting of a single line, those that comprise whole episodes of narrative or exposition, and even entire books or poems. Indeed, in my view, Vergil's allusiveness must be understood as a unified aspect of his style. Imitation at the level of individual words and phrases cannot in fact be separated from imitation on a larger scale, but can only be understood properly if we consider the place of these elements in a larger allusive context. Most critics, I believe, would subscribe to this proposition or to something very like it. A parallel as brief as the one we have been discussing is frequently not considered a true allusion unless the original context is in some way relevant to the imitative one. Because this particular allusion occurs near the end of Vergil's "Aetiology of *Labor*" (*G.* 1.118–159), the fact that *acervum* replaces Lucretius' *laborem* is no coincidence, but a pointed commentary on the different outlooks espoused by each author. But the contextual relationships indicated by Vergilian allusion are seldom simple, and this passage is no exception. Even if one feels that he understands the relationship between Vergil and Lucretius at this point, the reader must also reckon with the fact that the episode that this allusion closes is usually considered an imitation of Hesiod's "Myth of Ages" (*WD* 109–201). One can hardly avoid taking this Parthian shot as a strong signal that we must revise our familiar assumptions and reread this Hesiodic passage with Lucretius in mind. And if, as we ought, we reject the notion that Vergil works with only one source or model in mind at a time, we are then forced to consider not only the poetic, but also the literary-historical meaning of a passage that begins and sustains itself as an apparently full-blooded imitation of a Hesiodic set piece, and then concludes with an almost ostentatious reference to Lucretius.

All such passages raise difficult questions about Vergil's allusive style, but the questions are not unanswerable. As for the evidence itself, a vast number of small-scale allusions, involving single lines or half-lines, have been collected over the years and are easily accessible. What is needed is to understand the relationship between the contexts in which these allusions occur. Because so much depends on context, let us turn our attention to a more extensive allusion. Here it should be possible to learn more about Vergil's allusive style by examining how closely he adhered to a model in extended imitation and in what respects he felt free to depart from it.

"*The Farmer's* Arma"

In extended allusive episodes, the same basic techniques of imitation and variation discussed above can be observed on a larger scale. The principles of analysis that have been used to describe Vergil's allusive essays in the dramatic and narrative modes of the *Eclogues* and the *Aeneid* can be applied with equal success to the expository mode of the *Georgics*. Instead of defining an allusive episode as, let us say, a speech or a sequence of action, we need only recognize the parallelism between two lines of argument to establish a context of allusion, which may then be broken down into its components and studied in comparison with its model. The episode of *Georgics* 1 that has been labeled "The Farmer's *Arma*" (160–175) shows a clear and direct relationship of versecraft, structure, and theme to Hesiod's passage on "Farm Equipment" (*WD* 414–436). The relationship between these passages has been discussed so often, and is in most respects so clear, that I will confine my remarks to the bare minimum—as indeed Vergil does. The passage is one of his tersest imitations; the relationship rests on a few details only, most of them varied imitations of Hesiodic originals. These qualities make it a good passage with which to begin discussing the ways in which Vergil develops contextual imitations.

> ἦμος δὴ λήγει μένος ὀξέος ἠελίοιο
> καύματος ἰδαλίμου, μετοπωρινὸν ὀμβρήσαντος 415
> Ζηνὸς ἐρισθενέος, μετὰ δὲ τρέπεται βρότεος χρὼς
> πολλὸν ἐλαφρότερος· δὴ γὰρ τότε Σείριος ἀστὴρ
> βαιὸν ὑπὲρ κεφαλῆς κηριτρεφέων ἀνθρώπων
> ἔρχεται ἠμάτιος, πλεῖον δέ τε νυκτὸς ἐπαυρεῖ·
> τῆμος ἀδηκτοτάτη πέλεται τμηθεῖσα σιδήρῳ 420
> ὕλη, φύλλα δ᾽ ἔραζε χέει, πτόρθοιό τε λήγει·
> τῆμος ἄρ᾽ ὑλοτομεῖν μεμνημένος ὥρια ἔργα·
> ὅλμον μὲν τριπόδην τάμνειν, ὕπερον δὲ τρίπηχυν,
> ἄξονα δ᾽ ἑπταπόδην· μάλα γάρ νύ τοι ἄρμενον οὕτω·
> εἰ δέ κεν ὀκταπόδην, ἀπὸ καὶ σφῦράν κε τάμοιο. 425
> τρισπίθαμον δ᾽ ἄψιν τάμνειν δεκαδώρῳ ἀμάξῃ,
> πόλλ᾽ ἐπικαμπύλα κᾶλα· φέρειν δὲ γύην, ὅτ᾽ ἂν εὕρῃς,
> εἰς οἶκον, κατ᾽ ὄρος διζήμενος ἢ κατ᾽ ἄρουραν,
> πρίνινον· ὃς γὰρ βουσὶν ἀροῦν ὀχυρώτατός ἐστιν,
> εὖτ᾽ ἂν Ἀθηναίης δμῶος ἐν ἐλύματι πήξας 430
> γόμφοισιν πελάσας προσαρήρεται ἱστοβοῆι.
> δοιὰ δὲ θέσθαι ἄροτρα, πονησάμενος κατὰ οἶκον,

αὐτόγυον καὶ πηκτόν, ἐπεὶ πολὺ λώϊον οὕτω·
εἴ χ' ἕτερον [γ'] ἄξαις, ἕτερόν κ' ἐπὶ βουσὶ βάλοιο.
δάφνης δ' ἢ πτελέης ἀκιώτατοι ἱστοβοῆες. 435
δρυὸς ἔλυμα, πρίνου δὲ γύην. βόε δ' ἐνναετήρω
ἄρσενε κεκτῆσθαι· [τῶν γὰρ σθένος οὐκ ἀλαπαδνόν·
ἥβης μέτρον ἔχοντε·] τὼ ἐργάζεσθαι ἀρίστω.
οὐκ ἂν τώ γ' ἐρίσαντε ἐν αὔλακι κὰμ μὲν ἄροτρον
ἄξειαν, τὸ δὲ ἔργον ἐτώσιον αὖθι λίποιεν.

Works and Days 414–440

dicendum et quae sint duris agrestibus arma,
quis sine nec potuere seri nec surgere messes:
vomis et inflexi primum grave robur aratri,
tardaque Eleusinae matris volventia plaustra,
tribulaque traheaeque et iniquo pondere rastri;
virgea praeterea Celei vilisque supellex, 165
arbuteae crates et mystica vannus Iacchi;
omnia quae multo ante memor provisa repones,
si te digna manet divini gloria ruris.
continuo in silvis magna vi flexa domatur
in burim et curvi formam accipit ulmus aratri. 170
huic a stirpe pedes temo protentus in octo,
binae aures, duplici aptantur dentalia dorso.
caeditur et tilia ante iugo levis altaque fagus
stivaque, quae currus a tergo torqueat imos,
et suspensa focis explorat robora fumus.

Georgics 1.160–175

Consider first the structure of the two passages. Hesiod begins in leisurely fashion, taking nine full lines (414–422) to specify fall as the proper time of year for woodcutting. In the next four and a half lines (423–427a) we learn that he refers not just to firewood, but to boughs and branches of suitable shapes to serve as mortar and pestle, cart axle, mallet head, and felloe, or wheel rim. At 427b the poet mentions a plow-tree, and shifts his attention to the construction of a plow until line 435a (six lines); he then specifies the best age of oxen (435b–440) and plowman (441–447, not quoted above) to use. There follows a new paragraph (448–457) advising the listener to provide himself with the proper equipment well before plowing season begins. Vergil reduces this delightful ramble to his usual tight, meticulous order: two lines of introduction announcing the topic of farm equipment (*dicendum et quae sint duris agrestibus arma* | *quis sine nec potuere seri nec surgere messes* 160–161) are followed by a five-line list of miscellaneous tools (162–166) and a two-line gnome (167–168). The rest of the passage (169–175) concerns the various parts of the plow. Vergil has followed Hesiod's basic sequence of topics, with a few exceptions. First, two of Hesiod's topics (oxen and plowman) have been completely eliminated; second, another topic (the injunction to have all the tools ready well before they are needed) has been greatly compressed and shifted from last place in Hesiod to third in Vergil. The relative organization of the two passages is best illustrated by a chart:

Hesiod	Vergil
A. Introduction 414–422: Cut wood in the fall (9 lines)	A. Introduction 160–161: The farmer needs these tools 2 lines)
B. List of Tools 423–427a (4½ lines)	B. List of Tools 162–166 (5 lines)
	F. Transition 167–168: Have all these things ready well before plowing season begins (2 lines)
C. The Plow 427b–435a (6 lines)	C. The Plow 169–175 (7 lines)
D. The Proper Oxen 435b–440 (5½ lines)	
E. The Proper Plowman 441–447 (7 lines)	
F. Conclusion 448–465: Have all these things ready well before plowing season begins (18 lines)	

As a whole Vergil's paragraph is not only more tightly organized than Hesiod's (note the doubly symmetrical arrangement of material: A + B + F = 2 + 5 + 2, and AB + F + C = 7 + 2 + 7), it is much shorter and more focused as well. Hesiod's nine-line introduction, for instance, is cut to two, and instead of woodcutting, Vergil announces the real topic of what follows as "tools" (*arma* 160). Hesiod's eighteen-line observation on the importance of having these tools ready before the start of plowing season is represented through synecdoche by the single line that sums it up: τῶν πρόσθεν μελέτην ἐχέμεν οἰκήϊα θέσθαι (457) ≈ *omnia quae multo ante memor provisa repones* (167). Vergil then "pads" this line with a second in order to balance his two-line introduction. But the two main sections of Vergil's paragraph (B and C, 5 and 7 lines respectively) are slightly longer than their Hesiodic models (4½ and 6 lines). In fact, they contain more information as well. In section B, Hesiod bids his listener cut a three-foot mortar, a pestle of three cubits, a seven-foot axle—or, if he finds one of eight feet, to cut a foot off the end for a mallet head—and a three-span felloe for a ten-palm cart. Where Hesiod mentions five tools, Vergil lists nine. Plowshare and plow take pride of place, looking forward to the second half of the paragraph; then come carts (*plaustra* 163 ≈ ἀμάξῃ 426), threshing sledges, drags, mattocks (*rastra* 164 ≈ σφῦραν 425),[12] baskets, harrows or hurdles, and finally a winnowing fan. Vergil's imitation not only gives a fuller list of necessary equipment, but also embellishes its model. Thus Hesiod's plain carts become "the slow-rolling carts of the Eleusinian mother" (163), and Vergil continues in the same vein with two non-Hesiodic tools, "Celeus' humble rushwork implement...and Bacchus' mystic winnowing fan" (165–166). The *heuretes* motif, at once learned and religious, adds an elevated dignity to Hesiod's very matter-of-fact account.

Then follows the plow. Hesiod says to *look* for a plow tree, i.e. a suitably shaped bough of tough holm oak, either in the woods or in the field, and then to take it home (φέρειν δὲ γύην, ὅτ' ἂν εὕρῃς | εἰς οἶκον, κατ' ὄρος διζήμενος ἢ κατ' ἄρουραν, | πρίνινον 427–429). Vergil too begins with the plow tree, but says that a living[13] elm—a substitution mentioned by Plutarch[14]—must be *bent* into the shape required for a plow stock, not at home or in the field but right in the forest (*continuo in silvis magna vi flexa domatur | in burim et curvi formam accipit ulmus aratri*

12. The heavy (*iniquo pondere*) rastrum was used, like Hesiod's mallet, to break up clods (see White 1967.52–56), and is therefore its equivalent here.

13. Does Vergil recommend bending a live tree or a piece of timber? The Latin perhaps most naturally suggests the former: an elm is bent with great force and trained to grow in the shape of a *buris*. Servius Danielis (*ad G.* 1.170 [Thilo 1887.172]), however, seems to read the passage differently: *buris enim ut curvetur, ante igni domatur, id est amburitur; unde et quae naturaliter inveniuntur curvae, ita dicuntur.* This etymology would seem to have nothing to do with the process of shaping a live tree; on the other hand, it is directed more to the word itself than to the procedure that Vergil describes.

14. In his commentary on the *Works and Days* (*Moralia* fr. 64 [Sandbach 1967.44]).

169–170).[15] The alteration is in line with Vergil's general emphasis in *Georgics* 1 on the violence and artifice involved in working the soil.[16] The two poets mention the other parts of the plow in different orders. Hesiod is brief, naming just the stock or share beam and the yoke pole; only much later and in a different context does he mention the tail or stilt. Vergil describes a slightly different plow and does so more fully. Again he is more orderly, working from the top of the plow down and from front to back. He starts with the yoke pole and proceeds to the "ears" or mold boards (*aures* 172, not mentioned by Hesiod), and then to the *two* share beams instead of Hesiod's one (*dentalia* 172 ≈ ἐλύματι 430, ἔλυμα 436); finally he adds the two pieces that provide the plow with power and control, the yoke (*iugo* 173, also not mentioned by Hesiod) and the stilt (*stiva* 174 ≈ ἐχέτλης 467). In discussing the form and types of wood suited to these different parts, Vergil simply fills out Hesiod's account. Besides the plow tree, Hesiod mentions the best materials for the yoke pole (myrtle or elm 435) and the share beam (oak 436); Vergil, silent on these points, supplements Hesiod's advice by specifying linden for the yoke and beech for the stilt (173–174). Similarly, when Vergil measures the yoke pole at eight feet (171), he gives his reader another piece of instruction not found in *Works and Days;* but his inspiration is probably all those measurements that Hesiod does give in section B.[17] Again, Vergil closes his imitation with an injunction to season the plow by hanging it in the hearth (175). This point is not in the original; but we may see it as a substitute for Hesiod's opening concern that the wood be gathered at the proper time of year, *when it is least susceptible to rot* (ἀδηκτοτάτη 420). Furthermore, Hesiod twice in quite different contexts (*WD* 45, 629) advises hanging rudders in the hearth when they are not in use, also with a view to preventing rot.[18]

It is easy to see in this contextual allusion the same principles of variation that were observed in our survey of imitative versecraft. Vergil on the whole retains the internal structure of Hesiod's account (the relative positions of sections A, B, and C), eliminating two sections (D and E) and transposing a third (F) while greatly condensing the whole in order to satisfy his own aesthetic requirements. We saw something very similar in the one-line paraphrase of Lucretius discussed above; on a larger

15. Cf. Putnam 1979.37.

16. Cf. Putnam 1979.39; Ross 1987.79, 83, 236–238.

17. An eight-foot axle is in fact specifically mentioned by Hesiod (424), but Vergil's allusion is probably representative rather than specific.

18. See West 1978.154 *ad WD* 45.

scale, compare the funeral games of *Aeneid* 5 with those of *Iliad* 23.[19] Other parallels are not far to seek. Indeed, the formal relationship between these passages indicates some of the most typical procedures of Vergilian allusion. First of all, I call attention to the relative importance of imitation and variation. Even this relatively straightforward example of contextual allusion entails a serious reworking and condensation of its original. Most critics have seen here a close imitation of the plow passage only. This is of course the core of the allusion; but in fact Vergil casts his net much wider. Much is omitted, and yet enough is done by incorporating a few specific details (the *length* of the yoke pole; the stilt; the injunction to have all the equipment ready well before it is needed) extraneous to Hesiod's plow passage to indicate the true limits of textual congruity. At the same time, Vergil includes these extraneous details in such a way as to redefine the structure of the episode. Here I accept the traditional view that finds in such Vergilian reworkings evidence of a classical aesthetic emphasizing balance, order, proportion, and similar qualities. Certainly there is a further element of *aemulatio* both in Vergil's list of auxiliary equipment, which is longer than Hesiod's, and in the design of his plow, which is more complex than Hesiod's. It should be obvious that none of the points I have made will be apparent to a reader who does not consult the original and spend some time tracking down the sources of individual Vergilian motifs. But the painstaking intertextual craftsmanship of such allusions goes beyond the spirit of mere rivalry. It seems clear that Vergil conceived of his imitation and its original as interdependent in some respects. For example, one discovers the various proper materials for all the parts of the plow only by reading both passages together.

A further point: Wilkinson has well observed that neither Hesiod's nor Vergil's lines amount to useful instruction. In particular, the evidence is against any notion that Vergil's contemporaries made their own plows instead of buying them. Rather, "from the literary point of view the Roman Hesiod was concerned to work in enough Hesiodic matter to recall the original."[20] While Wilkinson begs an important question with the phrase "Roman Hesiod," he is absolutely right that the entire Vergilian passage is purely literary in nature. Useless to the would-be farmer, the passage speaks eloquently to the poetically aware reader, establishing an intertextual relationship between the *Georgics* and an earlier poem, here *Works and Days*. This too is a characteristic common

19. Heinze 1915.145–170.
20. Wilkinson 1969.58.

to all the passages we will be discussing; and while it is reasonably apparent here, in a context involving didactic epos, elsewhere the affinities between the *Georgics* and other poems of the epic tradition will necessitate a reassessment of Vergil's didactic stance.

Finally, the intertextual relationship between these two passages sharpens our appreciation of Vergil's argument at the level of theme. The issues raised here go to the poem's heart: the need to accept the recurring cycles of human life; the idea of preparation for predictable change; the violent nature of civilized existence. What is essential to grasp is that the issues are raised as elements in the relationship between two texts in such a way as to define a poetic tradition. Hesiod's instructions maintain the familiar stance of hard-headed, fatalistic practicality that animates the entire *Works and Days.* Vergil accentuates and enriches Hesiod's pessimism and gives it his own stamp, using specific images inspired by nothing in his model. To Hesiod, farming is hard work. To Vergil, throughout *Georgics* 1, it is warfare: thus he discusses not the farmer's *instrumentum*,[21] but his *arma* (160). His farmer is not only general, but priest, presiding with sacred implements over rites founded by Celeus, Iacchus, and the goddess of Eleusis (164–166). The land itself is sacred, and the object of its cultivation not mere sustenance, but glory (168). These details are not pieces of decorative appliqué intended to enliven a spare model, but evidence of the metaphorical character of Vergil's didactic exposition. A comparison between the two plows reinforces this point. The plow that attracts most of Hesiod's attention is natural thing, a found object; Vergil's is unambiguously a product of craft and force. Even so, Hesiod's straightforward machine remains a lifeless object; Vergil's is, if not humanized, then certainly animate, showing us its feet, ears, teeth, and back (170–171).[22] Vergil's farmer needs monsters such as this in his arsenal, to counter those that Earth bears (*quae plurima terrae | monstra ferunt* 184–185), the subject of the paragraph that follows.

From such details certain conclusions emerge clearly. A passage such as this (and there are many in the *Georgics*) is obviously a tour de force; as such, it threatens to disrupt the fabric of Vergil's poem, allowing us to treat it as no more than a signpost of Vergil's intent to become "the Roman Hesiod." If we read it properly, however, we see clearly how, even in such an obsessively careful imitation of a famous Hesiodic passage, Vergil animates his borrowed material with a sensibility that is

21. See, e.g., Varro *DRR* 1.17.1.
22. Putnam 1979.37

altogether new, and integrates a thoroughly foreign passage into the imagistic and thematic structure of his own poem.

Because allusion is so important an element of Vergil's style, it cannot be adequately understood from one or two examples only. Indeed, the very pervasiveness of the phenomenon ultimately requires us to search the poem exhaustively from end to end in order to understand how different allusive passages fit together in a unified whole. At the same time, more might be said to illuminate different aspects of this particular imitation; but the points I have raised with respect to "The Farmer's *Arma*" will serve to introduce the basic features that we shall find at work, sometimes in slightly different forms, in a number of other important contexts of allusion.

Imitation and Variation in Contextual Allusion: Five Passages

The *Georgics*, as noted above, is rich in episodes that are closely based on similar passages from other poems. It is no accident that the episodes in question are noteworthy for more than their allusive character. In addition—and despite the integrative thematic features which we have seen in "The Farmer's *Arma*"—most of them have traditionally been regarded as digressions from Vergil's main argument. Their significance has been variously assessed, but there is now a broad consensus that, despite their formal status as digressions, such passages are fully integrated into the thematics of the poem.

By examining a series of these episodes we will deepen our understanding of the imitative variations that Vergil favors in contextual allusion and gain insight into his treatment of a few specific models in the *Georgics*. The most commonly cited passages and their respective models are:

1.160–175	"The Farmer's *Arma*" ≈ *WD* 427–435 "Farm Equipment"
1.276–286	"Lucky and Unlucky Days" ≈ *WD* 802–813 "Days"
1.351–464	"Weather Signs" ≈ *Ph.* 733–1154 "*Diosemiae*"
2.315–345	"Praise of Spring" ≈ *DRN* 5.680–770 "The Procession of the Seasons" + 1.248–264, 2.991–1022 "*Hieros Gamos*"
3.478–566	"Plague of Noricum" ≈ *DRN* 6.1090–1286 "Plague of Athens"
4.315–558	"Aristaeus" ≈ *Iliad* 1.345–427 and 18.22–137 Achilles and Thetis" + *Odyssey* 4.351–572 "Menelaus and Proteus"

These imitations, as noted above, have received a good deal of critical attention, and my debt to previous scholarship will be apparent. The results of such activity, however, have not been fully integrated into a general interpretation of the poem; nor have the different passages been compared with one another for the purpose of illuminating Vergil's allusive style per se. Such comparison would have revealed the interesting fact that the different episodes, while they individually bear noticeably different relationships to their respective models, taken together define a single, coherent allusive style. The distinctions among these relationships consist primarily in the different types of *variatio* that can be observed in each case—the different techniques that define the overall style. Not that Vergil's treatment of each individual episode relies on a single technique. One might illustrate all or most of the salient aspects of his imitative style by examining one of these episodes exhaustively. My previous remarks on "The Farmer's *Arma*" may serve as an extremely rough approximation of this ideal. But in the scrutiny of only one passage an important perspective is inevitably lost. Besides the various facets of Vergil's imitative style, I want to illustrate his methodical application of it as a coherent whole to a variety of authors in the four books of *Georgics*. In discussing each of these passages, I will focus my attention on what seems to me the most important aspect of his imitative style—e.g. structural allusion, contextual *contaminatio*, etc.—at work in that passage, while paying less attention to other concerns.[23] Similarly, I will restrict myself at this time from far-reaching interpretive statements. These are better saved for a later chapter, as a superstructure to be erected on the foundation of formal analysis.

It would be possible to place the episodes discussed below along a single continuum according to differences in their allusive technique. In fact, the sequence in which I shall discuss them defines one such continuum: each successive episode relies more than its predecessors on the technique of *contaminatio*, or the combination of models. I chose this criterion because contextual *contaminatio* will prove to be an especially important element in my reading of the poem as a whole. But the differences between these episodes might be measured by other criteria, and an imitation that looks simple and straightforward in one respect

23. I do not pretend that my discussion covers every possible type of imitative technique that may be found in the *Georgics*, let alone in Vergil's entire *oeuvre* or in Latin poetry generally. This (in my view) overly ambitious claim is made by Thomas (1987.198) in his useful study of what he calls Vergil's "art of reference."

will prove to be complex and elaborate in others. I therefore do not wish to imply anything by the order in which I discuss the passages; the following sequence was chosen mainly with a view to convenience and clarity.

"Weather Signs"

In "The Farmer's *Arma*" we see a close structural relationship between imitation and original. While variations do occur, the essential sequence of topics is maintained. Only one element, Hesiod's concluding injunction to have all the necessary equipment ready well before it is needed, is transferred to a new position: a corresponding injunction falls in the middle of Vergil's imitation. This kind of variation through the transposition of imitated elements often takes place on a much larger scale. In the *Aeneid*, for instance, narrative structures are rearranged in a sweeping manner to create new episodes of contextual allusion with complex relationships to their originals.[24] We see the poet following the same procedures in his allusion to Aratus' *"Diosemiae"* (*Ph.* 733–1154) in *Georgics* 1.351–464, the "Weather Signs."[25] The structural relationship between these passages is fairly direct; major sequences of topics are left largely intact. But if we consider Vergil's imitation as a whole, we notice significant instances of editing and transposition.

Aratus' sequence of topics is ordered as follows. After an introduction (733–777), his first section (778–908) tells how to predict the weather by observing the heavenly bodies, proceeding upward through the spheres from moon to sun to stars. Each stage of the progression is clearly marked by anaphora:

Lunar signs 778–831:
σκέπτεο δὲ πρῶτον κεράων ἑκάτερθε σελήνην 778

Solar signs 832–891:
σκέπτεο δ' εἴ κε τοι αὐγαὶ ὑπείκωσ' ἠελίοιο 832

Sidereal signs 892–908:
σκέπτεο καὶ Φάτνην· ἣ μέν τ' ὀλίγῃ εἰκυῖα 892

24. The most outstanding examples are probably the "Funeral Games" of Book 5 (on which see the paradigmatic explication of Heinze 1915.145–170) and the "Catabasis" of Book 6 (exhaustively analyzed by Knauer 1964a.107–147).

25. Essential literature on this passage includes Regel 1907, Beede 1936, and Cole 1979; cf. Otis 1972.47–50.

The second major section (909–1012) deals with non-astronomical signs, usually based on animal behavior, of various weather conditions, including wind (909–932), rain (933–987), fair weather (988–998), and storms (999–1141). The last thirteen lines of the poem (1142–1154) form an epilogue.

Just as in "The Farmer's *Arma*," here again Vergil discards a fair amount of source material. The opening section of the "*Diosemiae*" (733–777) contributes practically nothing to the "Weather Signs."[26] Most of the signs given by the stars and constellations (*Ph.* 892–908, 995–998) are simply omitted, as is any mention of halos around the sun or moon (*Ph.* 811–817, 862–889, 941) and most of the miscellaneous information that concludes the "*Diosemiae*" (1012–1154). The core of Vergil's imitation consists of Aratus' two central sections on celestial and non-celestial signs; but Vergil treats of these topics in reverse order and drastically reduces their bulk.

Correspondences between major sections of Aratus' exposition and Vergil's imitations are marked by close translations or paraphrases. After a brief introduction (*G.* 1.351–355), Vergil begins his discussion of terrestrial signs (356–424) with a clear reference to Aratus:

> *σῆμα δέ τοι ἀνέμοιο καὶ οἰδαίνουσα θάλασσα*
> *γινέσθω καὶ μακρὸν ἐπ' αἰγιαλοὶ βοόωντες*
> *ἀκταί τ' εἰνάλιοι ὁπότ' εὔδιοι ἠχήεσσαι,*
> *γίνωνται, κορυφαί τε βοώμεναι οὔρεος ἄκραι.*
>
> *Phaenomena* 909–912

> continuo ventis surgentibus aut freta ponti
> incipiunt agitata tumescere et aridus altis
> montibus audiri fragor, aut resonantia longe
> litora misceri et nemorum increbrescere murmur.
>
> *Georgics* 1.356–359

This close paraphrase signals a correspondence between Vergil's first section (351–369) and Aratus' comments on signs of windy weather (909–932). Vergil's next section, which concerns signs of rain (370–392), similarly begins with a close rendering of the corresponding passage in Aratus (933–987):

26. Scholars once compared with this section Vergil's introductory statement, but see Beede 1936.11. Cole (1979.41) observes that the words *ipse pater statuit* (353) parallel Aratus' assertion that ἐκ Διὸς ἤδη πάντα πεφασμένα πάντοθι κεῖται (*Ph.* 743).

αὐτὰρ ὅτ' ἐξ Εὔροιο καὶ ἐκ Νότου ἀστράπτῃσιν,
ἄλλοτε δ' ἐκ Ζεφύροιο, καὶ ἄλλοτε πὰρ Βορέαο,
δὴ τότε τις πελάγει ἔνι δείδιε ναυτίλος ἀνήρ,
μή μιν τῇ μὲν ἔχῃ πέλαγος, τῇ δ' ἐκ Διὸς ὕδωρ·
ὕδατι γὰρ τοσσαίδε περὶ στεροπαὶ φορέονται.

Phaenomena 933–937

at Boreae de parte trucis cum fulminat et cum
Eurique Zephyrique tonat domus, omnia pleni
rura natant fossis, atque omnis navita ponto
umida vela legit.

Georgics 1.370–373

The opening statements of these sections show clearly that Vergil's argument here closely parallels that of Aratus. We may say that the expository structure *Georgics* 1.351–423 imitates that of *Phaenomena* 909–998, dealing first with signs of wind, then signs of rain, and finally signs of fair weather.[27]

Vergil's next paragraph, however, goes on to discuss the moon and then the sun as weather indicators. This entire section (*G.* 1.424–514) follows the structure of Aratus' *first* section (*Ph.* 778–891) on celestial signs. Here again a clear point of reference is established by close imitation; but in addition to elements of diction, imagery, and folklore Vergil borrows from Aratus the idea of gracing with an acrostic his paragraph on the moon as a predictor of weather:[28]

σκέπτεο δὲ πρῶτον κεράων ἑκάτερθε σελήνην
ἄλλοτε γάρ τ' ἄλλῃ μιν ἐπιγράφει ἕσπερος αἴγλῃ,
ἄλλοτε δ' ἀλλοῖαι μορφαὶ κερόωσι σελήνην
εὐθὺς ἀεξομένην, αἱ μὲν τρίτῃ, αἱ δὲ τετάρτῃ·
τάων καὶ περὶ μηνὸς ἐφεσταότος κε πύθοιο.

27. Aratus' comments on foul weather (*Ph.* 999–1141) are not paralleled by a distinct section in Vergil. Instead, a few signs from this section are imitated in other parts of the "Weather Signs" on an ad hoc basis.

28. Aratus' acrostic was discovered by Jacques 1960, Vergil's imitation by Brown 1963.96–114; important subsequent discussions include Ross 1975.28–29 and Bing 1990. Several scholars in their enthusiasm for word games have adduced some fairly dubious examples of acrostics, anagrams, and so forth (and their ranks include some truly eminent names: the correspondence between Saussure and Meillet, documented by Starobinski 1979.127, is especially sobering). Many who refuse to believe that Vergil, at least, would have had time for what they consider such nonsense seize upon these bogus examples in an effort to discredit the whole idea; but the evidence is pretty clear. On acrostics in general, see Vogt 1967; Higgins 1987.

ΛΕΠΤΗ μὲν **καθαρή** τε περὶ τρίτον ἦμαρ ἐοῦσα
Εὔδιός κ' εἴη, **λεπτὴ** δὲ καὶ εὖ μάλ' ἐρευθὴς
Πνευματίη· παχίων δὲ καὶ ἀμβλείῃσι κεραίαις
Τέτρατον ἐκ τριτάτοιο φόως ἀμενηνὸν ἔχουσα
'**Η**ὲ νότου ἀμβλύνετ' ἢ ὕδατος ἐγγὺς ἐόντος.
εἰ δέ κ' ἀπ' ἀμφοτέρων κεράων τρίτον ἦμαρ ἄγουσα
μήτ' ἐπινευστάζῃ μήθ' ὑπτιόωσα φαείνῃ,
ἀλλ' ὀρθαὶ ἑκάτερθε περιγνάμπτωσι κεραῖαι,
ἑσπέριοί κ' ἄνεμοι κείνην μετὰ νύκτα φέροιντο.

Phaenomena 778–791

luna revertentis cum primum colligit ignis,
si nigrum obscuro comprenderit aëra cornu,
MAximus agricolis pelagoque parabitur imber;
at si virgineum suffuderit ore ruborem,
VEntus erit: vento semper rubet aurea Phoebe.
sin ortu quarto (namque is certissimus auctor)
PUra neque obtunsis per caelum cornibus ibit,
totus et ille dies et qui nascentur ab illo
exactum ad mensem pluvia ventisque carebunt,
votaque servati solvent in litore nautae
Glauco et Panopeae et Inoo Melicertae.

Georgics 1.427–437

In each case the acrostic marks a sequence of five lines; in Aratus, the first word of the first line is λεπτή, which is also read vertically starting from that point. Vergil, as usual, varies his imitation. He does not translate λεπτή either in his text or his acrostic; instead *pura* (433) picks up Aratus' καθαρή (783),[29] the word that follows λεπτή in the first line of the original. We notice a second variation in the fact that *pura* in the last line of the imitation picks up καθαρή in the first line of the original. This is a signal: Vergil's acrostic must be read backwards. We then find that Vergil employs a slightly more complicated form of acrostic than Aratus. Rather than reading the initial letters of five consecutive lines as a single word, Vergil introduces the beginning, middle, and final lines of his acrostic with words that suggest, by homoeocatarchton, three other words, PUblius VErgilius MAro. This remarkable signature reveals both Vergil's detailed familiarity with Aratus' text and the lighthearted, eminently Alexandrian spirit of imaginative ζήλωσις with which he approached his models.

29. Ross 1975.29 n. 1.

Everyone, I think, realizes that Vergil transposed Aratus' celestial and terrestrial signs so that his "Weather Signs" might build more effectively to the conclusion of *Georgics* 1 on the cosmic signs that attended the murder of Julius Caesar.[30] It would be wrong to speak of "improvement" in this connection. Aratus' poem is chiefly devoted to describing the heavens, but concludes with a treatise on prognostication. The signs that Aratus discusses involve the sun and the moon, thunder and lightning, animal behavior, and so forth, and he takes them in an order that moves down from the heavens through the atmosphere to the earth. This "descending" sequence of topics fits the requirements of Aratus' poem just as well as Vergil's "ascending" sequence does his.[31] Yet Vergil's structural variation on Aratus is paralleled in other contextual allusions, and several scholars have used such comparisons to illustrate a general principle of Vergilian style. Otis, for instance, sees a desire for climax or crescendo as a crucial difference between Vergil and most of his models;[32] and certainly Vergil's restructuring of Homeric episodes in the *Aeneid*, as I have already noted,[33] bears this notion out. Still, I would not suggest that crescendo is the only, or even the chief, motive behind this type of allusive variation. The most important lesson to be learned from comparing the "Weather Signs" with the *Aeneid* passages mentioned above is that the imitative technique that Vergil was eventually to use in adapting narrative episodes already informs extended contextual allusions in the *Georgics*.[34] Indeed, essentially similar techniques can be observed even in the *Eclogues*.[35] In the case of the "Weather Signs" the transposed elements are as clearly defined as the single words and phrases that are often transposed within individual lines. This is not always the case. We have already seen that variation in allusive versecraft takes many forms, and the same is true of contextual allusion as well. What should be clear from these comparisons is that the principle of transposition was fixed in Vergil's mind as an important tool of allusive variation, both at the level of versecraft and that of large contextual units.

30. This suggestion was first made, I believe, by Regel (1907.29).
31. This point is well made by Beede 1936.10.
32. Otis 1972.47–50, 1976.18–19.
33. See p. 79 n. 24.
34. Otis 1976 makes this point.
35. I plan to demonstrate this point in a paper on imitation in *Eclogue* 3.

"The Plague of Noricum"

In "The Farmer's *Arma*" and the "Weather Signs" allusive variation takes a particular form. Both imitations involve an exclusively internal rearrangement of structural elements borrowed from their respective models. A more direct structural relationship exists between *Georgics* 3.478–566 ("The Plague of Noricum") and *De Rerum Natura* 6.1090–1286 ("The Plague of Athens").[36] Here, however, we encounter a new form of variation in which material from a secondary passage is, so to speak, grafted onto an imitation of the primary model.

To some extent the theme of plague itself and the ostensibly scientific manner in which Vergil approaches it determine the structure of his account. Indeed, because Lucretius himself adapted his "Plague" from Thucydides, the traditional and generic aspects of these passages stand out more clearly than usual.[37] Furthermore, just as Thucydides' account shows affinities with the Hippocratic Corpus,[38] so Lucretius and Vergil go outside their immediate sources to actual medical writings for supplementary material. A notable instance concerns signs of imminent death, which Lucretius lists at 6.1182–1196. His signs correspond to nothing in Thucydides, but in fact find their precedents in "Hippocrates."[39] Similarly, where Lucretius makes sweat a general symptom of his plague (6.1187), Vergil specifies *cold* sweat as a sign of imminent death (*G.* 3.500–501), following Hippocrates *Aphorism* 4.37 and in keeping with his own application of humoral theory.[40] Just as Lucretius imitates Thucydides' consultation of actual medical treatises, so Vergil imitates Lucretius' manner of "improving" on his source by

36. The following is a development of Farrell 1983.29–31 and 119–125. Other studies of the material include Klepl 1940.52–101, Liebeschuetz 1965 and 1967–1968, Grimm 1965.55–61, Nethercut 1973.47–48, Raabe 1974.39–57, Harrison 1979, and West 1979. I have tried to indicate specific debts to these studies, not to catalogue every point of agreement or disagreement. It will be clear that my view of the relationship between Vergil and Lucretius differs from most of these in important ways (although I find myself in general sympathy with Liebeschuetz).

37. All three authors address first the aetiology of the plague, next its symptomatology, and last its epidemiology, employing the same arrangement that we find in proper medical treatises. The Hippocratic *Epidemics*, which report on a number of plagues that struck various ancient communities, establish a generic paradigm for dealing with such a topic. On the literary treatment of plagues in general see Grimm 1965.

38. See Gomme 1956.149–150. Rusten 1989.179–180, however, denies that there is such a connection.

39. See Munro 1886.2.394–395 *ad DRN* 6.1180–1192.

40. Hippocrates: West 1979.221–222 n. 6; humoral theory: Ross 1987.55–74, 177–183.

incorporating material from the technical tradition into an essentially literary account.

Formally, Vergil's "Plague" looks almost like a literary version of some lost treatise on a real occurrence; at least one commentator, in fact, credulously takes it "not only as a work of art, but as a noteworthy testimonium of veterinary history."[41] But as West has shown, judged by the twin criteria of clinical accuracy and poetic license, Vergil's "Plague" stands in relation to its model much as that of Lucretius does to Thucydides—and, to complete the analogy, as Thucydides' account does to treatises like the Hippocratic *Epidemics*. All of these authors use the formulae of clinical observation and medical analysis; but they can easily be placed on a scale, beginning with "Hippocrates" and ending with Vergil, of increasing emotional affect and, in Vergil's case at least, decreasing clinical accuracy.[42] Consider how Vergil skews the reader's attention away from questions about the plague's origin and characteristics in favor of its effects. Thucydides and Lucretius both devote considerable space to each of these three topics (aetiology 2.47–48 ≈ 6.1090–1144, symptomatology 2.49 ≈ 6.1145–1214, epidemiology 2.50–54 ≈ 6.1215–1286). Vergil however gives the first two topics very short shrift—four lines each (causes 3.478–481, symptoms 482–485) as compared with a full sixty-two lines concerning effects (486–547)—no doubt because they offer much less scope for the emotional element that he wishes to emphasize. Moreover, he has abandoned the precision with which Thucydides and Lucretius attempt to describe the plague of Athens; but it is here that his debt to Lucretius emerges most clearly. The details of Vergil's description seem to derive from accurate clinical observation, though they make little sense *as he presents them*. For example, in *Georgics* 3.541–543 we learn that the plague affected not only birds and beasts, but creatures of the sea as well: Vergil shows us their "bodies washed up on shore as if after a shipwreck" (542). But, one might ask, what shore? This plague is in Noricum, a landlocked territory.[43]

A detail such as this is clearly motivated by a desire to describe the plague as a universal affliction, attacking every sphere of animate life,

41. Richter 1957.320. Scholarship on the question of the plague's historicity is conveniently summarized by Harrison 1979.4–6.

42. We might extend the scale to include Ovid's "Plague of Aegina" (*Met.* 7.523–613), a still more exclusively literary set piece lacking any pretension to science (literature: Bömer 1976.332–334), and to other Silver Latin set pieces (e.g. Lucan *BC* 6.78–117, Silius Italicus *BP* 14.580–640).

43. West 1979.84.

and here Vergil is on his own. At bottom, however, his "Plague" is almost a cento of Lucretian ideas, images, and diction. This is true especially in those passages where imitation takes the form of pointed inversion, or *oppositio in imitando*.[44] To cite one of the more obvious examples, Lucretius in a famous passage notes that medicine was helpless against the Athenian plague:

> nec requies erat ulla mali: defessa iacebant
> corpora, mussabat tacito medicina timore.
>
> *De Rerum Natura* 6.1178–1179

In Vergil, impersonal medicine becomes a specific mythological image:

> quaesitaeque nocent artes, cessere magistri
> Phillyrides Chiron Amathaoniusque Melampus.
> saevit et in lucem Stygiis emissa tenebris
> pallida Tisiphone Morbos agit ante Metumque,
> inque dies avidum surgens caput altius effert.
>
> *Georgics* 3.549–553

Here Vergil is not merely exaggerating Lucretius, but actually inverting a basic principle of Epicurean philosophy. Not only does he associate medicine with the mythical Chiron and Melampus, but he depicts the opponent of these creatures not as a conventional plague explicable by the atomic theory, but as something that rises out of hell, driven by a Fury and identified by the twin personifications of Disease and Fear. What fear Vergil means is poignantly clear; for the essential message of Lucretius' poem is that all phenomena, including the terrifying disasters studied in Book 6, can be explained without recourse to the religious superstition that breeds fear of the gods. Tisiphone's skyward growth surely recalls Lucretian *religio*, looming above a cowering humankind prior to Epicurus' triumph of enlightenment.[45] Thus Vergil brings out his characteristically religious stance through pointed variation in alluding to Lucretius.[46] But even in passages where there is no parallelism of content between the two "Plagues" we find Vergil plundering Lucretius

44. Cf. Harrison 1979.30–31.

45. *DRN* 1.62–79; this point is made by Conington–Nettleship 1898.336 *ad G.* 3.553.

46. His purpose in doing so, however, has more in common with Lucretius than is usually perceived; I will return to this point on p. 102.

for his diction.[47] West's summary of the relationship between these episodes says it well: "In general then Virgil has eschewed the clinical detail of Lucretius, but has put together words and concepts culled from Lucretius' richly detailed and coherent symptomology to produce a simplified picture of a disease in two phases which is emotionally effective, rhetorically arresting by means of paradox and pathos, and has no regard for historical or scientific truth."[48]

Such an imitation may well strike the modern reader as strange. Vergil borrows a multitude of rather minute details from Lucretius, and the traditional view that he was intimately acquainted with "The Plague of Athens" is surely right. West himself adopts this view, but seems to suggest that Vergil, despite his familiarity with the passage, did not fully understand it.[49] I would not go so far. Vergil's interest in Lucretius' poem may not include all aspects of its scientific content, and his ancient reputation as a polymath need not count for much, even if he is credited with a knowledge of medicine.[50] But most modern readers find "The Plague of Athens" remarkably clear, even without any special knowledge of ancient or modern medicine. Is it likely that Vergil really misunderstood so badly a passage to which he alludes at such length and in such detail?

To put this question in perspective, there is good reason to believe that Vergil was seriously interested in philosophy throughout his life; and yet his most overtly "philosophical" passage—Anchises' description of the cycle of metensomatosis in *Aeneid* 6—presents the reader with an almost unintelligibly confused and eclectic eschatology. Such passages, which give the *appearance* of a detailed concern with technical knowledge while presenting a seriously confused or even deliberately wrong account, are in fact typical of Vergil; and because he ignores or distorts the content of his sources so frequently, one is forced to conclude that he was either an exceptionally inattentive reader, or that he deliberately and habitually emptied his sources of their original content, focusing on formal considerations. This is certainly what he has done with Lucretius'

47. Examples: West 1979.81 ("And yet even here where Virgil is far from Lucretius he uses Lucretian words").

48. West 1979.79.

49. "Virgil seems to have supposed for no good reason that deer would be immune" (1979.85); "...a tendency to write for effect, and lose sight of truth" (88).

50. *Vita Donati* 15 (Hardie 1966.9). Such a notice could be mere scholiastic inference deriving ultimately from this very passage.

"Plague." Lucretius' clinical detachment and scientific clarity, his wish to describe with great accuracy the historical plague that actually did beset Athens almost four hundred years before he wrote, are of no interest to Vergil, who turns the account into emotional rhetoric with only the appearance of technical accuracy. And yet the language of the imitation remains largely that of the original. What, one asks, is the purpose of alluding to Lucretius here at all?

The answer, I think, is that Vergil's concern is with the thematic rather than the "scientific" content of Lucretius' "Plague" and with the emotionalism that fairly throbs just below its surface. The emotional element of Vergil's imitation is no simple essay in *inventio*. It would be more useful to recognize it as the result of Vergil's effort to make explicit what Lucretius merely implies. On this view, Vergil empties Lucretius' clinical language of its rational content because this element does not fit in with the thematic agenda of his imitation.

This effort to bring out the hidden qualities of his model becomes particularly clear when we realize that the sources of Vergil's imitation include Lucretian passages other than "The Plague of Athens." By using *contaminatio* to vary his imitation of Lucretius' "Plague," Vergil is able to use Lucretius' words in the service of his own thematic goals. The following passage, for example, looks very much like a reference to one in *De Rerum Natura* 3:[51]

mox erat hoc ipsum exitio, furiisque refecti
ardebant, ipsique suos iam morte sub aegra
(di meliora piis, erroremque hostibus illum!)
A2 discissos nudis laniabant dentibus **artus**.
ecce autem duro fumans sub vomere taurus
A1 **concidit et mixtum spumis** vomit *ore* cruorem
B **ex**tremosque ciet **gemitus**.

Georgics 3.511–517

quin etiam subito vi morbi saepe coactus
ante oculos aliquis nostros, ut fulminis ictu,
A **concidit** et **spumas** agit, ingemit et tremit **artus**,[52]
desipit, extentat nervos, torquetur, anhelat
inconstanter, et in iactando membra fatigat.
. . . .

51. Cf. Harrison 1979.16
52. Cf. *tremit artus* | *G*. 3.84.

> B **exprimitur** porro **gemitus**, quia membra dolore
> adficiuntur et omnino quod semina vocis
> eiciuntur et *ore* foras glomerata feruntur
> qua quasi consuerunt et sunt munita viai.
>
> *De Rerum Natura* 3.487–498

The symptomatology of Vergil's plague, being based on that of Lucretius, presents an opportunity to exploit another Lucretian medical passage. The context of this source passage—it is one of Lucretius' twenty-nine proofs of the soul's mortality[53]—is not immediately relevant to the Vergilian context. Lucretius argues that the soul, like the body, is subject to diseases such as epilepsy; and that it therefore follows that the soul, like the body, is equally subject to death. Lucretius' implicit moral argument has an ambivalent relationship to Vergil's imitation. On the one hand, Vergil protests the futility of living the good life that Lucretius advocates, but which is nevertheless subject to the horrifying reality of death (*G.* 3.551–553). On the other hand Vergil, too, praises this way of life in the finale of Book 2. Lest we miss this irony, he provides the reader with an obvious cross-reference.[54] A further point. In Lucretius, epilepsy is a human affliction. In Vergil, the Lucretian symptoms of epilepsy become those of an animal plague. Here Vergil underscores the fact that his animal plague is based on Lucretius' description of a human one. He thus transforms two Lucretian passages; but in doing so, he observes a thoroughly Lucretian precept, that the fundamental nature of all forms of life—indeed, of all matter—is identical. This principle pervades both the poetry and the philosophy of *De Rerum Natura*, and Vergil's adherence to it is pronounced, as we shall see.[55]

Lucretius' epilepsy symptoms blend easily into Vergil's "Plague." A slightly greater imaginative leap was required to assimilate *De Rerum Natura* 2.352–367 into this context.[56] While this Lucretian passage, like the one discussed above, has nothing to do with the "Plague" per se, nor even with disease, it can plausibly be incorporated into Vergil's imitation as a study of the effect of a loved one's death upon his or her survivors:

53. Number six according to Kenney 1971.143 *ad DRN* 3.487–509.
54. Cf. *G.* 2.458–474 ≈ *G.* 3.520–530.
55. I shall develop this point further on pp. 197ff.
56. Klepl (1940.54 and 61 n. 104) notes a similarity between *Georgics* 3.486 and *DRN* 2.352–353, but says nothing more.

> ecce autem duro fumans sub vomere taurus
> concidit et mixtum spumis vomit ore cruorem
> extremosque ciet gemitus. it tristis arator
> maerentem abiungens fraterna morte iuvencum,
> atque opere in medio defixa reliquit aratra.
> non umbrae altorum nemorum, non mollia possunt
> prata movere animum, non qui per saxa volutus
> purior electro campum petit amnis; at ima
> solvuntur latera, atque oculos stupor urget inertis
> ad terramque fluit devexo pondere cervix.
> quid labor aut benefacta iuvant? quid vomere terras
> invertisse gravis? atqui non Massica Bacchi
> munera, non illis epulae nocuere repostae:
> frondibus et victu pascuntur simplicis herbae,
> pocula sunt fontes liquidi atque exercita cursu
> flumina, nec somnos abrumpit cura salubris.
>
> *Georgics* 3.515–530

> nam saepe ante deum vitulus delubra decora
> turicremas propter mactatus concidit aras
> sanguinis exspirans calidum de pectore flumen.
> at mater viridis saltus orbata peragrans
> quaerit humi pedibus vestigia pressa bisulcis,
> omnia convisens oculis loca si queat usquam
> conspicere amissum fetum, completque querelis
> frondiferum nemus adsistens et crebra revisit
> ad stabulum desiderio perfixa iuvenci,
> nec tenerae salices atque herbae rore vigentes
> fluminaque illa queunt summis labentia ripis
> oblectare animum subitamque avertere curam,
> nec vitulorum aliae species per pabula laeta
> derivare queunt animum curaque levare:
> usque adeo quiddam proprium notumque requirit.
>
> *De Rerum Natura* 2.352–366

This Lucretian passage has very little to do with "The Plague of Athens," Vergil's primary model for the end of *Georgics* 3. Lucretius' point in this passage is, however, very much a part of Vergil's imitation. Much more is at stake here than the simple motif of the grieving animal. Vergil's animal plague is modeled on Lucretius' human one. He endows it with the Lucretian symptoms of a human disease. His *iuvencus*, even more than its master (the *tristis arator* of 517) elicits a peculiarly human sympathy, while Lucretius avoids overt anthropopathism in depicting the animal's suffering.[57] Yet Lucretius, too, makes clear the analogy between the human and the animal world. By way of introducing this vignette, he observes that all creatures, human and animal, despite their basic uniformity within species, nevertheless are clearly differentiated as individuals: otherwise mother and child would be unable to recognize each other, as they clearly do, animals no less than humans (*nec minus atque **homines** inter se nota cluere, DRN* 2.351). Lucretius thus makes two crucial points: that atoms, despite their common properties, are nevertheless differentiated, and that this atomic principle dictates the structure of nature in the animal and human world as well.[58] But what serves as illustration in *De Rerum Natura* becomes the essential point in the *Georgics.* Lucretius certainly argues for, or even assumes, a basic similarity between beasts and humans; Vergil in his imitation emphasizes this idea, in characteristic fashion, making it one of the thematic keynotes of his poem. *Georgics* 3, after all, is not really about animal husbandry. It is about those internal passions and social forces that both enrich and threaten to destroy human life as well.[59]

Thus the basic motif of Lucretius' human plague is fortified with additions from other passages of *De Rerum Natura* that point to both the human (epilepsy *DRN* 3.487–498 ≈ *G.* 3.511–517) and the animal (mother/plowmate *DRN* 2.352–367 ≈ *G.* 3.515–530) sphere. The interaction between these illustrations is so great that the difference between the two worlds becomes blurred. Here Vergil emphasizes intergeneric similarity by focusing on a single quality, the unselfish love of one being for another, which transcends the boundaries of genus and species. The human and animal worlds become as one.

A third reference in Vergil's "Plague" carries this point even further. Vergil's passage on the death of the plow animal differs from its model in two respects: first, it concerns the reaction not of the beast's mother, but of its yokemate;[60] and second, the animal's death is caused not by sacrifice, but by disease. Reminded, perhaps, by the circumstances of the Lucretian exemplar, Vergil also incorporates language from a second

57. West 1979.82–83.

58. The passage tacitly makes a further ethical point condemning anthropocentrism, something that no doubt made an impression on Vergil.

59. Cf. Liebeschuetz 1965. The correspondence is most apparent in the discourse on sex: see Putnam 1979.191–201.

60. The idea of a blood relationship is retained, however, in Vergil's phrase *fraterna morte* (*G.* 3.518).

passage in *De Rerum Natura* where a child is lost through sacrifice. Vergil describes the futile efforts of priests to expiate the plague by sacrificing a calf.[61] The language at several points recalls Lucretius' pathetic vignette of "The Sacrifice of Iphianassa" (*DRN* 1.80–101).[62] The scene that Vergil describes was repeated frequently (*saepe* 486); Lucretius argues that his doctrines will not breed impiety, but that this has more frequently (*saepius* 82) been true of traditional religion. Vergil stations his victim at the altar (*stans hostia ad aram* 486) adorned with the traditional fillet (*circumdatur infula* 487) until it falls amidst the attendants (*ministros* 488). He further notes that after the death blow (*ferro mactaverat* 489) the victim was found to have hardly enough blood even to stain the sacrificial knives (*vix suppositi tinguntur sanguine cultri* 492). All these details find their parallels in Lucretius. Iphianassa too is a victim (*hostia* 99) standing before an altar (*ante aras adstaret* 89) adorned with the fillet (*infula...circumdata* 87), fallen (*concideret* 99) among attendants (*ministros* 90) and ritually slaughtered (*mactata* 99); her blood, however, abundantly stains the sacred place (*Iphianassai turparunt sanguine foede | ductores Danaum delecti, prima virorum* 85–86). The wealth of detail and congruence of language and imagery secure the reminiscence and confirm the thematic linkage between the two passages.

In Lucretius' sacrifice, the comparison between the human and the animal world is a travesty. Human sacrifice is so obviously repugnant that the superstitious motives behind it seem criminally absurd. The polemic against religion gains strength with the pathetic sacrifice of the calf in *De Rerum Natura* 2. The same idea recurs in many connections throughout the poem until its final appearance in "The Plague of Athens":

> omnia denique sancta deum delubra replerat[63]
> corporibus mors exanimis onerataque passim
> cuncta cadaveribus caelestum templa manebant,
> hospitibus loca quae complerant aedituentes.
> nec iam religio divum nec numina magni
> pendebantur enim; praesens dolor exsuperabat.
>
> *De Rerum Natura* 6.1272–1277

61. I presume, on the basis of the various Lucretian parallels, that this is the purpose of the sacrifice.

62. Again Klepl notes parallelisms between *Georgics* 3.486 and *DRN* 1.87 (1940.54 and 61 n. 104) and between *Georgics* 3.492–493 and *DRN* 1.85–86 (1940.62 n. 106), but makes nothing interesting of them. Harrison (1979.10) has some useful remarks on the overall imitation. On Vergil's recurrence to this passage in the *Aeneid* see Hardie 1984.

63. Cf. *DRN* 2.352 *nam saepe ante deum vitulus delubra decora....*

Religion is powerless to alleviate the plague. This is also the point of Vergil's sacrifice motif. Indeed, Vergil follows the imagistic development of this idea in *De Rerum Natura* right down to the cadavers that litter the plague-ridden area:

> iamque catervatim dat stragem atque aggerat ipsis
> in stabulis turpi dilapsa cadavera tabo
>
> *Georgics* 3.556–557

In the georgic world, it is the *stabulis* rather than the *deum delubra* and *caelestum templa* (*DRN* 6.1272 and 1274) that are filled with bodies, but the effect and the intent are identical.

In fine, the essential difference between Vergil's "Plague" and that of Lucretius consists in the mythic personification of science on the one hand and death on the other. I have already mentioned this passage as an example of *oppositio in imitando*. While Lucretius indulges in occasional mythic personification, he generally explains himself after the fact in more prosaic terms, and never uses the technique, as Vergil does here, to intensify the horrific aspect of death, the very thing which his poem is intended to allay.[64] Vergil's use of the technique is therefore surprising in this carefully constructed, overwhelmingly Lucretian context. In this terrifying image of hellish forces, the reason that guides man's relationship with nature dissolves before the final irrationality of death. Because this personification and the use to which it is put seem so alien to Lucretius' philosophical message, most readers have taken the passage as another element of *anti-Lucrèce chez Virgile*. One would do well to avoid leaping to this conclusion before examining Vergil's use of Lucretius' poem as a whole. Increasingly scholars have learned to appreciate the *Georgics* as a form of philosophical discourse.[65] Accordingly, we may wish to read allusion to *De Rerum Natura* as a method of philosophical debate. But the model of discourse that we should bear in mind is cooperative dialectic rather than crudely polemical doxography. In particular, we must put off the old assumptions of specious psychobiography that lurk in so much *Georgics* scholarship—that Vergil's early Epicureanism gave way to a maturing Augustanism, and that the *Georgics* enacts the poet's exorcism of the spirits that animated his early work by

64. The most explicit example of mythic personification and an attendant disclaimer is the "*Magna Mater*" episode (*DRN* 2.600–660). On Lucretius' use of myth in general see Ackermann 1978.

65. Most outstandingly Ross 1987.

simply rewriting the work of an early influence, Lucretius.[66] Such a simplistic view of the matter does Vergil no justice; nor does the idea of a quasi-Oedipal rejection of Lucretius' Epicureanism mesh well with recent thinking about the poet's "maturing Augustanism." Above all, we should remember that we are dealing here with a philosophical debate carried out by poets using the tools of poets—image, gesture, and evocative diction, not technical terms and syllogisms. When we consider Vergil's concluding image of medical futility against hellish force in the light of our overall discussion of the "Plague," one point seems clear: that both poets in these passages regard religion as useless at best, and potentially harmful. Indeed, Vergil develops his position on this point with repeated allusion to Lucretius. By adopting a more overtly theistic stance than his predecessor, on one level he parts company with Lucretius; but his view of religion throughout the "Plague of Noricum" is as grim and dissatisfied as anything in *De Rerum Natura*.[67]

"The Praise of Spring"

Contaminatio, the process of combining or conflating passages by multiple allusion, is obviously a compositional tool of great potential, and Vergil thoroughly explored its many facets. Some of its applications are fairly simple, even simpler than what we observed in "The Plague of Noricum." At the most basic level, *contaminatio* is a convenient way of condensing a source. While such a motive may seem hardly worth mentioning, one should remember that practically all contextual allusions in Vergil are markedly briefer than their models, so that it is hard to avoid the conclusion that he regarded allusive condensation as an end in itself. The phenomenon is most readily illustrated by those quotations that combine elements of adjacent verses into a single imitative line:

> ἔνθ' ἄρ' ἔην Γλαύκη τε **Θάλειά τε Κυμοδόκη τε,**
> **Νησαίη Σπειώ τε Θ**όη θ' Ἁλίη τε βοῶπις
>
> *Iliad* 18.39–40
>
> **Nisaee Spioque Thaliaque Cymodoceque**[68]
>
> *Aeneid* 5.826

66. This view is most succinctly expressed by Farrington 1963.
67. For a different view of Vergil's religiosity in this passage see Harrison 1979.
68. Editors rightly bracket this line at *Georgics* 4.338 as an interpolation from *Aeneid* 5.826.

temptabant **fructusque feros m**ansuescere terra
cernebant indulgendo blandequ**e colendo**[69]

De Rerum Natura 5.1368–1369

agricolae, **fructusque feros m**ollit**e colendo**

Georgics 2.36

et quid quaeque qu**eant** per foedera naturai
quid *porro nequeant*, sancitum quandoquidem exstat

De Rerum Natura 1.586–587

et quid quaeque ferat regio et ***quid*** *quaeque recuset*

Georgics 1.53

Similarly in contextual allusion *contaminatio* is used purely or essentially to condense a source. The battle scenes of *Aeneid* 9–12 provide abundant examples.[70] Here the poet solves the problem of how to summarize the much more extensive battle scenes of *Iliad* 3–22 by composing only a few individual scenes out of elements borrowed from several Homeric episodes. We may thus say that Vergil's much briefer narrative imitates or encapsulates that of Homer not merely in part, but as a whole. But cases like this are few. A need for condensation per se is seldom an adequate explanation of contextual *contaminatio*. Certainly it is correct to say that "The Farmer's *Arma*" condenses its Hesiodic model; but at the same time, Vergil expands upon the original in some respects and even introduces into his imitation extraneous Hesiodic ideas drawn from other contexts. This process appears even more clearly in "The Plague of Noricum," where Vergil actually subverts the "scientific" intention of the Lucretian original in order to emphasize through contextual *contaminatio* its latent emotionalism. In such cases we may regard contextual *contaminatio* as a means of creating allusive contexts that contain cross-references, so to speak, between related passages of a given source. These contexts cannot usefully be considered simply as reworkings of rhetorically attractive material. Vergil's "Plague," for instance, must be read not simply against Lucretius' "Plague," but against the several passages to which it alludes, passages which in combination define an important Lucretian and Vergilian theme. Imitations of this kind reveal

69. Another echo of this Lucretian passage occurs at *Georgics* 2.239: *frugibus infelix ea, nec mansuescit arando*. Parallels of this sort have traditionally been taken to indicate a motive of stylistic revision. For Vergil's tendency to compress Lucretian borrowings, see Merrill 1929.403.

70. Knauer 1964a.317–318, 335; cf. 1964b.65 (=1981.875).

the dynamism of Vergilian allusion, which begins with the poet's imaginative, analytical reading of a source and proceeds by editing, rewriting, and finally outlining a web of correspondences within that source. The allusive passage serves rather as a map of how Vergil reads his models.

The "contaminated" verses listed above illustrate another important point. In no case is it possible to distinguish a primary from a secondary model; the Vergilian line alludes to both in equal measure. The same is often true in contextual *contaminatio*. Unlike "The Plague of Noricum," in which it is possible to speak of a primary model with cross-references to secondary models, many Vergilian allusions create a web of references that involve a number of equivalent models. In the case of *Eclogue* 3, for instance, it is useless to distinguish Theocritus' first, fourth, fifth, sixth, eighth or ninth *Idyll* as its primary model: rather, the Vergilian poem imitates parts of all, as well as something that all of them have in common, the idea of alternation and exchange. The same holds for the action of *Aeneid* 1. The hero's arrival on the coast of Libya and subsequent adventures allude at various points to Odysseus' landings on Scheria, Aeaea, and Ithaca.[71] It is impossible to say that any one of these correspondences truly predominates: the multiple allusion signals to the reader that Aeneas' experience at this point is highly ambivalent. Like Odysseus, he has escaped from storm at sea and arrived in the land of a friendly people; he will encounter a woman who will entrance him and attempt to keep him with her forever; and, by wishful thinking, he will come to think or pretend that he has found a new home. Unlike Odysseus, he finds himself in these different circumstances not at three different times and places, but upon his arrival in a single place, Carthage. Such is the power of allusive *contaminatio*.

A similar allusive process is apparent in the central "digression" of *Georgics* 2, the Lucretian "Praise of Spring" (*G.* 2.315–345).[72] It is in many ways the quintessential Vergil that speaks these lines: one finds in them that unique combination of childlike delight in the beauty of nature with an almost superhuman pity for all living, dying things. But as W. Y. Sellar noted long ago, "the passage...is thoroughly Lucretian in feeling, idea, and expression."[73] It should be no surprise to find that Vergil is most himself when he is being most Lucretian. Imitation is the

71. Knauer 1964a.148–180; cf. 1964b.68–72 (= 1981.878–881). On these allusions see p. 268 n. 127 of the present work.

72. The following discussion is based on Farrell 1983.98–108. See also Klepl 1940. 11–52.

73. Sellar 1897.246.

very cornerstone of his craft, and Lucretius is one of his most important models. To conclude his explanation of the uniformity and rhythm that govern celestial mechanics (*DRN* 5.509–770), Lucretius refers the reader to the analogy of the fixed and definite progression of the seasons:

> it **ver** et **Venus**, et Veneris praenuntius ante
> pennatus graditur, **Zephyri** vestigia propter
> Flora quibus mater praespargens ante viai
> cuncta coloribus egregiis et odoribus opplet.
> inde loci sequitur Calor aridus et comes una
> pulverulenta Ceres <et> etesia flabra **Aquilonum**.
>
> *De Rerum Natura* 5.737–742

In the *Georgics* passage, the progression of thematic elements—*ver* (2. 322–323), *Venus* (329), *Zephyri* (330), *Aquilonibus* (335)[74]—is very close to that of Lucretius. Indeed, Vergil gratuitously incorporates the of idea seasonal regularity (*certis diebus* 329) from the Lucretian passage (*certo tempore* 5.750), where it is much more relevant, simply as a token of indebtedness. It would seem to go without saying that the general picture of spring as a season of clemency is the same in both passages. The point is well worth making, however, because this congruency with the Lucretian passage dramatically reverses a position taken up earlier in the *Georgics*. In a discussion of "Vernal and Autumnal Storms" (*G.* 1. 311–350), spring and fall appear as times of uncertain or even dangerous weather; but in the passage quoted above, Vergil portrays both these seasons, and especially spring, as periods of almost preternatural clemency.[75]

At least one important element of the Vergilian passage, however, is not found in the immediate Lucretian context. This is the motif of the *hieros gamos*.[76] Although some form of this motif appears in nearly all primitive mythology, Vergil's use of it here is without doubt a specifically Lucretian allusion. It is a familiar part of Lucretius' argumentative style

74. Note that Vergil varies Lucretius' learned reference to Aquilones as prevailing Summer winds (cf. Seneca *NQ* 5.10–11) by alluding to Aquilo's more commonly attested (and in Vergil invariable) association with winter storms.

75. Cf. Ross 1987.88–89, 119–121.

76. For a general discussion of this motif and its prevalence see Frazer 1911.120–170, especially pp. 129–155. In surviving classical literature before Lucretius, the motif is most fully developed in two tragic fragments (Aeschylus fr. 44 [Radt 1985.159–160] from the lost *Danaids*, and Euripides fr. inc. 898.10ff. [Nauck 1964.648–649]).

to conclude a section of closely reasoned scientific theory with an effusive "digression" in which he elaborates an easily grasped intuitive truism. His purpose, no doubt, is to use the familiar conviction as a way of making his argument seem more plausible than it otherwise might have, and perhaps also to provide a reward for the diligent student.[77] By using the *hieros gamos* motif he makes his closest approach to an authentically primitive religious outlook.

The appearance of this motif at *Georgics* 2.324–327 reflects Vergil's appreciation of the way in which Lucretius advances his argument by developing imagery and theme. Lucretius twice uses the *hieros gamos* motif to underline one of the most important tenets of his system. At *De Rerum Natura* 1.248–264 he introduces the motif as an emotional climax to the poem's first major proof. Nothing, Lucretius argues, can be reduced to nothingness; rather all things at their death are resolved into atoms (248–249). Rain, for example, is sent down from heaven by *pater aether* into the lap of *terra mater*, the result being field crops and fruit trees, which in turn feed beasts and humans: thus can we speak almost literally of fertile cities "aflower" with young (*laetas urbis pueris **florere** videmus* (255), knowing that the propagation of all species depends on this principle (250–261). The poet concludes by restating his main point, that things do not perish utterly, but rather provide material for the creation of new things (262–264). As often, however, his final point hints at an idea that he will develop more fully later: *nec ullam | rem gigni patitur nisi morte adiuta aliena* (263–264). Here as elsewhere, then, Lucretius uses a highly allegorical mythic *exemplum* to explain a perfectly natural phenomenon.[78] The phenomenon in turn illustrates an analogical principle of Epicurean physics. In this case, the primitive mythic conception of a cosmic union between the sky-god/father and the earth-goddess/mother illustrates the functioning of something as prosaic as the water cycle. This phenomenon serves in turn as an analog of what occurs at the atomic level when a thing is born or dies. As the "death" of rains makes possible the birth of plants, so does the dissolution of one atomic structure make available material from which a new structure can be created.

The same idea and illustrations occur again at *De Rerum Natura* 2.991–1022, a closely related context. Here, too, Lucretius is concerned

77. An aspect of the "honeyed cup" image (*DRN* 1.936–950 ≈ 4.11–25).

78. Again we may compare the "*Magna Mater*" passage (*DRN* 2.581–660), where the rationale behind this technique is explained.

with explaining the atomic basis of Epicurean physics. He has already demonstrated the existence of *corpora prima*, indivisible and unalterable particles lacking any of the characteristics with which sensible objects are endowed, but which combine with one another to produce such objects (*DRN* 1.146–634, 2.62–990). The paragraph in question stands as the conclusion to this argument. Following as it does a long series of more formal, dispassionate proofs, the emotional impact of the passage is especially great:

> denique caelesti sumus omnes semine oriundi;
> omnibus ille idem pater est, unde alma liquentis
> umoris guttas mater cum terra recepit,
> feta parit nitidas fruges arbustaque laeta
> et genus humanum, parit omnia saecla ferarum,
> pabula cum praebet quibus omnes corpora pascunt
> et dulcem ducunt vitam prolemque propagant;
> quapropter merito maternum nomen adepta est.
>
> *De Rerum Natura* 2.991–998

As in *De Rerum Natura* 1.248–264, this motif is immediately related to the idea of balance between birth and death. Recalling the point with which the earlier passage concludes, Lucretius now focuses more intently on the idea of death. What was of earth goes back to earth, what of the sky back to the sky: death does not destroy the fundamental particles of matter that make up all things, but merely releases them from *coetus*, enabling them to recombine in new shapes, colors, and so forth (*DRN* 2.999–1006). What matters is how these atoms are combined, with what other atoms and in what position, how they move relative to one another. Death is simply the rearrangement of these factors; the atoms are eternal (*DRN* 2.1007–1012).

The relationship between the two Lucretian passages is obvious: they are companion pieces written deliberately to recall one another. Vergil clearly realized this and wrote *Georgics* 2.324–331 with both passages in mind. Many traces of the two *loci* are clearly visible in his "Praise of Spring." There at the beginning of spring we find *pater Aether* (325: cf. *DRN* 1.250, 2.992, 1000) sending his rain—i.e. *genitalia semina* (324–325: cf. *DRN* 1.250, 2.991)—into the lap of his *coniunx* (326: cf. *DRN* 2.1004 *coniungit*). This event nourishes (*alit* 327, *almus* 330: cf. *DRN* 1.254 *alitur*, 2.992 *alma*) all their offspring (*omnis alit fetus* 326–327: cf. *DRN* 1.248 *hinc alitur porro nostrum genus atque ferarum*,

2.991–995). Further, it results in parturition (*parturit almus ager* 330: cf. *DRN* 2.993–995 *mater terra feta parit*). Vergil's debt extends even to such details as the songbirds mentioned among the offspring of this cosmic union. A line such as:

> avia tum resonant avibus virgulta canoris
>
> *Georgics* 2.328

not only takes into account a specific element in one of the immediate source passages:

> frondiferasque novis avibus canere undique silvas
>
> *De Rerum Natura* 1.256

but also incorporates a *figura etymologica* in the manner of Lucretius to illustrate an ontological relationship between *trackless* wilderness (*avia)* and birds (*avibus*).[79] This bare list of verbal similarities cannot express the precision of Vergil's imitation in versecraft or its thematic depth. Nevertheless, it will suffice to show that Vergil drew both on elements common to the two Lucretian passages, and on elements unique to each in constructing his double allusion.

Vergil's allusiveness ramifies in a way that reflects the interconnectedness of Lucretius' major themes. In the "*Hieros Gamos*" passage and in the arguments that precede it (*DRN* 1.215–264), Lucretius seems to echo the language of the proem of *De Rerum Natura* 1:

> praeterea quaecumque vetustate amovet aetas,
> si penitus peremit consumens materiem omnem,
> unde *animale genus* generatim in ***lumina*** *vitae*
> redducit *Venus*, aut redductum **daedala tellus**
> unde alit atque auget generatim ***pabula*** *praebens?*
>
> *De Rerum Natura* 1.225–229

> alma *Venus*, caeli subter labentia signa
> quae mare navigerum, quae terras frugiferentis
> concelebras, per te quoniam *genus* omne *animantum*
> concipitur visitque exortum ***lumina*** *solis:*
> te dea, te fugiunt venti, te nubila caeli

79. A similar pun on *volucris* is found at *DRN* 2.145; cf. *avium* 5.1379/*avia* 5.1386. For the general principle, see Friedländer 1941 and Snyder 1980.

adventumque tuum, tibi suavis **daedala tellus**
summittit flores...

De Rerum Natura 1.2–8

hinc **fessae pecudes p**ingui **per pabula laeta**

De Rerum Natura 1.257

unde **ferae pecudes per**sultant **pabula laeta**

De Rerum Natura 1.14

ludit lacte mero *mentis* ***perculsa*** novellas

De Rerum Natura 1.261

significant initum ***perculsa****e corda* tua vi

De Rerum Natura 1.13

These associations center upon the figure of Venus. It is through Venus especially that one finds the Lucretian link between the *hieros gamos* and the other ideas with which Vergil associates it in the "Praise of Spring." This link leads the reader back to *De Rerum Natura* 5.737–742, the procession of Spring, Venus, Zephyr, and the rest. The crucial association of spring with Venus occurs in three of the four Lucretian passages, the *hieros gamos* in two. Moreover, the three earlier passages—*DRN* 1.2–8, 1.215–264, and 2.991–1022—clearly prepare the reader for the last. Just as the atomism of *De Rerum Natura* 1 and 2 is logically prior to the history and sociology of Book 5, so is the attendant imagery of the earlier books poetically prior, as it were, to that of the later ones. Thus the arguments and motifs presented in *De Rerum Natura* 5 are the culmination of a long and elaborate exposition from which they cannot easily be separated. In studying *De Rerum Natura*, Vergil perceived these connections as a vital part of the poem's thematic structure. Consequently, by way of honestly representing the stage of argument that Lucretius reaches in *De Rerum Natura* 5, he took care to maintain the same nexus of related motifs in the elegant imitation that he created in the *Georgics*.

The rigor of Lucretius' thinking is expressed in the directedness of his poem's thematic development. Despite the shifting nature of the associations which he draws between different thematic elements, his symbolic discourse on nature leads to a satisfying and seemingly necessary conclusion. Thus the arguments presented in *De Rerum Natura* 5 are the focus both of Lucretius' poem and of Vergil's allusions. "The

Procession of the Seasons" (*DRN* 5.737–742) exists as a symbol of the ever-present cycle of birth and death. Lucretius never loses sight of this truth, that *Venus*, or *terra mater*, or *natura creatrix*, in order to bring new things to life, requires the death of things that already exist. Vergil suppresses or ignores this point, and many have taken this departure from his model as another element of anti-Lucretian polemic. In fact, this procedure is merely the complement of what was observed in the "Plague of Noricum." There Vergil intensifies the dread with which the reader experiences his "Plague" by personifying Lucretius' relatively colorless opposing forces of medicine and disease as Chiron and Tisiphone. Here he intensifies the joy with which we contemplate the miracle of life by ignoring the connection, explicitly drawn by Lucretius, between birth and death. If we consider these allusions together, the variations appear not so much departures from Lucretius as references to the thematic emphasis of the two main source passages. Vergil has, in a way, simplified his source. Where Lucretius takes pains to show his reader that life is never so very far from death, and that death itself, no matter how horrible, is fundamentally nothing more than a rearrangement of atoms, Vergil seizes on the same images and motifs to create a strong emotional contrast. But the contrast is one that is unquestionably found in the original, and one, to anticipate the argument of Chapter 5, that is central to the thematic structure of *De Rerum Natura*.

Lucretius' "Procession of the Seasons" concludes one exposition, but introduces another. Having completed his account of celestial mechanics, Lucretius goes on to discuss the beginnings of life on earth in the period that he calls the *novitas mundi:*

> nunc redeo ad mundi novitatem et mollia terrae
> arva, novo fetu quid primum in luminis oras
> tollere et incertis crerint committere ventis.[80]
>
> *De Rerum Natura* 5.780–782

The words *nunc redeo* (780) characterize the entire preceding section on astronomy (509–770) as a digression from Lucretius' main topic, the birth of the world and of all its creatures. It is equally clear that Vergil's interest is in this topic, and not at all in astronomy; for it is primarily to

80. Several have seen here a parallel to *Georgics* 2.332–333 (*inque novos soles audent se gramina tuto | credere*): see, e.g., Merrill 1918a.198 (#1186); Conington–Nettleship 1898.260 *ad G.* 2.332 ff.

Lucretius' account of the *novitas mundi* that he looks in developing his "Praise of Spring":

> non alios prima crescentis origine mundi
> inluxisse dies aliumve habuisse tenorem
> crediderim: ver illud erat, ver magnus agebat
> orbis et hibernis parcebant flatibus Euri,
> cum primae lucem pecudes hausere, virumque
> terrea progenies duris caput extulit arvis,
> immissaeque ferae silvis et sidera caelo.
> nec res hunc tenerae possent perferre laborem,
> si non tanta quies iret frigusque caloremque
> inter, et exciperet caeli indulgentia terras.
>
> *Georgics* 2.336–345

Vergil's assertion that life began in an epoch of springlike clemency is more emphatic than its Lucretian inspiration (*DRN* 5.801–804), and it is made in more general terms than Lucretius uses, but the debt is no less clear. Other correspondences between the passages confirm that debt.[81] The Lucretian passage is moreover rich with cross-references to other parts of *De Rerum Natura*. The most important of these is the "earth mother" theme:

> linquitur ut merito maternum nomen adepta
> terra sit, e terra quoniam sunt cuncta creata
>
> *De Rerum Natura* 5.795–796

> quare etiam atque etiam maternum nomen adepta
> terra tenet merito, quoniam genus ipsa creavit
> humanum atque animal....
>
> *De Rerum Natura* 5.821–823

The phraseology here of course recalls

> quapropter merito maternum nomen adepta est.
> cedit item retro, de terra quod fuit ante,
> in terras et quod missumst ex aetheris oris,
> id rursum in caeli rellatum templa receptant.
>
> *De Rerum Natura* 2.998–1001

81. Cf. *G.* 2.331 ≈ *DRN* 5.806; *G.* 2.344–345 ≈ *DRN* 5.818–819.

and this final link completes the circle of associations on which the Vergilian passage stands.

While here as in the other passages that we have considered it is possible and even useful to speak of particular sources, the "Praise of Spring" reads like an allusion not to one or a few limited passages in *De Rerum Natura*, but to a complex of Lucretian themes. While the idea of multiple allusion as cross-reference remains valid and important, in passages such as this the process of tracing a multitude of cross-references leads ultimately to the impression that Vergil's "source" is actually a single, unified idea. Lucretius develops the idea gradually throughout his poem, above all in the passages that Vergil imitates in "The Praise of Spring." The allusion itself, however, takes for granted Lucretius' lengthy, careful development of the idea and presents it as a whole encapsulated in a brief passage of compelling force and resonance. More and more we see that allusion does not merely establish relationships between individual passages of limited extent. Here and elsewhere Vergil employs allusion as a compendious way of including in his own poem important thematic and imagistic concerns that are more fully developed in previous literature. As in "The Praise of Spring," the previous development of these ideas may have originally spanned five lengthy books of hexameters; or, as in the following example, it may involve two entirely different poems of even greater length.

"*Aristaeus*"

In the final episode of Book 4 we find one of the most fascinating examples of contextual allusion in the *Georgics*. The episode consists of two narratives, a frame telling how the hero Aristaeus discovered the *bugonia*, and a second tale enclosed by the first, concerning the death of Eurydice and the catabasis of Orpheus. What interests me here is the way in which Vergil has modeled the introductory "Aristaeus" narrative on a pair of Homeric episodes. In some respects, this narrative may be regarded as parallel to "The Praise of Spring," a contextual allusion through several passages to a nexus of important themes in the source. In at least one respect, however, the two allusions are very different: the Homeric episodes, unlike the Lucretian ones discussed above, are found in different poems. Both the *Iliad* and the *Odyssey* contribute to the "Aristaeus."

Many of the developed techniques of narrative allusion described by Knauer in his study of the *Aeneid* occur already in this episode of the

Georgics. For instance, in Aristaeus' complaint and subsequent interview with his mother Cyrene we find an episode that is clearly modeled on similar, but quite distinct episodes of the *Iliad.* Twice in Homer we find Achilles bewailing his fate on the seashore as a prelude to an interview with his mother Thetis. The "Aristaeus" begins by following the first of these passages (*Iliad* 1.345–427), in which Achilles makes an articulate complaint about his situation to his mother, who comes out of the sea to visit him. Aristaeus' complaint, like that of Achilles, concerns what he sees as a mark on his honor. Standing at the source of the river Peneus, he reproaches his mother for the loss of his bees, referring bitterly to his lofty parentage (*praeclara stirpe deorum* 322), to his immortal destiny (*caelum sperare* 325), to his exalted position among men (*vitae mortalis honorem* 326), and to her evident unconcern for his reputation (*taedia laudis* 332). Just so Achilles complains that, since Thetis bore him to be a man of short life (μινυνθάδιόν περ 1.352), Zeus should at least grant him honor (τίμην 353). Indeed, Aristaeus by comparison with Achilles even amplifies the theme of honor. Thus it is clear at the beginning that the passage from *Iliad* 1 is Vergil's primary model, especially since, in the second passage (*Iliad* 18.22–137), Achilles makes no formal complaint. He simply cries out in anguish, and his mother hears him (σμερδαλέον δ' ᾤμωξεν· ἄκουσε δὲ πότνια μήτηρ 18.35). But conversely, in the earlier Iliadic passage, when Thetis comes to her son, she comes alone, while in the later one, she is accompanied by her sister Νηρηΐδες (38 and 49), whom Homer names in a catalogue (39–48); and in Vergil too we find Cyrene surrounded by other nymphs[82] who are also named in a catalogue (336–344). Thus we detect an allusive shift from Book 1 to Book 18 of the *Iliad.*[83]

Such combinations of scenes are of course common in the *Aeneid,* where Vergil does not limit his allusions to only one of the Homeric poems at a time. Indeed, one of his chief principles of allusion in the *Aeneid* is to search for points of similarity between the *Iliad* and the *Odyssey* and to make these similarities the basis of narrative, thematic and

82. Specifically Oceanids, although only two (the *sorores* Clio and Beroe 341) are explicitly designated as such.

83. The shift occurs perhaps as early as *Georgics* 4.333, where the word *sonitum* aptly denotes what Cyrene actually hears at the river bottom, the sound of her child's voice; but from the reader's point of view, the word refers rather oddly to Aristaeus' articulate and highly rhetorical speech (321–332). If we bear in mind Vergil's models, we may appropriately recall Achilles' inarticulate groan in *Iliad* 18.

structural characteristics of his own poem. This principle informs the "Aristaeus" as well. Thetis in the *Iliad*, after each of her interviews with Achilles, goes to enlist the help of a god—in Book 1 Zeus, in Book 18 Hephaestus—to defend her son's τίμη. In the *Georgics*, Cyrene informs Aristaeus that he must enlist the aid of Proteus if he wishes to learn why he has lost his bees and how he may acquire a new swarm (387–414). Both this scene and the hero's subsequent encounter with the vatic sea god are based on a new model, *Odyssey* 4.351–572. There the nymph Eidothea advises Menelaus that he may discover why his homecoming has been so long delayed and how he might finally achieve it by capturing and then consulting her father—the very same Proteus, in fact. In both the *Odyssey* and the *Georgics*, a hero receives from a goddess explicit instructions on how to overpower this god when he takes his rest from shepherding his seals at midday;[84] and in both poems the hero succeeds in holding on to Proteus as he turns himself into various shapes in an effort to escape, but finally agrees to disclose the information that the hero desires.

In combining elements of three distinct episodes into one allusive context, Vergil introduces another technique that appears frequently in the *Aeneid*. We have in Aristaeus a very clear example of a "conflated" character.[85] In his plaint by the side of the river Peneus to his mother

84. Here an element of Vergilian *variatio* may be noted. In Homer, Eidothea disguises Menelaus and two of his men by draping them in seal hides, and when they complain of the stench, she smears ambrosia under their noses. Vergil eliminates the disguise and along with it the stench. One might compare this revision to that of *Aeneid* 5.327–330, where Nisus slips in a pool of sacrificial blood instead of a pile of dung, as Ajax does in Homer (*Iliad* 23.773–777). But Vergil's squeamishness and lofty seriousness may not be the only factors at work here. Schlunk (1974.14–15) has shown that in the footrace episode Vergil alters a detail that had been censured by ancient critics of Homer. Something similar may be at work in the *Georgics*, where we find that Vergil retains Homer's ambrosia, but gives it a nobler purpose. Cyrene applies it not to Aristaeus' nostrils, but rather anoints his whole body with it; and when he inhales its fragrance, his strength is enhanced (*habilis membris venit vigor* 418). This treatment of the ambrosia may correspond to the efforts of ancient critics to interpret such undignified elements in Homer allegorically; see, for example, Σ E *in Odysseam* 4.445 (Dindorf 1855.1.213): ἀλληγορικῶς ἀμβροσίην τὴν εὐελπιστίαν τοῦ ἀποτελέσματος. ὑπέμεινε γὰρ τὴν δυσωδίαν διὰ τὸ μέλλειν κατορθῶσαι τὸ ἑαυτοῦ συμφέρον. In Chapter 6 I will argue that Vergil relied heavily on allegorical exegesis of the Homeric poems in composing his "Aristaeus." Note too (with Richter 1957.387 *ad G.* 415 ff.) the use of ambrosia for similar purposes at *Iliad* 19.170, *Odyssey* 10.287 and 14.229.

85. See Knauer 1964a.542 s.v. "Kontamination: Vereinigung mehrerer homer. Gestalten in," and cf. Knauer 1964b.66 = 1981.875.

Cyrene, Aristaeus "is" Achilles, who twice stands on the seashore and addresses complaints to his mother Thetis, first bemoaning his ill-treatment at the hands of Agamemnon, and second lamenting the death of his friend Patroclus. Similarly, by extracting vital information from the sea god Proteus, Aristaeus "becomes" Menelaus, who in Homer performs the very same feat. Cyrene too is a conflation: her role combines that of Thetis in the *Iliad* with that of Eidothea, Menelaus' patroness in the corresponding episode of the *Odyssey*. These general relationships are enriched and complicated by further correspondences involving still other characters. For instance, in both Iliadic passages Thetis comes to visit Achilles, while in the *Georgics* Aristaeus' underwater journey to the home of his mother, his amazement at its beauty, and the heroic welcome that he receives there, are described in great detail (*G.* 4.359–379). This passage corresponds to nothing in the *Iliad*, nor to any part of Menelaus' adventure; but it finds its parallel in the wonder of Telemachus and Pisistratus on their arrival at Menelaus' palace and in the heroic reception accorded them there (*Odyssey* 4.36–58).[86] Thus Aristaeus is in some respects "derived" from Telemachus as well as from Achilles and Menelaus.

In general we may regard the main characters of Vergil's "Aristaeus"—with the exception of Proteus, who is nearly identical with the Homeric prototype—as conflations, and the narrative itself as a sequential combination of several specific Homeric scenes. To be sure, this sequence is interrupted by passing allusions to a number of other contexts; but these will not concern us here.[87] It is the main elements of the allusion that I want to discuss. The formal procedures that Vergil followed are easy to identify. The seams between those sections of his narrative where he shifts from one model to another indicate the points of similarity or difference between them, the features that made it possible to combine and conflate the elements of several Homeric passages into a unified allusive context. Thus the overall similarity between Achilles' interviews with Thetis made it possible to allude to each of them in successive parts of one episode; the fact that Achilles and Menelaus are in all of these scenes the protégés of sea goddesses facilitated the composition of a single episode alluding to all three. But necessary as it is to

86. Norden 1934.638 suggests that Bacchylides 17 ("Theseus") offered Vergil a collateral model for the hero's underwater voyage, citing lines 100–109 in particular; but despite the parallel situation, it is difficult to see how Bacchylides influenced Vergil in any significant way.

87. I shall deal with these other allusions in Chapter 6.

identify these features, merely identifying them cannot tell us much. Why does Vergil allude to Homer here in the first place, and why to these particular scenes—not the most obvious ones if Vergil's interest was merely to provide his epyllion with a bit of Homeric color? So careful an allusion must be informed by some specific purpose, and our interpretation should be guided by an attempt to discern it.

I believe that Knauer is surely right in holding that the "Aristaeus" is a clear and direct anticipation of the analytical approach to Homer that informs the *Aeneid*.[88] By selecting particular episodes from both Homeric poems to be the basis of his allusive epyllion, Vergil follows the same basic procedure that would later produce his final masterpiece. His selection of episodes makes sense only if we assume that it was informed by an interpretive analysis of his twin exemplars. I would go so far as to say that, like the *Aeneid*, the "Aristaeus" combines the *Iliad* and the *Odyssey* in a very ambitious sense: that the scenes chosen for imitation are regarded as exemplary—not, to be sure, of the Homeric epics in their entirety, but of an important trait that they both share. From this point of view, the "Aristaeus" and the *Aeneid* differ mainly in degree rather than in kind. In the *Aeneid*, despite the fact that the poem is only one-third the length of the two Homeric poems, it was possible to develop a much more intricate and therefore complete system of allusion to nearly every aspect of both models. In the very brief "Aristaeus" it would have been necessary to be even more selective, to concentrate only on those elements that spoke to Vergil's immediate purpose.

If this view of Vergil's working method is correct, it should be possible to read his "Aristaeus" as an allusion to some essential thematic element of both Homeric poems. Even this is asking a lot of a very short and idiosyncratic narrative that is more often tantalizing than explicit in its handling of essential issues. Yet I think that this conclusion is justified by the obvious correspondences in allusive technique between the "Aristaeus" and the *Aeneid* and also by the conclusions to be drawn from Vergil's selection of models. The episodes in question, as I have noted, are hardly typical, nor do they appear to address the main themes of the two poems in a direct way. To represent Achilles' anger, one might first have chosen a scene such as his battle with the river Xanthus in *Iliad* 21 or his abuse of Hector's body in 23. Why would Vergil have chosen the scenes that he did if his purpose was to allude to what Homer

88. Knauer 1981.910–914.

himself announces as the main theme of the *Iliad*? As far as the *Odyssey* is concerned, there is the even more perplexing question of why that poem's hero is completely absent from the passage to which Vergil alludes. But these questions arise only if we attempt to make Vergil's interpretation of Homer in the *Georgics* conform to our own. His purpose here is not necessarily to advance a convincing interpretation of Homer's main themes, but to find in the *Iliad* and the *Odyssey* elements that shed light on the foremost concerns of the *Georgics*—and, more specifically, of the "Aristaeus." If, therefore, we ask what themes the episodes in question define, and, more important, if we attempt to answer this question with regard to their redeployment in the "Aristaeus," we may find that Vergil's allusion does in fact outline a coherent interpretation of Homer and, like the allusions to Hesiod, Aratus, and Lucretius that I have already discussed, establishes the place of the *Georgics* in a particular tradition of epic poetry—in this case narrative and heroic rather than expository and didactic.

The Iliadic episodes in question mark decisive stages in the poem's narrative. It is in these scenes that Achilles gives voice to his anger and acts on it, by appealing to his mother for help, in such a way as to guarantee that it will have tragic results. In Book 1, because he has lost his war prize Briseis, Achilles determines that he must stop fighting the Trojans, abandon his former allies, and live a long and inglorious life. In Book 18, because he has lost Patroclus, he determines that he must reenter the fray, avenge his friend's death, and live a short but glorious life. This is Achilles' fate; his decision to opt for a long and inglorious life begins the first stage of his magnification as a hero, a dramatic illustration of his importance through the suffering of the Greek army in his absence. This decision creates the circumstances that bring about the death of Achilles' surrogate Patroclus, the event that precipitates Achilles' decision to return to the war and his second meeting with Thetis; and this episode launches the second phase of Achilles' heroic magnification. In this sense, we may say that the loss of Briseis and Patroclus set in motion the two main actions of the *Iliad* and so cause the fulfillment of Achilles' anger.

The Odyssean episode indicated by the "Aristaeus" bears a similar relationship to that poem's central theme. On a priori grounds, one might have expected Vergil to choose Odysseus' meeting with his mother Anticleia in the "*Nekyia*" as a formal parallel to Achilles' meetings with his mother Thetis. Instead, he chooses a different episode in which the hero does not even appear. I have already noted that Aristaeus at one point "is" Telemachus, and at another Menelaus; but he never represents Odysseus. Thus Vergil's Iliadic allusion directly involves that poem's hero, but his Odyssean allusion does not. We must ask ourselves why Vergil chose this particular episode to represent the *Odyssey* if his purpose was to fashion a synecdochic allusion to both Homeric poems.

The solution to this problem is to be found in its most puzzling feature, the absence of the hero. The journey to Sparta is the last major episode of the "Telemachy." Vergil's allusion is to this entire episode, for he incorporates elements of both its beginning (Telemachus' and Pisistratus' arrival) and its climax (the story of Menelaus' *nostos*). If we inquire into the purpose of this episode, its contribution to Homer's narrative in the *Odyssey*, I believe that most readers will give a similar answer. Telemachus is acting out of a sense of separation from his father, of loss if you will. His purpose in leaving Ithaca is to get news either of his father's death or of his homecoming. In Menelaus he encounters the last of the Greeks to return from Troy, and from him he hears a fantastic tale that anticipates in many respects Odysseus' tale of his own *nostos* in Books 9–12. While Telemachus can take back with him to Ithaca no definite word about his father, he finds in Menelaus' adventures a reason to hope that Odysseus may yet return, that the loss of his father may be overcome. He also learns that those who knew his father recognize Odysseus in him.

The Iliadic and Odyssean episodes, then, are functionally identical. All three are liminal scenes. Achilles' first interview with Thetis marks the point at which he abandons the war, and its result is that Thetis successfully petitions Zeus to make the Greeks suffer in Achilles' absence; while the second interview marks the hero's decision to return to battle, and from this point on his presence dominates the narrative. In just the same way, Telemachus' visit to Sparta concludes the "Telemachy," the only section of the *Odyssey* that is characterized primarily by the absence of the hero, the very function of which is to make the reader (or listener) feel Odysseus' absence by showing how much Telemachus suffers in his absence. Up to the point of Telemachus' journey, Odysseus himself has not yet appeared in the *Odyssey;* but this he will shortly do, and from the point when he does to the end of the poem, his presence will dominate the narrative.

The episodes are similar in another sense as well: all three represent a character's reaction to loss. Achilles, like Telemachus, acts from a sense of loss, first of Briseis, then of Patroclus. All of his actions in the *Iliad* are direct responses to these losses. His fate, however, is different from that of Telemachus. Achilles will learn that his loss is irreparable, that his attempt to avenge the loss of Briseis and, especially, Patroclus, will lead inexorably to his own death. Telemachus finally makes good his loss by aiding Odysseus in his return to Ithaca and life. Aristaeus' situation conflates these experiences. In the first place, the hero complains that he has lost his honor, just as Achilles does after losing Briseis to Agamemnon. But Aristaeus' loss involves death, the death of his bees, and so resembles Achilles' loss not of Briseis, but of Patroclus. Now Patroclus in the *Iliad* is Achilles' doppelgänger: his death prefigures that of Achilles, and in fact sets in motion the chain of events that leads to the death of Achilles, which is formally ἐξὼ τοῦ δράματος.[89] By the same token, Menelaus and Telemachus are, in different ways, surrogates for the absent Odysseus. Just as the death of Patroclus heralds that of Achilles, so do the adventures of Telemachus and Menelaus, not to mention Homer's characterization of them, herald the homecoming, the "return to life," of Odysseus. The Iliadic and Odyssean episodes are complementary in this sense. Aristaeus represents Achilles in his bitterness over the loss that he has suffered; but in other respects he resembles Telemachus and Menelaus.

A further connection: In both of his interviews with Thetis, Achilles mentions with emotion his father, Peleus. Aristaeus, too, when he complains to Cyrene about the loss of his bees, refers to his father. But Aristaeus' reference is given more prominence than those of Achilles, and is given a different turn: what good is my divine parentage, he asks,—if, as you maintain, my father really is Apollo of Thymbra (323). This allusive variation to Achilles' speech anticipates the course of the episode in assimilating Aristaeus' situation to that of Telemachus. When Athena in the guise of Mentes the Taphian asks Telemachus whether he is really the son of Odysseus, his reply is diffident: "My mother says that I am his, but I don't know; for no one ever has personal knowledge of his own birth" (*Odyssey* 1.215–216). The sentiment is conventional, but for Telemachus the problem is real. His task in the *Odyssey* is to come to

89. Cf. Knauer 1981.912: Vergil "may have grasped while reading Homer...that the double loss suffered by Achilles could be welded into one much shorter scene. In the 'Iliad' it is just these two steps, the loss of Briseis (Il. 1) and the loss of Patroclus (Il. 16 and 18), which finally open the way to the predestined death of Achilles and consequently the fall of Troy."

terms, in one way or another, with his father's absence and so to acquire a sense of himself. By journeying to Sparta and hearing the *nostos* of Menelaus, he makes great progress toward this goal and prefigures its ultimate realization later in the poem. Aristaeus, in the end, will not succumb to death, but will overcome it; he will not "be" Achilles, but Odysseus.

In conclusion, I will note that there is strong external evidence that Vergil interpreted the Homeric episodes in just this way, viewing the theme of the hero's absence as an important similarity between them. In *Aeneid* 9 and 10, it is clear that Aeneas plays the roles of Telemachus, Odysseus, and Achilles.[90] In his voyage up the Tiber to the site of Rome (Book 8), he "is" Telemachus journeying to the courts of Nestor and Menelaus. The story of Cacus, like Menelaus' tale of homecoming, contains a special message for the one who hears it. At the same time, in his absence from his son Ascanius and from the Trojan army he resembles Odysseus and Achilles respectively, as we see in Book 9. Again in Book 8, Aeneas correctly interprets a heavenly sign, as Helen does at the close of the "Telemachy," and receives from his mother arms made by Vulcan, just as Thetis had given Achilles armor made by Hephaestus; the correspondence is made memorable by the twin shield ecphrases. Thus possessed of divine armor, both Aeneas and Achilles are properly equipped to return to battle. It is clear from these correspondences that Vergil considered the theme of the hero's absence an important parallel between the Homeric poems; and from our analysis of his Homeric allusion in the "Aristaeus," I think it fair to say that the poet of the *Georgics* anticipated himself not only technically, but with respect to his interpretation of Homer in the *Aeneid* as well.

There is much more to say about this allusive context. For now, it is sufficient to note that the Vergil's "Aristaeus" narrative could have been created only through a process of allusive *contaminatio* that depended on a careful analysis of the relevant Homeric episodes. I believe it is clear how closely the allusive process we can observe in this episode resembles that of the *Aeneid;* only the historical dimension, as Knauer has noted, is lacking.[91] I disagree with Knauer, however, on a related point. Although it is true that the "Aristaeus" differs markedly in texture, tone, and style from any other episode of the *Georgics*, I do not believe that

90. The following basically summarizes Knauer 1964a.239–266; cf. 1964b.76–81 = 1981.884–888.

91. Knauer 1981.912.

this difference is due to the introduction of a new allusive technique. It is very reasonable to assume that this episode both differs from what precedes it and resembles the *Aeneid* because the character of its chief model, Homer, despite all of Vergil's *variatio*, shows through. But I do not find Vergil's method of alluding to Homer in the Aristaeus essentially different from the one he applies to Hesiod, Aratus, and Lucretius in other parts of the *Georgics*—or, for that matter, to Theocritus in the *Eclogues*. In all of the contexts that I have examined, it is clear that Vergil's allusiveness uses the same essential techniques of *imitatio* and *variatio* to take account of the same essential elements—diction, meter, structure, theme, and so on—that inform his allusion to Homer in the "Aristaeus." It is also clear that the care and precision with which he combines and conflates different passages must be based on a thorough, analytical study of his models. If the relationship between Vergil's "Aristaeus" and its models seems richer in some respects, this is for two reasons. First, the "Aristaeus"—unlike almost any other episode of the *Georgics*—is a complete and coherent narrative.[92] In the case of the other allusive contexts that we have examined, final judgment depends on factors that I have not yet addressed. Second, the passage that we have just examined, along with the Orpheus and Eurydice narrative embedded in it, is the conclusion of the *Georgics*. It has been interpreted many ways, but most convincingly as a symbolic coda summing up the main thematic issues that are developed throughout the poem. In it Vergil attains an exquisite balance that is in marked contrast to the dialectical extremes that precede it. Its self-sufficiency and its privileged position suggest that its purpose is in every way to sum up what has gone before. Those contexts of allusion that precede it are every bit as technically sophisticated; but our interpretation of those earlier episodes depends on our understanding of their relationship both to one another and to their climax, the "Aristaeus."

This interpretation will be the subject of Part 2. Before we can proceed to it, we must examine one last context of allusion, one that will introduce that aspect of Vergilian allusion which will be most important in our study of Vergil's overall allusive plan in the *Georgics*—a plan based on a model that is neither a single poem nor a single author but involves a variety of participants in the epic tradition.

92. Only the "Orpheus," with which it is associated, is comparable; and here no model can be identified.

"Lucky and Unlucky Days"

Most of the passages discussed in this chapter involve imitative episodes of an extent more limited than but still roughly comparable to that of their originals. Lucretius' "Plague of Athens" occupies 196 lines, Vergil's "Plague of Noricum" 126; Aratus' *"Diosemiae"* spans 422 lines, Vergil's "Weather Signs" 153;[93] Of course, when one inspects more carefully the relationship between imitation and exemplar, it sometimes becomes necessary to redefine the boundaries of the source passage. According to most scholars, Vergil's six lines on the parts of plow imitate nine lines of Hesiod; but my analysis suggests that Vergil has really fashioned a context of sixteen lines (*G.* 1.160–175) based on a Hesiodic passage of forty-four lines (*WD* 414–457). Moreover, the formal relationship between these passages appears to be much more intricate than was previously thought. Where *contaminatio* creates multiple allusions, the matter becomes still more complex: we may still define with precision the boundaries of an imitative passage, but to count the lines of its sources becomes difficult and rather absurd. Vergil's "Praise of Spring," as we saw, occupies only a relatively few lines (*G.* 2.315–345). It nevertheless involves imitation of several relatively lengthy and widely separated passages of *De Rerum Natura.* In my discussion of this allusion, I concluded that it seems more sensible to regard Vergil's model as a complex of Lucretian themes rather than as a collection of particular verses. Similarly in the case of the "Aristaeus," we could perhaps tally up the lines of the several episodes to which Vergil alludes; but, if my analysis is correct, by doing so we would miss the point that the Homeric passages were chosen for their representative value. The imitated episodes are evidently meant to stand for important themes that inform the entire *Iliad* and *Odyssey.*

In any such imitation the principle of selectivity is obvious. As a conclusion to this survey of contextual allusion in the *Georgics,* I wish to adduce what I believe is among the most highly selective, representative, and intricate imitations in the poem. It is also one of the most taken for granted. In *Georgics* 1.176–186, some scholars have alleged that Vergil's treatment of his model (*WD* 765–828, especially lines 800–813) is cursory, that the allusion is confined to the level of versecraft—really to only a couple of borrowed words and motifs—and has

93. Or 113, depending on whether one counts *Georgics* 1.311–350 as part of the episode; see pp. 157–158.

little contextual importance.[94] A careful examination of the passages in question will show that Vergil has done much more than merely select a few easily recognized motifs from his source. Instead, he has used his normal methods of *contaminatio* to produce an extremely dense allusion of only a few lines to a much longer passage of *Works and Days;* and within this context he has managed an important secondary allusion to an apparently unrelated passage not only of a different poem, but of a different author.

It is true that general similarities between Vergil's calendar and that of Hesiod need not betoken an important allusive relationship. Italian farmers in Vergil's time, like their counterparts in seventh-century Boeotia, liked to coordinate certain activities with the phases of the moon. In general, growth in any sense was felt to be favored by a waxing moon; therefore planting would if possible be done early in the lunar month, so that the seeds would emulate the moon as it grew to fullness at midmonth.[95] Conversely, tasks such as clearing land of trees and uprooting stumps were best performed under a waning moon to inhibit brush from reclaiming the hard-won farmland.[96] Vergil's educated Roman readers would have been familiar with these superstitions, and probably believed in them. Such attitudes were not confined to the peasantry and their rustic activities: even the emperor Tiberius visited the barber religiously in the dark of the moon, to ensure that his hair would grow back with the new moon![97]

Along with this general superstition about the moon's phases, a good deal of folklore attached itself to specific days of the lunar month. Divine birthdays explain the character of certain days. The fourth and seventh days of the new moon, for instance, were widely held to be lucky, the former because it was the birthday of Hermes,[98] the latter

94. Cumont 1933 went too far in denying that Hesiod was Vergil's direct source for the calendar, pointing instead to Hellenistic astrological sources. Wilkinson 1969.58 calls the imitation "perfunctory" and notes, "it is clear that he attached no importance to the subject except as a reminder of Hesiod." Putnam similarly suggests, "it may not be incidental that Hesiod allots some sixty lines (*W. and D.* 765–828) to a subject Virgil scans in 11 (276–286), as if to signal an uncertainty about the accuracy of his instruction" (1979.46 n. 36). M. L. West (1978.355 *ad WD* 779) comically reprimands Vergil for "his cursory and irresponsible imitation of the Days"!

95. Palladius 1.6.12, 3.4, 13.1; Pliny *NH* 18.321 ff.

96. Cato *Agr*. 31.2, 37.4; Columella *RR* 11.2.52.

97. Pliny *NH* 18.321; cf. Varro *DRR* 1.37.4.

98. E.g. [Homer] *Hymn*. 4.19.

that of Apollo.[99] Some surviving calendars incorporate such traditions as these into comprehensive surveys of lunar days, linking each day of the month with a tutelary figure who determines, or at least explains, its character.[100]

Thus the kind of information contained in Vergil's calendar was common "wisdom" available to him from many sources. A few obvious parallels with Hesiod's account have long been recognized, but there are divergences as well. Some scholars have therefore held that Vergil follows some other lunar calendar than the one found in *Works and Days*. Nevertheless, it seems to me essential that we recognize Hesiod as the primary model of this passage. When we see it for what it really is, a mere part of Vergil's much more thoroughgoing imitation of *Works and Days*, the absurdity of denying this relationship will become apparent. This will be the business of the next chapter. For now it will be useful to examine in detail the correspondences between Vergil's calendar and Hesiod's, without regard to context.

In fashioning his "Days" imitation, Vergil as usual condenses his source. How drastically he condenses here has not, however, been appreciated. When we study this passage with care, it quickly becomes clear that "Lucky and Unlucky Days" is one of Vergil's most ambitious and successful essays in concentrated allusion. The peculiarities of Hesiod's archaic calendar and his tendency to go back over old material in a new context offered Vergil a number of ways to condense his source by alluding to several days or groups of days at once. Therefore, instead of merely selecting a few particular days from his source and imitating them obliquely, Vergil has constructed a new, tersely allusive calendar encapsulating that of his more prolix source. If the imitation seems to represent Vergil's immediate model inadequately, as many critics have charged, the reason is not far to seek. The poet's concern is not to follow Hesiod slavishly, nor is he willing to devote much space to this episode. Nevertheless, in the first line of it, Vergil clearly indicates the extent of his reference:

> ipsa dies[a] alios alio[b] dedit ordine[c] Luna[d]
> felicis[e] operum.[f]
>
> *Georgics* 1.276–277

99. Σ *vet. in Hesiodi Opera* 770a (Pertusi 1955.237–238); Apollodorus 244 F 37; Plutarch *QC* 717d; Σ *in Aeschyli Septem* 800–802a and b, 800b, e, f, h–j (Smith 1982.342–343); Martini–Bassi 1901.33.

100. Cumont 1933.

> **ἤματα**[a] δ' ἐκ Διόθεν πεφυλαγμένος εὖ **κατὰ μοῖραν**[c]
> πεφραδέμεν δμώεσσι τριηκάδα **μηνὸς**[d] **ἀρίστην**[e]
> **ἔργα**[f] τ' ἐποπτεύειν ἠδ' ἁρμαλιὴν δατέασθαι.
>
> *Works and Days* 765–768

> αἵδε μὲν **ἡμέραι**[a] εἰσὶν ἐπιχθονίοις μέγ' ὄνειαρ·
> αἱ δ' ἄλλαι μετάδουποι, ἀκήριοι, οὔ τι φέρουσαι.
> ἄλλος δ' **ἀλλοίην**[b] αἰνεῖ, παῦροι δ' ἴσασιν.
>
> *Works and Days* 822–824

These purely verbal echoes, I believe, are offered as boundary markers. They define the beginning and the end of the Hesiodic "Days," and so indicate the intended scope of Vergil's imitation. While its correspondences with *Works and Days* 802–813 are close, Vergil's calendar alludes not just to this limited Hesiodic passage; rather, the ways in which Vergil varies Hesiod's account must be seen as an effort to summarize *all* of the "Days" (*WD* 765–828) in a few brief lines. That Vergil manages to compress over sixty lines of poetry into a mere eleven should be no surprise in view of his normal method of imitation through conflation and combination.

Such a compressed imitation could be managed only by alluding to representative classes of Hesiodic days, rather than to specific, individual days. The point is easily illustrated. At *Works and Days* 802–813, the immediate model of *Georgics* 1.276–286, Hesiod mentions five days: the fifth, the seventeenth, the fourth, the nineteenth, and the ninth. Vergil has only three days: the fifth, the seventeenth, and the ninth. None of these days corresponds exactly to the Hesiodic day of the same name. For example, in *Works and Days* the ninth is a day "entirely devoid of trouble" (παναπήμων 811); but the character of Vergil's *nona* is mixed. We find in Hesiod several days of this kind, days that are good for one activity but bad for another, or better at one time than at another. Such a day is the nineteenth, or middle-ninth, which is "*better* (λώϊον 810) before evening." Vergil's ninth too is mixed, although in a different way, being "*better* (*melior* 286) for running away, but unfavorable for thievery." We may therefore wonder whether Vergil's *nona* is in fact to be taken as the middle ninth, i.e. as *nona* [*post decimam*] following *septima post decimam* (284).[101] If so, then Vergil's reason for eliminating Hesiod's fourth becomes clear: the result is an orderly chronological progression

101. On this odd expression, see p. 119.

through the month, from the fifth to the seventeenth to the nineteenth, in contrast to Hesiod's achronological progression from seventeenth to fourth to nineteenth to ninth. To this "rationalization" of Hesiod we may compare various features of the plow imitation.[102] But even by referring to Hesiod, one cannot be sure which "ninth" Vergil's *nona* really is: it corresponds exactly to neither. Here another factor presents itself. Hesiod speaks not of two "ninths" only, but of a third as well, all of them coming in an associative sequence forming an ascending tricolon (εἰνὰς ἡ μέσση 810; πρωτίστη δ' εἰνάς 811–813; τρισεινάδα 814–818). The "thrice-ninth" too is relevant to Vergil's *nona*, because *few call it by its real name* (παῦροι δέ τ' ἀληθέα κικλήσκουσιν 818). It thus appears that Vergil's *nona* is an amalgam of Hesiod's three adjacent "ninths": its name most resembles that of the first-ninth; its character is ambivalent, like that of the middle-ninth; and its real identity is therefore uncertain, like that of the thrice-ninth.

The case of the ninth day in fact can teach us a lot about the principles of analytical imitation that Vergil brought to bear on Hesiod's *Days*. Clearly more than just the names of individual days is at stake; and more days than those found in Vergil's proximate model (*WD* 802–813) may be involved in his imitation. I mentioned that Hesiod's nineteenth is only one of several days that are not wholly good or bad, but possess a mixed character. I do not think it fortuitous that of the three days in Vergil's account, one is lucky, one unlucky, and one mixed. The range of characters attributed to these days is surely meant to represent the three general types that Vergil found in his source. To put it differently, each of Vergil's days represents not just a single Hesiodic day, but rather the three Hesiodic classes of lucky, unlucky, and mixed days. For example, whereas Hesiod says to "avoid" ([ἐξ]αλέασθαι 780, 802; ἀλεύασθαι 798) four days, Vergil says to avoid (*fuge* 277) only one, the fifth (≈ *WD* 802), which comes to stand for the entire class. Similarly, the seventeenth is a lucky day; and its representative character is especially clear, because it is a composite of several Hesiodic days:

> **Μέσσῃ δ' ἑβδομάτῃ** Δημήτερος ἱερὸν ἀκτὴν
> εὖ μάλ' ὀπιπεύοντα ἐυτροχάλῳ ἐν ἀλωῇ
> βάλλειν, ὑλοτόμον τε ταμεῖν θαλαμήϊα δοῦρα
> νήϊά τε ξύλα πολλά, τά τ' ἄρμενα νηυσὶ πέλονται.
>
> *Works and Days* 805–808

102. See pp. 70–77.

> **Septima post decimam** felix et ponere vitem
> et prensos domitare boves et licia telae
> addere.
>
> *Georgics* 1.276–286

In both poems this day follows the fifth. Clearly, too, Vergil's odd and un-Latin way of designating the seventeenth as *septima post decimam*, which caused ancient commentators some trouble,[103] imitates Hesiod's μέσσῃ δ' ἑβδομάτῃ, which occupies the same *sedes*. Both poets also call this day good for more than one kind of activity; yet the tasks that the two poets assign to this day are very different. Hesiod says to thresh and to cut timber for building houses and ships, both reasonable jobs for the second half of the month.[104] Vergil remarkably says to plant vine shoots, to tame wild bulls, and to weave. It is unlikely that Vergil found this advice in another source. Planting anything is no task for the seventeenth, but should normally be done earlier in the month, i.e. under a waxing moon.[105] Similarly Hesiod recommends weaving (779) and training animals (794–797) just *before* the full moon, on the twelfth and fourteenth respectively. Why Vergil transfers these tasks to a day just *after* the full moon is unclear; but as in the case of planting, we can say that in assigning these tasks to the seventeenth, he not only lacks authority, but actually flouts the basic principles followed by most of our sources.[106] But this flagrant departure from received wisdom at the level of content

103. Servius *ad loc.* (Thilo 1887.196) thinks that Vergil means to indicate that the seventeenth is lucky, but suggests that the words might also mean that the seventh is lucky, but not so lucky as the tenth; and he records the opinion of "others" that the day in question is the fourteenth or *septima duplicata, id est cuius numerus post decimam invenitur.* The other scholia show similar uncertainties.

104. Pliny (*NH* 18.322) recommends threshing *circa extremam lunam*, i.e. during the last quarter, which is even later than Hesiod; and all authorities are agreed that timber is best cut in the second half of the month, because with the full moon the trees had stopped growing, i.e. the sap had stopped flowing, so that the timber is thus less liable to rot or to be devoured by termites. The waning moon itself was thought to have a preservative effect by advancing the process of drying, curing, or seasoning: see Tavenner 1918.71–72.

105. Hesiod assigns this task to the thirteenth, which is too late for sowing, but just right for transplanting (780–781), and notes that the sixteenth—the first day of the waning moon—is very unfavorable for this activity (μάλ' ἀσύμφορός ἐστι φυτοῖσιν 782). Tavenner (1918.69–70) cites *Georgics* 1.284 as the only passage in ancient literature that recommends planting under the waning moon, and comments, "I am at a loss to explain the apparent departure from an otherwise uniformly attested custom of planting exclusively during the waxing moon or just before."

106. West 1978.354 *ad WD* 775 notes that "the end of the growing period was appropriate for the jobs here mentioned," including weaving. Livestock should be broken on

throws into relief the purely formal principle that led Vergil to connect these several tasks with a single day. Hesiod's seventeenth is only one of several days associated with more than one job. Three of these days are the very ones from which Vergil borrows the tasks that he assigns to his seventeenth. Thus planting vine shoots is borrowed from Hesiod's thirteenth (bad for sowing but good for transplanting: φυτὰ δ' ἐνθρέψασθαι ἀρίστη 781 ≈ *felix et ponere vitem* 284), taming wild cattle from the middle-fourth[107] (good for the birth of a girl as well as for taming or training sheep, cattle, dogs, and mules 795–797 ≈ *prensos domitare boves* 285), and weaving from the twelfth (good for shearing sheep and harvesting 775, as well as for weaving 779 ≈ *licia telae* | *addere* 285–286). The tasks that Hesiod assigns to his own middle-seventh are ignored, a clear indication that Vergil's allusion is to no single Hesiodic day, but to a class.

Where Hesiod's advice about individual days largely agrees with what we know of ancient moon lore, Vergil's is willfully out of step. To a great extent, his imitation of the "Days" is simply a game of forms. His redaction is based chiefly on structural aspects of Hesiod's calendar, rather than on its underlying themes. Hesiod, by way of introducing his account, stresses the notion that the days are "from Zeus" (ἐκ Διόθεν 765; Διὸς παρὰ μητιόεντος 769), and that one must keep close track of their character (πεφυλαγμένος εὖ κατὰ μοῖραν 765). In contrast, Vergil makes no mention of Zeus/Jupiter; rather, the moon itself has determined the respective characters of the various days (*ipsa dies alios alio dedit ordine luna* 276). Vergil's *ordine* here picks up Hesiod's κατὰ μοῖραν, but in a different sense: Hesiod's farmer is to mark the days well, each according to its μοῖρα, while Vergil's moon has established each of the days in a different *ordo*. Clearly Vergil means his reader to take *ordo* in a sense similar to μοῖρα, but the two ideas are not quite congruent. In Hesiod, a day's μοῖρα is the character that Zeus has appointed for it; in Vergil, its *ordo* or place in the sequence of days is what determines its character.

This difference is significant for what follows. Hesiod of course discusses the individual days in his own idiosyncratic order. He begins with a chronological sequence. The first day that he mentions is the thirtieth (766); then come the first, the fourth and seventh (770), eighth

---

the eighth or tenth (Martini–Bassi 1901.34) or on the sixteenth (Delatte 1924.197).

107. Solmsen 1970.84 brackets line 798, making 799 refer to the fourth of the waning month, which is more in line with Vergil's recommendation.

and ninth (772), eleventh and twelfth (774), the thirteenth (780), and the sixteenth (782). This sequence, though not all-inclusive, is chronological. But at this point the poet introduces a new sequence which is largely associative. From the sixteenth, or middle-sixth, he goes back to the first-sixth (785), and similarly from the middle-fourth (794–795) to the fourths of the waxing and waning moon (798), and again from the middle-ninth (810), as we have seen, to the first-ninth (811) to the thrice-ninth (814). The section that Vergil takes for his immediate model shows elements of both ordering principles: the sequence given is fifth, middle-seventh, fourth, middle-ninth, and ninth, with thrice-ninth to follow. Two of these days are obviously out of chronological sequence. Vergil, as I have argued above, eliminates the fourth, leaving the fifth, the seventeenth and the ninth. By retaining these particular days, Vergil redacts Hesiod's unruly (by classical standards) sequence according to an abstract ordering principle, in fact the one suggested at the beginning of his imitation: chronology. The ninth, which in Vergil as in Hesiod follows the seventeenth, apparently tells against this interpretation; but Vergil's *nona* has traits of both Hesiod's nineteenth and twenty-seventh as well as his ninth. The allusive ambivalence of this conflated day in fact calls the reader's attention to Vergil's manner of departing from his original.

Still, in selecting a few from the many tasks that Hesiod mentions, Vergil confines himself chiefly to those that have a thematic relevance to his own poem. The tasks that he assigns to the seventeenth illustrate this point. In general Vergil borrows these tasks from days that Hesiod calls good for several tasks, as I have shown. Hesiod's thirteenth, being a day of mixed character, may therefore seem an odd source of material for Vergil's middle-seventh. I believe that we can detect here a conflict between the poet's formal and thematic goals. Planting vine shoots obviously anticipates the subject of *Georgics* 2, just as taming bulls (borrowed from Hesiod's middle-fourth) does that of Book 3. Thus the thematic importance of viticulture led Vergil to borrow this task from a Hesiodic day that did not meet the formal requirements of his own middle-seventh.[108] But while a slight formal inconcinnity is allowed, discordant thematic material is rigorously excluded. For example, Vergil replaces any reference to nautical activities with other kinds of work, and

108. Still, Hesiod's thirteenth is suitable for a transitional role: he says to avoid sowing on that day (the text is *τρισκαιδεκάτην ἀλέασθαι* | *σπέρματος ἄρξασθαι* 780–781); and in Vergil's imitation the previous day, the fifth, is also a day to avoid (see p. 119).

refers not at all to Hesiod's fourth, a day wholly given over to shipbuilding. By the same token he does not adopt any of Hesiod's references to gelding, a normal farm operation that is never mentioned in the *Georgics*, no doubt because sexuality, along with mortality, is one of the two essential characteristics of farm animals in Book 3. Had Vergil allowed his husbandmen the simple expedient of gelding, his observations on the force of sexual passion would have been a good deal less effective.

Normally, though, Vergil manages to satisfy both his formal and thematic requirements. In *Works and Days*, the eleventh and twelfth days are both good, but the twelfth is better because on that day the spider spins its web and the ant piles up its hoard; therefore on that day a woman should "set up the loom and get on with her work" (777–779). Weaving, as noted above, is the final task that Vergil assigns to the middle-seventh (285–286). Its inclusion here completes a progression of tasks involving the vegetable (planting vines), animal (taming bulls), and human (weaving) worlds. This progression, of course, mirrors the thematic development of the *Georgics*. I have already drawn a thematic connection between the tasks of planting vines and taming bulls and the subject matter of *Georgics* 2 and 3 respectively. Weaving in Hesiod connects the activity of insects (ants and especially spiders)[109] with that of humans by means of an *aetion;* and this is exactly what Vergil will do in *Georgics* 4. There the bee state implies a detailed parallelism between the lives of humans and bees, and the "Aristaeus," juxtaposed with the story of Orpheus and Eurydice, extends this parallelism to include the religious and social issue of death and rebirth by way of relating the *aetion* of the *bugonia*.

But this is to anticipate. The question at this point is, if these are the factors that led Vergil to follow Hesiod in assigning weaving to his day of many tasks, then why does he neglect to allude explicitly to the weaving

109. Aristotle, the most advanced entomologist of the ancient world, in one passage defines insects as creatures possessing three body segments and three pairs of legs (*HA* 531b.26–30); but even here he classifies creatures that are long and many-footed as insects too. Elsewhere he speaks explicitly of spiders (*HA* 550b.31, *GA* 721a.4–5) and other arachnids (*HA* 532a.16–17, 542a.11, 552a.16) as insects, to say nothing of such creatures as worms and centipedes; and indeed Linnaeus 1735 does the same. It therefore seems a reasonable inference that most ancient Greeks and Romans regarded spiders and insects as essentially similar. On this subject in general, see Meyer 1855.197; Byl 1980.325–326; Davies–Kathirithamby 1986.19.

aetion—an important element in the equation suggested above—that he found in his source? The answer is that to do so would have been otiose. Hesiod, as is the case with many other formal elements in the "Days," gives several *aetia;* Vergil, as usual, gives only one. Besides the weaving *aetion,* Hesiod explains the lucky character of the seventh day by making it the birthday of Apollo (771) and the unlucky character of the fifth by making it the birthday of Ὅρκος (803–804). One of these birthday *aetia,* that of Ὅρκος/Orcus, becomes the basis of the only *aetion* related in Vergil's "Days," where it is again connected with the fifth day of the month (277–278); but Hesiod's Apollo *aetion* remains important. Indeed, Vergil's remarkable handling of his sources makes it possible to say that he has not only preferred to include the ominous *aetion,* but has actually suppressed the propitious one, going to significant lengths to make the reader feel its absence. In constructing his lone birthday *aetion,* Vergil departs from Hesiod in a number of small but meaningful details:

> Πέμπτας δ' ἐξαλέασθαι, ἐπεὶ χαλεπαί τε καὶ αἰναί·
> ἐν πέμπτῃ γάρ φασιν Ἐρινύας ἀμφιπολεύειν
> Ὅρκον γεινόμενον, τὸν Ἔρις τέκε πῆμ' ἐπιόρκοις.
>
> *Works and Days* 802–804

> Quintam fuge: pallidus Orcus
> Eumenidesque satae; tum partu Terra nefando
> Coeumque Iapetumque creat saevumque Typhoea
> et coniuratos caelum rescindere fratres.
> Ter sunt conati imponere Pelio Ossam
> scilicet atque Ossae frondosum involvere Olympum;
> ter pater exstructos disiecit fulmine montis.
>
> *Georgics* 1.277–283

Both poets agree that the fifth is to be shunned; but Vergil's departures from Hesiod have convinced unimaginative critics that he must have been following another source.[110] The "translation" of Ὅρκος as Orcus is clearly, as most scholars have realized, a familiar type of Vergilian pun. Similarly, substituting the apotropaic euphemism *Eumenides* for Ἐρινύας requires no specific source; and while Vergil may have had an authority for the belief that the Furies and Orcus were conceived on the same day, it was certainly not beyond him to concoct such a statement out of

110. Cumont 1933.

Hesiod's assertion that these goddesses served as midwives at Horcus' birth. Furthermore, Hesiod's participle *γεινόμενον* agrees with the usual practice of explaining the character of a day by the belief that some divinity was *born* on it, rather than conceived.[111] Vergil's *satae* is therefore an arbitrary departure, but one that introduces a metaphor consonant with the dominant conceit of the *Georgics.*

Thus far there is no need to look past Hesiod's fifth day for the model of Vergil's fifth, and we may say that *Georgics* 1.277–278 closely follows *Works and Days* 803–804. But in lines 278–283 Vergil makes the fifth the birthday of Coeus and Iapetus, of Typhoeus, and of Otus and Ephialtes, something of which there is no hint in Hesiod. Here one may well look for another source; but the Titans and Typhoeus are not linked with the Aloidae before Vergil, nor are the birthdays of these figures ever mentioned in selenodromic literature. Here too Vergil's departure from his primary model looks like an act of imagination. But, as often, Vergil's imagination expresses itself through literary reference. By mentioning the birth of Coeus, Iapetus, and Typhoeus, the poet turns his attention from *Works and Days* to *Theogony:*

tum[a] **partu**[b] **Terra**[c] nefando
Coeumque[d] **Iapetumque**[e] **creat**[f] saevumque T**yphoea**[g]

Georgics 1.278–279

ἧ δὲ [sc. **Γαῖα**[c] (126)] *καὶ ἀτρύγετον πέλαγος* **τέκεν**[f], *οἴδματι θυῖον,*
Πόντον, ἄτερ φιλότητος ἐφιμέρου· αὐτὰρ **ἔπειτα**[a]
Οὐρανῷ εὐνηθεῖσα **τέκ'**[f] *'Ωκεανὸν βαθυδίνην*
Κοῖόν τε[d] *Κρεῖόν θ' Ὑπερίονά τ'* **'Ιαπετόν τε**[e]

Theogony 131–134

ὁπλότατον **τέκε**[f] **παῖδα**[b] **Τυφωέα**[g] **Γαῖα**[c] *πελώρῃ*[112]

Theogony 821

The transition is unforced: a passage on the birth of gods even when encountered in *Works and Days* might easily remind a reader of Hesiod's earlier work devoted to this theme. The following transition, however, is more surprising. On the same day, Vergil says, Earth bore the twin brothers Otus and Ephialtes, who thrice attempted to pile Mt. Ossa on

111. Not that I wish to press the point; *satus* is, of course, a frequent poetic synonym for *natus* when the letter is metrically inconvenient. But I generally find arguments based on metrical exigency unconvincing where Vergil is concerned.

112. Note too the similarity between the clausulae *Γαῖα πελώρῃ* ≈ *Terra nefando.*

Mt. Pelion and Olympus on Ossa in an effort to invade the heavens; but thrice Jupiter blasted apart their work with his thunderbolt (280–283). This story may have been told by Hesiod;[113] but most scholars have felt, correctly I believe, that Vergil's model here is Homer.

Several factors may have led Vergil to connect the Homeric and Hesiodic passages. For instance, Homer's chief account of Otus and Ephialtes is related as part of Odysseus' "*Nekyia*," and Vergil, having altered Hesiod's Ὅρκον to *Orcus*, clearly has the underworld in mind here.[114] If Hesiod did mention the Aloidae in the *Catalogue of Women*, then Vergil may have noted a contextual similarity between this passage and *Odyssey* 11: Homer tells the story of the twins in order to identify their mother, Iphimedeia, one of the many heroines whom Odysseus sees in the land of the shades. Furthermore, both the Titans, here represented by Coeus and Iapetus, and Typhoeus, like the Titans a child of Earth, appear in *Theogony* 617–868 as opponents of Zeus's regime who bring force against him. They are therefore typologically quite similar to Homer's enormous twins. Of course there are differences between Vergil's account and Homer's. Most obviously, in the *Odyssey* Otus and Ephialtes are the offspring of Iphimedeia by Poseidon; in the *Georgics* they are, like Coeus, Iapetus, and Typhoeus, *terrigenae*. It seems certain that Vergil willfully altered the Homeric tradition in this respect to facilitate the introduction of the Aloidae here.[115] Other departures are more closely based on Homer's text. Putnam, for instance, notes that Vergil, by changing the order in which Homer names the three mountains (Ὄσσαν, Οὐλύμπῳ, Ὄσσῃ, Πήλιον 315–316 ≈ *Pelio, Ossam, Ossae, Olympum* 281–282), "follows the order of the event itself" and "puts Olympus in the place of honor and gives to it the epithet Homer allots Pelion."[116] Similarly, Vergil's statement that the brothers made three attempts on heaven and that Jupiter thrice repelled them is not in

113. The Aloidae were indeed mentioned by Hesiod according to Σ *in Apollonii Argonautica* 1.482a (Wendel 1935.42), and the *Suda* 5.2221 s.v. Ἐπιάλτης (Adler 2.348.20) attests his use of the unaspirated form of Ἐφιάλτης. Merkelbach–West 1967.12 ascribe these passages to the *Catalogue of Women* (frr. 19 and 20); but there is no basis for assuming that Hesiod told the story of Pelion and Ossa unless one also accepts Merkelbach's adventurous supplement of P. Oxy. 2075 fr. 6 (ἵν' οὐρανὸς] ἀμβατὸ[ς εἴη; cf. *Odyssey* 11.316), which he and West include as fr. 21 of the *Catalogue*.

114. This idea is just barely present in Hesiod's mention of the Erinyae. Vergil's amplification of it is his own contribution.

115. He may, however, have had a precedent in Eratosthenes, according to Σ *in Apollonii Argonautica* 1.482a (Wendel 1935.42).

116. Putnam 1979.46.

the *Odyssey;* yet Vergil's anaphora *ter...ter* (281–283) may have been inspired by Homer:

> **ἐννέωροι** γὰρ τοί γε καὶ **ἐννεαπήχεες** ἦσαν
> εὖρος, ἀτὰρ μῆκός γε γενέσθην **ἐννεόργυιοι.**
>
> *Odyssey* 11.311–312

One further departure from Homer's treatment of the myth bears mentioning. Whereas Vergil states that Jupiter himself opposes and defeats the monstrous brothers, in the *Odyssey* it is not Zeus but Apollo who plays this role.[117] Here too Vergil's language appears to play off that of Homer. In the *Georgics*, Jupiter is referred to tersely as *pater* (1.283); in the *Odyssey*, Apollo is designated in the fulsome epic style as Διὸς **υἱός**, ὃν ἠύκομος τέκε Λητώ (11.318). Thus in Vergil the father replaces the son. This displacement of Apollo from his role in Homer's myth parallels Vergil's selection of the Horcus *aetion* from *Works and Days*, instead of the similar Apollo *aetion*, to represent Hesiod's manner of explaining the character of certain days. This parallelism is all the clearer because Homer too refers explicitly to the birth of Apollo and to his mother Leto, just as Hesiod does in explaining the lucky character of seventh days.

We may ask, why does Vergil insert this Homeric reference into an otherwise clearly Hesiodic context? The question is interesting and important, but we as yet lack the data to answer it. We can, however, make one observation. The analysis of Vergil's imitation here reveals the fundamentally pessimistic nature of his account.[118] Where Hesiod's "Days" cling to conventional folk wisdom as a means of lending order to the randomness of life, Vergil goes out of his way to reduce conventional thinking about the lunar calendar to nonsense. In grafting Homeric material onto this account, he expands upon one of Hesiod's more frightening *aetia* and; and, by excluding Apollo from his normal role in the Homeric passage, reminds us that he has eschewed the more cheerful Hesiodic *aetion* concerning Apollo's birthday. Vergil depicts a far more baleful universe under the sway of a Jupiter jealous of his power and ruthless in the suppression of revolt. This is an important element in the tragic world view elaborated in the first book of *Georgics*.

117. The normal version; sometimes Artemis helped (e.g. Σ *in Iliadem* 5.385b [Erbse 1969–1988.2.60]).

118. As noted by Putnam 1979.45 note.

With this episode we have come a long way from where we began. "The Farmer's *Arma*," as we have seen, is a relatively simple reworking of a single passage from *Works and Days*. Its companion piece, "Lucky and Unlucky Days," is an elaborate and involved condensation of another Hesiodic passage, but it is more than this. By alluding so openly and so definitely to Homer in this emphatically Hesiodic context Vergil in a way surpasses even the extensive double allusion to the *Iliad* and the *Odyssey* found in the "Aristaeus." I have tried in the case of "The Plague of Noricum" and "The Praise of Spring" to explain multiple allusion or *contaminatio* as a kind of index to the thematic structure of a given source. In the case of "Aristaeus" and "Lucky and Unlucky Days," it becomes clear that the process cannot be confined to internal cross-reference; instead, allusion becomes a way of finding common ground between distinct poems and even distinct authors. Usually we think of allusion rather simply as a relationship between two poems. Here Vergil employs allusion to create a relationship between his own poem and, not a single poem or even a single author's *oeuvre*, but the epic tradition within which both Hesiod and Homer—and, for that matter, Aratus and Lucretius and a multitude of others—worked. We have not answered all the questions about allusion in the *Georgics;* but we have, I think, made progress towards our goal of understanding the poem, if only in microcosm, as an allusive whole, and towards defining the object of its allusiveness—its "model." Clearly all the different models that we have examined in this chapter are involved; and through this examination of allusive *contaminatio*, it has become possible to begin to see how they may be related. A fuller exploration of these relationships will be the business of the following chapters.

PART II

Vergil's Program of Allusion in the Georgics

CHAPTER 4

Hesiod and Aratus

Preliminaries

I HAVE ALREADY REMARKED upon a crucial difference between the models of the *Eclogues* and the *Aeneid* on the one hand and those of the *Georgics* on the other. In the case of Vergil's first and last works, it is correct to speak of a single, dominant model; in the *Georgics*, we must recognize a number of models, the influence of each one being comparatively restricted. Whereas Theocritus exerts a pervasive influence on the *Eclogues*, as does Homer on the *Aeneid*, the various models of the *Georgics* exert an essentially local influence on individual parts of the poem. We have now reviewed some of the most important passages in which such influence is felt, and have gained a measure of insight into Vergil's method of using these models. It remains to ask how these passages contribute to the poetic unity of the *Georgics*.

Until fairly recently there was no ready way to answer this question. Before Knauer, even the *Aeneid*, with its extensive debt to Homer, was not fully recognized as a continuous allusion to both the *Iliad* and the *Odyssey*. But if we work out in detail the system of imitations by which Vergil combines and restructures both Homeric epics, the poem becomes a rigorous examination of heroism, ethics, and political ideas.[1] Although the *Eclogues* still await a similar treatment,[2] it is clear that they too stand

1. Knauer himself sees Vergil's achievement essentially as a re-creation, *mutatis mutandis*, of Homer by way of emulative tour de force (sc. the combination of two epics into one). Most subsequent readers (myself included) have tended to read the poem as a searching, even, in passages, disapproving criticism of the heroic code; see, outstandingly, Putnam 1965 and Johnson 1976.

2. Posch 1969 and Thill 1979 are the two most detailed treatments to date of allusion in the *Eclogues*, but neither is a comprehensive study of Vergil's overall allusive program.

in a complex intertextual relationship to their model, comprising in their entirety a carefully planned imaginative re-creation of Theocritus' bucolic world.

In both his first and last works, then, it is evident that Vergil followed an extensive and elaborate program of allusion: imitation is used not as an occasional decorative device, nor yet to create an occasional *aperçu* between a Vergilian passage and some other poem. Rather, it serves in both the *Eclogues* and the *Aeneid* as a fundamental element of poetic unity. It would be strange if Vergil had temporarily abandoned this working method in his intervening work, the *Georgics;* and it is even stranger that no one has yet attempted to discern in the *Georgics* a unifying program of allusion, similar to those found in Vergil's earlier and later works, underlying the diversity of source material that appears from book to book.

This diversity of source material is not as serious an obstacle to unity as one might initially have supposed. Perhaps the most salient characteristic of Vergilian allusion is its unifying tendency. In the *Eclogues* this fact is not very pronounced, because the models of, for instance, *Eclogue* 3 share many obvious points of similarity.[3] But in the *Aeneid* one must remember that Vergil's program of Homeric allusion depends upon the poet's critical ability to discover points of similarity between what most readers regard as very different poems, the *Iliad* and the *Odyssey.* A careful study of verbal borrowings and narrative patterns shows quite unmistakably that Vergil derives an action such as that involving Aeneas, Lavinia and Turnus from the similar Homeric action involving Odysseus, Penelope and the suitors on the one hand, but also from the one involving Menelaus, Helen, and Paris on the other.[4] An important effect of such allusive *contaminatio* is to reveal points of similarity between distinct narratives, points that the reader may never before have grasped. In such a case, the previously unrecognized similarity between the two Homeric epics becomes the basis of a single, unified action in a new poem. This process is repeated many times within the program of Homeric allusion that informs the *Aeneid,* to say nothing of the extensive secondary allusions with which Vergil adorns his poem, allusions

3. The poem's chief models share important formal characteristics, such as an amoebean structure (*Idylls* 5, 8) and thematic concerns, particularly the motifs of contest (4, 5, 8, 6, 9) and exchange (1, 9). I hope to explore these relationships in detail in a future study.

4. Knauer 1964a.239–266; cf. Knauer 1964b.76–78 = Knauer 1981.884–885.

involving authors such as Apollonius, Callimachus, Naevius, Ennius, and others.

My working hypothesis is that the *Georgics*, too, is informed by an allusive program that finds elements of unity among its diverse models. Unlike the *Aeneid*, the *Georgics* does not base this unity on the supposed identity of a single major source author. For his final masterpiece, it is possible to state Vergil's allusive program as a simple theme, such as "the fundamental unity (moral, religious and political) of the Homeric epics." For the *Georgics*, at least at this point, no such obvious theme presents itself. Yet our study of Vergil's allusive style encourages us to try to identify some such theme and suggests how best to do so. We know, in the first place, what Vergil's chief models in this poem are. We know also where their influence tends most to be felt. In the previous chapter I discussed six passages of obvious contextual allusion. Three of these passages occur in *Georgics* 1: two, "The Farmer's *Arma*" and "Lucky and Unlucky Days," primarily involve Hesiod's *Works and Days*, while the third, the "Weather Signs," concerns Aratus' *Phaenomena*. Two of the remaining three passages involve Lucretius' *De Rerum Natura*. These occur in *Georgics* 2 ("The Praise of Spring") and 3 ("The Plague of Noricum") respectively. The final passage, the "Aristeaus," concludes *Georgics* 4 and derives chiefly from Homer. Of course it is not yet possible to maintain that the general allusive character of the various books corresponds in some way to that of their major allusive contexts—that Book 4 is "Homeric," Books 2 and 3 "Lucretian," and Book 1 either "Hesiodic," "Aratean," or both. On the other hand, it has been claimed that the overall pattern of Book 1 frequently alludes to Hesiod and Aratus, and that the same is true of Lucretius in Books 2–3 and of Homer in Book 4.[5] The situation is not dissimilar to the old view that the *Aeneid* is composed of "Odyssean" and "Iliadic" halves.[6] We now know, thanks to Knauer, that the situation is more complex;[7] and the process by which the old view came to be discarded in favor of a more subtle and complete understanding of the involved relationship between the *Aeneid* and the Homeric poems may serve as a model for how to proceed in this case. Let us assume provisionally that we are dealing with a tripartite poem as determined by pattern of allusion, that we have in Book 1 a "Hesiodic/Aratean" *Georgics*, in Books 2–3 a "Lucretian"

5. Wender 1979, unjustly criticized by Griffin 1981.29–30.
6. Otis 1964 et al.
7. See also the interesting observations of Cairns 1989.177–214.

and in Book 4 a "Homeric" one. Our task is to examine the poem book by book in order to test this assumption—i.e. both to see in what respects it is true and, where it is not, to modify it as the evidence dictates. In this way it may be possible to develop a more satisfactory conception of Vergil's allusive program.

Vergil's *Works and Days*

In Chapter 2 I suggested that Vergil's acknowledgement of Hesiod in the phrase *Ascraeum carmen* (*G.* 2.176) signals the end of Hesiodic influence on the *Georgics*. It has often been observed that Hesiodic influence is much more prominent in Book 1 than in the later books.[8] One or two have gone so far as to assert that *Georgics* 1 was originally conceived as a self-contained poem based on that of Hesiod, and that *Georgics* 2–4 are the result of fresh inspiration at a later date.[9] While this view has properly won little acceptance, it is true that the structure of *Georgics* 1 is strikingly congruent with that of *Works and Days*. The very title of Hesiod's poem suggests its structure in broad outline: it begins by illustrating the general necessity to work, mentioning most of the farmer's basic tasks, and ends by stating the best days to do various jobs. Servius may be right in detecting in *G.* 1.1 an allusion to the title of Hesiod's poem: *quid* ***faciat*** (≈ ἔργα) *laetas segetes,* ***quo sidere*** (≈ ἡμέραι) *terram* | *vertere*. Certainly Vergil structures *Georgics* 1 according to this dual theme: lines 43–203 concern the farmer's tasks and so correspond generally to Hesiod's ἔργα, while lines 204–514 discuss the best times to perform these tasks, as do the Hesiodic ἡμέραι.[10] Thus Vergil's division of subject-matter into "Works" and "Days" indicates an overall structural congruity between *Georgics* 1 and its model.

This general congruity of theme and structure is greatly reinforced by Vergil's deployment of specific Hesiodic allusions. I have already discussed two well-known and very similar examples. Both "The Farmer's *Arma*" and "Lucky and Unlucky Days" are specific and unmistakable

8. Cf. the statement of Griffin 1986a cited on p. 30 n. 9.

9. Bayet 1930; Le Grelle 1949.

10. Most scholars would put the end of Vergil's ἡμέραι earlier in the book, either at line 310 or 350, on the theory that the very end of the book is modeled on Aratus' "*Diosemiae*" rather than on Hesiod's "Days." I will not yet comment on this point, which does not really affect the correspondence that I have outlined above. It remains clear that *Georgics* 1 as a whole borrows its structure from *Works and Days*, with or without the addition of an Aratean coda.

points of contact between *Georgics* 1 and *Works and Days.* Indeed, these have long stood out as the most overtly Hesiodic passages in the *Georgics.* Despite obvious divergences from their respective models, both passages are extremely Hesiodic: one feels this character especially in their apparently matter-of-fact purpose, the dispensation of homely wisdom. They are in a way emblems of Hesiod's poem in its simplest aspect, as a dissertation on how and when to perform the essential tasks of farm life. But they are also more. It would be wrong to think of these passages as isolated from one another or from any larger allusive program. Their relationship to one another is made obvious by their very directness. Only in these passages does Vergil follow Hesiod so closely for more than a line or two, and it is hardly possible to think of either passage without recalling the other. But if the two passages are deliberately paired, what is their function, beyond recalling Hesiod with such unmistakable clarity? Why does Vergil choose to recall these particular Hesiodic passages at just these points in his own exposition? Again the answer seems obvious: first, because the two passages to which he alludes clearly represent Hesiod's twin themes, ἔργα καὶ ἡμέραι; second, because the positioning of these allusions in their context resembles the positioning of the source passages in their context, and does so closely enough to establish a general structural and thematic correspondence between *Georgics* 1 and *Works and Days.* Just as a word or two repeated exactly from another context may establish an allusive relationship between two lines, and as a repeated line may indicate a more extensive contextual imitation, so do these paragraphs anchor the structure of *Georgics* 1 to that of *Works and Days*, organizing *Georgics* 1 into a discourse on *opera* and *dies* that closely parallels that of Hesiod. The twin allusions therefore draw the reader's attention not only to themselves as isolated imitations, but to an important aspect of Vergil's allusive program, inviting detailed comparison between *Georgics* 1 and *Works and Days* as a whole.

We will surely fail to understand Vergil's program unless we make this comparison. Let us begin by pressing the idea that the structure of both the original and the imitation falls into halves that can be labeled "Works" and "Days."

The Technical Discourse

The part of Hesiod's poem that deals chiefly with farm operations begins at line 383 and continues to the end; and we may indeed without undue distortion divide this lengthy exposition into roughly equal sections dealing with "Works" (383–617) and "Days" (618–828). The first section begins with an outline of the farmer's year (383–413), and proceeds by discussing the various annual operations in order.[11] The structure of the second section is partly chronological, partly associative.[12] As a test of how far the structure of *Georgics* 1 follows its prototype, let us compare it with the two poets' first technical passages.

One could hardly call the relationship close. The "Works" begins with a passage made famous by the *Certamen Homeri et Hesiodi*, where Hesiod chooses it as the most beautiful lines of all his poetry:

> Πληϊάδων Ἀτλαγενέων ἐπιτελλομενάων
> ἄρχεσθ' ἀμήτου, ἀρότοιο δὲ δυσομενάων.
> αἳ δή τοι νύκτας τε καὶ ἤματα τεσσαράκοντα 385
> κεκρύφαται, αὖτις δὲ περιπλομένου ἐνιαυτοῦ
> φαίνονται τὰ πρῶτα χαρασσομένοιο σιδήρου.
> οὗτός τοι πεδίων πέλεται νόμος οἵ τε θαλάσσης
> ἐγγύθι ναιετάουσ' οἵ τ' ἄγκεα βησσήεντα
> πόντου κυμαίνοντος ἀπόπροθι, πίονα χῶρον, 390
> ναίουσιν· γυμνὸν σπείρειν, γυμνὸν δὲ βοωτεῖν,
> γυμνὸν δ' ἀμάειν, εἴ χ' ὥρια πάντ' ἐθέλῃσθα
> ἔργα κομίζεσθαι Δημήτερος, ὥς τοι ἕκαστα
> ὥρι' ἀέξηται, μή πως τὰ μέταζε χατίζων
> πτώσσεις ἀλλοτρίους οἴκους καὶ μηδὲν ἀνύσσεις. 395
>
> *Works and Days* 383–395

Hesiod starts the farmer's year with the heliacal rising of the Pleiades in the spring,[13] saying that this is the time to begin reaping (383–384a). Their setting in the fall marks the time for plowing and for sowing the next year's crop (384b). The (approximately) forty days before their rising on which they are invisible is the time to prepare for the work ahead

11. Hesiod begins in fall with woodcutting and building equipment (414–457) and continues with plowing (458–492), inactivity in winter (493–563), pruning (564–570), harvest (571–581), inactivity in summer (582–596), threshing (597–608), and pressing grapes (609–617).

12. See in general the remarks of West 1978.346–350 *ad WD* 765–828 and, for a more penetrating analysis of the thematic organization of the "Days," Hamilton 1989.78–87.

13. Hence their Latin name *Vergiliae*.

(385–387). Vergil too begins in the spring (*vere novo* 43) and ends in the fall (*sub ipsum* | *Arcturum* 67–68); and like Hesiod, he measures time by the stars. But his specific operations and their schedule differ considerably from Hesiod's. He begins by noting that *early* spring—much earlier, I presume, than the rising of the Pleiades in May—is the normal season not to harvest, but to plow (*G.* 1.43–46). This plowing is different from the one mentioned by Hesiod. New or fallow land had to be broken up in the spring, and might be plowed several more times before the final plowing and sowing in fall, the one that Hesiod connects with the *setting* of the Pleiades (384b). Similarly, Vergil concludes with the supplementary observation that poor soil is better plowed lightly at the rising of Arcturus in early September (67–68)—that is, roughly, in early fall, but again considerably earlier than the setting of the Pleiades in late October.

This is the extent of parallelism between Hesiod's and Vergil's first paragraphs of technical instruction. The allusion, if we admit that one exists here, is very faint; and in fact, the passages basically differ from one another in ways that these oblique similarities can hardly conceal. Hesiod's passage reads like a brief calendar of the year's work, one that marks only the the first and last activities of the sidereal and agricultural year, the spring harvest and the fall sowing. It is characterized, in a way that the corresponding Vergilian passage is not, by fulsome eloquence (Πληϊάδων Ἀτλαγενέων ἐπιτελλομενάων 383) and proverbial terseness (*γυμνὸν σπείρειν, γυμνὸν δὲ βοωτεῖν,* | *γυμνὸν δ' ἀμάειν* 391–392). Its basic point is easily reduced to the blunt warning with which it concludes: get your work done on time while the weather allows it, if you want to prosper and not be reduced to begging (392–395). But the bulk of Vergil's much longer paragraph concerns not timing, but work itself. "Plow twice a year (47–49), but before plowing a new field, ascertain its character and capability (50–53); different kinds of soil are better suited to different crops (54–59, with a mythological *aetion* 60–63); but at any rate, plow early in the year and vigorously; or if the soil is poor, plow lightly in the Fall (63–70)." The basic thrust of the two passages is obviously quite different. In fact, this difference corresponds to a larger one between the two poems in general on which M. L. West has remarked:

> [Hesiod] assumes a general understanding of the purpose and method of ploughing, reaping, threshing, and so forth. What he has to say is

> concerned primarily with the right time for beginning each job, or, to look at it the other way, with the jobs to be done at each time in the year. It is therefore natural that he takes as a framework the succession of the year's seasons, and not a division of subjects such as we find in the four books of Virgil's *Georgics.*[14]

This is exactly what we see in the two passages we have been comparing. Hesiod does not trouble to explain why or how to plow, but rather contents himself with when. His first technical paragraph is in this sense programmatic, since it indicates the overall structural importance of the agricultural calendar to his discourse. For Vergil, the schedule is not of primary importance; one must find out what kind of soil he is dealing with, decide what to plant in it, and learn how best to prepare it. The farmer's activities are not presented as a list of maxims keyed to a calendar, but are analyzed, classified, separated and explained almost as departments of an intellectual discipline. Here we see clearly the influence of prose authors such as Cato, who wrote for senatorial investors uninterested in supervising the day-to-day operations of their estates. It goes without saying that nothing in Hesiod's first technical paragraph that corresponds to this approach.

On the other hand, we cannot conclude that Vergil's first technical paragraph owes nothing to Hesiod. In fact, while this passage basically ignores the beginning of Hesiod's technical exposition, it leans heavily on a different section of the "Works." The specific advice with which Vergil begins has much in common with what Hesiod has to say some seventy-five lines after the beginning of his "Works":

> εὖτ᾽ ἂν δὴ πρώτιστ᾽ ἄροτος θνητοῖσι φανήῃ,
> δὴ τότ᾽ ἐφορμηθῆναι, ὁμῶς δμῶές τε καὶ αὐτός,
> αὔην καὶ διερὴν ἀρόων ἀρότοιο καθ᾽ ὥρην,
> πρωὶ μάλα σπεύδων, ἵνα τοι πλήθωσιν ἄρουραι.
> ἔαρι πολεῖν· θέρεος δὲ νεωμένη οὔ σ᾽ ἀπατήσει·
> νειὸν δὲ σπείρειν ἔτι κουφίζουσαν ἄρουραν.
> νειὸς ἀλεξιάρη παίδων εὐκηλήτειρα.
>
> *Works and Days* 458–464

Here we find a pronounced contextual similarity between Hesiod and Vergil. In this passage Hesiod emphasizes the need to begin the year's

14. West 1978.52.

work "as soon as plowing season appears to mortals" (458) and to "get at it very early" (461). At length he specifies this time as spring (ἔαρι 462). This advice, if not explicitly at odds with the opening paragraph of the "Works," at least begins the farmer's year with different operations at a different time, plowing in early spring rather than harvest in late spring. More to the point, it agrees exactly with Vergil's advice in the opening paragraph of his "Works." The same is true of what follows. Hesiod goes on to stress the importance of letting the land lie fallow (*νεωμένη* 462, *νειόν* 463, *νειός* 464) if it is to be really productive; and this is what Vergil does as well ("It is plowland that has twice felt sun and chills that answers the greedy farmer's prayers" *G.* 1.47–48).[15] Hesiod tells his farmer to follow this course "so that your fields may be full" (461); as for the farmer who takes Vergil's advice, "His massive harvests burst his barns" (49). Finally, Page noted a further point of similarity between these passages in a curious idiom. Hesiod, speaking of the last plowing that is performed in conjunction with the fall sowing, says to sow fallow land that is just "lifted up" (*κουφίζουσαν* 463); Vergil, speaking at the end of his first technical passage of how to treat poor soil, says not to plow in the spring and let it lie fallow, but that that it is sufficient just to "lift it up" (*suspendere* 68) from the furrow at the rising of Arcturus.[16]

From these parallels it is clear that Vergil's first technical paragraph has a markedly Hesiodic background. Although it virtually ignores Hesiod's first technical paragraph (*WD* 383–404), it begins and ends by clearly imitating, with typical forms of variation, a different Hesiodic passage (*WD* 458–464). Thus in this case it has proved necessary to seek Vergil's model elsewhere than where we expected to find it. By the same token, we must be alert to Hesiodic influence on *Georgics* 1 in unexpected contexts. Just because the beginnings of the Hesiodic and Vergilian "Works" are dissimilar, we cannot assume that Vergil took no account whatsoever of Hesiod's first technical paragraph. As it happens, Vergil openly quotes from this same Hesiodic passage, not in the *Georgics* passage that corresponds to it structurally, but in an entirely different context:

15. Properly understood by Servius Danielis *ad G.* 1.48 (Thilo 1887.145).
16. Page 1898.190 *ad G.* 1.68.

> at rubicunda Ceres medio succiditur aestu
> et medio tostas aestu terit area fruges.
> **nudus ara, sere nudus;** hiems ignava colono.
> frigoribus parto agricolae plerumque fruuntur
> mutuaque inter se laeti convivia curant.
>
> *Georgics* 1.297–301

Not only is the quotation very close to *Works and Days* 391–392, but the point is the same: in both passages the farmer is advised that the warm season will be his busy time. This parallelism shows that Vergil did not simply ignore the beginning of Hesiod's argument; and it establishes *Works and Days* 383–404 as a model of *Georgics* 1.287–310. Why Vergil chooses to recall the beginning of Hesiod's "Works" at just this point in his own exposition is a question that must be asked. To answer it, we must step back from these individual points of contact to take a broader view of how Vergil alludes to Hesiod in *Georgics* 1.

I have already articulated a version of this broader view, suggesting that the structure of both *Georgics* 1 and *Works and Days* follows a bipartite division of subject matter into "Works" and "Days," and that Vergil's allusions to Hesiod conform to a program of imitation based on this structural and thematic congruity. This view is based primarily on two passages, "The Farmer's *Arma*" and "Lucky and Unlucky Days." One might expect other allusive passages to show a similarly high degree of structural and thematic correspondence with *Works and Days.* But this expectation is immediately frustrated. Contrary to our hypothesis, Vergil's opening paragraph of "technical instruction" alludes not to Hesiod's first bit of agricultural advice, but to a passage that is "out of sequence"—namely *WD* 458–464, which immediately follows *WD* 414–457, the model of "The Farmer's *Arma.*" Similarly, Vergil quotes from Hesiod's opening passage at *G.* 1.299, i.e. in a passage (*G.* 1.287–310) that immediately follows "Lucky and Unlucky Days" (*G.* 1.276–286). In view of Vergil's normal standards of imitation, we should be ill-advised to conclude that these "out-of-sequence" allusions vitiate our hypothesis about structural imitation in general. Rather, our experience with Vergilian allusion on a smaller scale encourages us to view the unsequential allusions as elements of variation within an established allusive pattern. In the case of allusion to Hesiodic elements in *Georgics* 1 we find the same principle applied on a larger scale. In other words, by alluding to these Hesiodic passages, Vergil has rearranged their original order as follows:

	Works and Days		*Georgics* 1
A.	First technical paragraph: γυμνὸν σπείρειν, κτλ (391–392)		
		C.	First technical paragraph: tasks (43–70)
B.	"Farm Equipment" (414–457)	B.	"The Farmer's *Arma*" (160–175)
C.	Tasks		
D.	"Days of the Moon" (765–828)	D.	"Lucky and Unlucky Days" (276–286)
		A.	*nudus ara*, etc. (299)

Of course, we need not and should not consider such variation random; and again the precision with which we have seen Vergil vary other allusions encourages us not to do so. Rather, we must look for some purpose in them, which means considering the relationship between imitation and model in all its aspects.

While both allusions seem out of place in terms of structural imitation, they are thematically very much at home. I have noted the varying emphasis on the farmer's yearly schedule as an important difference between the ways in which Hesiod and Vergil begin their expositions. Clearly this difference in emphasis reflects a decision on Vergil's part to focus on "Works" in his first technical paragraph more narrowly than Hesiod does in his. Rather than begin with a brief calendar outlining the essential dates in the farmer's year, Vergil concentrates on the various factors that one must take into account when beginning to plow. In doing so, he alludes to material found in another section of *Works and Days*—a section devoted more to operations than to their schedule. By the same token, Vergil does allude to Hesiod's opening paragraph, clearly and unmistakably; but he does so in a context that is concerned precisely with schedule instead of with operations. It is as if Vergil had, through allusion, modified the structure of Hesiod's poem in such a way as to strengthen the contrast between ἔργα καὶ ἡμέραι.

Here we have a second important aspect of Vergil's allusive program. Not content merely to borrow the general structure of *Georgics* 1 from *Works and Days*, Vergil actually intensifies his model's thematic dualism by structural revision. In opening his imitation of the Hesiodic "Works"

(*WD* 383–617 ≈ *G.* 1.43–203), the poet purges his source of material more in keeping with the Hesiodic "Days" (*WD* 618–828), replacing it with more strictly "Works" oriented material. When he finally does allude to Hesiod's opening technical paragraph, he does so in the thematically related context of his own "Days" imitation (*G.* 1.204–350). The result is that Vergil's imitation clarifies and intensifies the chief structural and thematic characteristics of his source. This practice conforms to what we observed in the case of "The Farmer's *Arma*" and "Lucky and Unlucky Days," namely, that Vergil redacts his archaic source in such a way as to give it a shape in keeping with the more classicizing aesthetic of his own poem. Finally, this allusive practice may be viewed as embodying a view of literary history. By subjecting his source to significant redaction and revision, Vergil seems to forestall the charge that his own poem is a derivative composition. Rather, he suggests that the *Works and Days* is a kind of prototype, an approximation of the formal perfection achieved only in the *Georgics*. The suggestion need not be taken as an impugnment of Hesiod, nor must the modern scholar accept such a tendentious literary-historical pronouncement at face value. Rather, we should recognize it as an important element of the rhetoric that the allusive poet uses to win himself a place in an established tradition—or, better perhaps, to establish the tradition within which he desires to take his place.

The Mythic Discourse

It is by now well agreed that *Works and Days* goes far beyond the dispensation of homely wisdom to farmers about what to do and when to do it.[17] A major goal of the poem is to ground humankind's common struggle to achieve and to maintain a dignified way of life upon a mythic conception of universal order. Increasingly scholars have identified similar concerns as among the most important themes of the *Georgics* as well. But while scholars have convincingly linked Vergil's mythic conception to that of Hesiod at several points, the differences between *Georgics* 1 and its original are equally apparent. As we have already seen, individual passages of concentrated allusion differ significantly from their Hesiodic exemplars. If we examine the structure of the *Georgics* against the paradigm offered by *Works and Days*, we find that the imita-

17. On the poem's thematic organization see Hamilton 1989.

tive poet has carefully adjusted the plan of his exemplar. This is the conclusion that we reached by assuming a bipartite structure for *Works and Days* as well as *Georgics* 1, an account of *opera* followed by one of *dies;* and clearly this structural and thematic division is an essential characteristic of both poems. But this is not a sufficient view either of Hesiod's poem or of Vergil's imitation.

In expressing the relationship between myth and praxis, Vergil differs markedly from his source. The structure of *Works and Days* suggests that the mythic material is in every way prior to any more overtly practical wisdom. Indeed, Hesiod addresses these larger concerns before settling down to the humbler topics of when to plow and reap, what sort of tools to use, and so forth. The first major section of the poem (lines 11–381 following the proem) concerns universal principals and their causes—ἔρις, the story of Pandora and Prometheus, the "Myth of Ages," δίκη, and the absolute necessity to work. Through this succession of myths Hesiod develops at great length the idea that work is humankind's most authentic response to and participation in the cosmic forces that shape his world. Structurally and thematically it is obvious that this introductory section stands apart from the following discussion of specific tasks in a tripartite scheme: part 1, the need to work (11–382); part 2, ἔργα (383–617); part 3, ἡμέραι (618–828).[18] We observe the fact that Hesiod's "preamble" on the need to work occupies a longer section of his poem than either the "Works" or the "Days," which are about equal in length.

The structure of *Georgics* 1 is very different. Vergil's imitation truly is bipartite, consisting of a section of *opera* (43–203) followed by one of *dies* (204–350): there is no third section corresponding to Hesiod's mythological introduction. Yet Vergil does not simply ignore this material. In developing the mythic background to his imitation of Hesiod, he does not, it is true, take over the Hesiodic material intact; rather, as we have now seen in so many similar cases, he analyzes his archaic original into its elementary components, discovers in these *disiecta membra* the germ of new poetic possibilities, and creates through imitation an original idea in harmony with the demands of a poem for a new age. In this particular case, we find that the mythic background of *Georgics* 1 is developed gradually, though not exclusively, through a series of formal

18. Even this analysis is too simplistic, but it takes account of all those sections of the poem that are relevant to Vergil's imitation.

digressions that extend from one end of the book to the other. Indeed, it would be more accurate to say that the series extends beyond Book 1 to provide an underlying thematic context for the entire poem; but this is to anticipate. Confining our purview for the moment to Vergil's program of Hesiodic allusion, it seems reasonable to suggest that Vergil's series of occasional mythic digressions corresponds to the mythic preliminaries of *Works and Days*, just as his reorganized discourse on *opera et dies* proper is based on that of Hesiod.

The contrast, both in *Works and Days* and in *Georgics* 1, between myth and "practical" instruction is essentially a formal one. If Vergil's means of expressing the relationship between myth and mundane existence, as represented by his "technical instruction," differ from those employed by Hesiod, nevertheless the relationship between the two spheres remains in place. By virtue of seeing in the most ordinary farm operations the principles of a universal order, Vergil resembles his model closely. A further difference is perhaps that where Hesiod proceeds deductively—i.e. since (he argues) certain principles exist, as the myths prove, to lend order to the universe, we should observe them and act in this way—Vergil fashions his "technical" passages in such a way as to produce by induction an awareness of universal principles.

A more specific comparison of the mythic elements in both poems will bear out this hypothesis. Hesiod, as I have noted, begins his poem with a brief proemium to the Muses (1–10) and then relates a series of *μύθοι* (11–382) that develop the cosmic importance of his subject. Vergil, after an elaborate proemium that owes little to Hesiod (1–42), launches immediately into advice on when and how often to plow (43–70). In this paragraph there is a passing reference to the myth of Pyrrha and Deucalion (*G.* 1.60–63). Thereafter, apart from one astronomical reference (*Arcturum* 68), one "decorative" epithet (*Lethaeo... somno* 78), and an instance of conventional metonymy (*neque illum | flava Ceres alto nequiquam spectat Olympo* 95–96), mythology is avoided until lines 118–159, the "Aetiology of *Labor*." This, the key mythic section of *Georgics* 1, corresponds in theme and in particular motifs to Hesiod's lengthy discourse on the need to work, and especially to his famous "Myth of Ages" (*WD* 109–201); and this passage, which has not yet figured in our analysis, is commonly adduced as the main source of Vergil's "Aetiolgy."

In accounting for the need to work, Vergil divides human history into two distinct periods, the earlier one free of *labor*, the later one governed and even defined by it. The "Aetiology" thus shows affinities with a common religious world view, one that explains humankind's current state as deficient in comparison with a previous, more blessed existence. While this myth is common in one form or other to many cultures, it is in classical civilization identified primarily with Hesiod; and that Vergil has Hesiod's account in mind is clear not only from thematic and structural parallels, but from specific verbal echoes as well. Above all, both authors agree that it is "Father Zeus" (*WD* 143 Ζεὺς δὲ πατήρ ≈ *G.* 1.121 *pater ipse*) who governs the transition from the earlier to the later condition; Hesiod's golden tribe had enjoyed the spontaneous produce of the earth (καρπὸν δ' ἔφερε ζείδωρος ἄρουρα | αὐτομάτη πολλόν τε καὶ ἄφθονον 117–118), just as all humankind originally had done in Vergil's conception (*ipsaque tellus* | *omnia liberius nullo poscente ferebat* 127–128); but these pristine, paradisiacal conditions give way to a harder life and the need to work. Of course, as in all imitations, Vergil varies his model considerably. The phrase *gravi...veterno* (124) is widely regarded as an echo of Hesiod's δειλόν | γῆρας (113–114), although the two phrases are used very differently. In fact *gravi veterno* would be a perfectly accurate translation of δειλὸν γῆρας under normal circumstances; but here the similarity of the two phrases underlines a difference in conception that stands out when we render the two passages in English. Hesiod states that the golden tribe did not suffer *grievous old age;* Vergil that Jupiter did away with the easeful conditions of the earlier age because he would not allow his kingdom to wallow *in oppressive sloth.* The two poets use similar diction to express incompatible views on humankind's original work-free condition. Hesiod presents this time as free of trouble in all its forms (νόσφιν ἄτερ τε πόνου καὶ ὀϊζύος 113), including δειλὸν γῆρας and the need to work. Vergil's Jupiter, however, equates this freedom from all forms of *labor* (≈ πόνος) precisely with *gravis veternus.* Thus Vergil, by a typical procedure of allusive dialectic, in essence uses Hesiod's own words to contradict what his model says about humankind's original state.

Allusive variation appears in other guises as well. By way of reworking Hesiod's "Myth of Ages," Vergil reserves some details for contexts other than "The Aetiology of *Labor.*" For instance, Hesiod's golden tribe is associated with the reign of Kronos (*WD* 111),[19] later tribes with Zeus (138, 143, 158, 168, 180); the decline of paradisiacal conditions for

19. West 1978.195 *ad WD* 173.

humankind is thus associated with the cosmic succession myth. In *Georgics* 1 we hear nothing of Kronos' Roman counterpart Saturnus; but his presence is felt from time to time in Book 2.[20] In the "*Laudes Italiae,*" which paints a picture of Italy as a land of almost supernatural abundance, Vergil salutes his homeland as *Saturnia tellus* (*G.* 2.173); and it is in this same passage, as we have seen, that the poet refers to his own creation as an *Ascraeum carmen* (176). Both phrases obviously complement "The Aetiology of *Labor*" by looking back to Hesiod's "Myth of Ages." A related reference occurs in the final episode of Book 2:

> ante etiam sceptrum Dictaei regis et ante
> impia quam caesis gens est epulata iuvencis,
> aureus hanc vitam in terris **Saturnus** agebat;
> necdum etiam audierant inflari classica, necdum
> impositos duris crepitare incudibus ensis.
>
> *Georgics* 2.536–540

Here Vergil clearly alludes to the succession myth, associating Saturnus with an earlier, idealized golden age (***aureus*** *Saturnus* 2.138) and Jupiter (*Dictaei regis* 536) with a postlapsarian period characterized by necrophagy (537) and warfare (539–540). Both these details correspond with elements of Hesiod's account: his bronze tribe is associated above all with a lust for warfare, and is also is the first not to eat bread—i.e. to eat meat:

> οἷσιν Ἄρηος
> ἔργ᾽ ἔμελεν στονόεντα καὶ ὕβριες, οὐδέ τι σῖτον
> ἤσθιον
>
> *Works and Days* 145–147

Finally, earlier in the finale of Book 2, Vergil inserts yet another detail from Hesiod. In praising the life of the Italian farmer, he notes that it was among them that Justice spent her last days on earth (*extrema per illos | Iustitia excedens terris vestigia fecit* 473–474). This is an allusion to, ultimately, the very end of "The Myth of Ages," where Hesiod predicts that the iron tribe will degenerate until finally "Aidos and Nemesis will abandon humankind and go off to join the tribe of the immortals, and woeful pains will be left for mortals, and there will be no defense against

20. Mention of the planet at line 336 (*Saturni...stella*) does not, however, seem apposite to the "Myth of Ages," nor does the myth alluded to at 3.93. On Saturnus in Vergil's conception of the Golden Age see Johnston 1980.62–89.

evil" (*WD* 199–201). Again, of course, Vergil departs significantly from Hesiod: Aidos and Nemesis have been combined into the single figure of Iustitia.[21] More important, Vergil's "golden Saturn" lives the life of a farmer on earth (*in terris* 2.538); here we may also note a passing reference to Saturnus' pruning hook (*G.* 2.406). This association of ideas is quite un-Hesiodic: in *Works and Days* Kronos rules in the sky (*οὐρανῷ ἐμβασίλευεν* 111), and the golden tribe is fed by the earth's spontaneous produce (117–118); farming is never mentioned.

Obviously Vergil develops the Golden Age theme in a more leisurely way than Hesiod does, integrating it with his argument at crucial points within Books 1 and 2 of the *Georgics*. It is also apparent that Vergil's conception of the Golden Age is in some respects at odds with that of Hesiod. Finally, although Hesiod's "Myth of Ages" is undoubtedly the main inspiration of the "Aetiology of *Labor*," Vergil—as often—"contaminates" his primary source with allusions from other parts of *Works and Days* that explore its internal thematic structure.

The "Myth of Ages" is not Hesiod's sole explanation of the need to work. The poet first accounts for this need by declaring that "the gods once hid the means of life from men, and they continue to hide it" (*κρύψαντες γὰρ ἔχουσι θεοὶ βίον ἀνθρώποισιν* *WD* 42). The story that he tells to justify this claim concerns Prometheus' theft of fire and Zeus's subsequent retribution, exacted through his unwitting agent, Pandora. "Zeus became angry at heart and he hid [the means of life] because crafty Prometheus had deceived him: this is why he devised baleful sorrows for humankind *and hid fire* (*κρύψε δὲ πῦρ* 50)...." In Vergil, too, Jupiter hides fire, but for a different reason. "He put evil venom into black serpents, bade wolves ravage and the sea toss, struck the honey from the leaves *and hid fire* (*ignemque removit*, *G.* 1.131), and checked the wine that ran in channels here and there, so that experience might with thought gradually hammer out the different crafts and seek the blade of grain in furrows, so that it might knock *the hidden fire* (*abstrusum...ignem* 135) out of veins of rock" (129–135). The precise verbal allusion recalls Hesiod's first explanation of the need to work, but brings the motif of hidden fire into line with Vergil's revised view of the Golden Age: it becomes just one more Jovian technique for shaking humankind out of its easeful torpor. Human craft is dissociated from Promethean deception and, ironically, becomes the (possibly unwelcome) gift of the very

21. Vergil borrows this conflation from Aratus: see Solmsen 1966.

same god whom Hesiod had previously portrayed as Prometheus' victim and opponent.

Again, the naming and charting of the constellations—a process essential to the postlapsarian craft of navigation—is marked by allusion to Hesiod:

> Pleïadas, Hyadas, claramque Lycaonis Arcton
>
> *Georgics* 1.138
>
> Πληϊάδες θ' Ὑάδες τε τό τε σθένος Ὠαρίωνος
>
> *Works and Days* 615

In *Works and Days*, of course, this line shortly anticipates Hesiod's advice about seafaring (*WD* 618–694);[22] Vergil thus spins another allusion into his web of thematic associations.

Finally, we may note that the "Aetiology" concludes with the characteristically Hesiodic observation that "unless you keep after the weeds with assiduous rakes and frighten the birds with noise and check with your pruning hook the shadows of the darkening farm and call rain with prayers, ah! in vain will you gaze on another's great pile, and shake the forest oak to appease your hunger" (*G.* 1.155–159); "strip to sow, strip to plow, strip to reap, if you want to accomplish all Demeter's tasks in season: thus will each thing grow for you in season, lest afterwards in need you go begging to other men's homes and accomplish nothing" (*WD* 391–395). It is this same passage to which Vergil alludes later in Book 1, as we have seen, with the words *nudus ara, sere nudus: hiems ignava colono* (*G.* 1.299). But note that he chooses to incorporate the moral lesson of this Hesiodic passage into his own mythic interlude on work, while the direct, unvarnished injunction takes its simple place within a matter-of-fact list of homely tasks. From such observations, we must conclude that Vergil's analytical approach to his model concerns not only the larger units of exposition, but extends even to the development of individual lines and phrases.

Structure and Selectivity

With this Hesiodic background established, it should be obvious that "The Aetiology of *Labor*" takes its place beside "The Farmer's *Arma*" and

22. Cf. *WD* 619 εὖτ' ἂν Πληϊάδες σθένος ὄβριμον Ὠρίωνος.

"Lucky and Unlucky Days" as the third pillar of Vergil's Hesiodic program. The three passages stand, allowing for the expected allusive variation, in substantially the same relationship to each another and to *Georgics* 1 as a whole as do their Hesiodic models—"The Myth of Ages" (109–201), "Farm Equipment" (414–457) and "The Lunar Calendar" (765–828, especially 800–813)—to one another and to *Works and Days* as a whole. Structurally and thematically, they represent the three main sections of Hesiod's poem—its mythic introduction (11–382), the ἔργα (383–617) and the ἡμέραι (618–828). Taken together, they are as it were the skeleton of Vergil's book-long allusion to *Works and Days*.

At the same time, we must recognize that "The Aetiology of *Labor*" differs from "The Farmer's *Arma*" and "Lucky and Unlucky Days" in this respect: the latter two passages define the bipartite structure of *Georgics* 1; and within this structure, the "Aetiology" is formally a digression. The Hesiodic passages on which Vergil draws in these opening paragraphs of *Georgics* 1 for the most part occur within a fairly short span of lines roughly corresponding to *WD* 414–478. The "Aetiology of *Labor*" draws chiefly on *WD* 11–382, especially 109–201 ("The Myth of Ages"). Vergil follows this mythic "digression" with more "practical" advice on farm equipment (*G.* 1.160–175), a passage heavily indebted, as I have shown, to *Works and Days* 426–431. This paragraph contains the last overt reminiscence of *Works and Days* in the ἔργα portion of *Georgics* 1; and again the Hesiodic material to which Vergil alludes is found in the same short span of lines that have provided so much of the "practical" material in this book.

The fact that so much of Vergil's Hesiodic material stems from a fairly restricted passage of *Works and Days* is interesting for two reasons. First, it suggests that Vergil viewed these lines as the didactic kernel of Hesiod's poem, its most essential repository of "practical" advice, or perhaps as the most characteristically Hesiodic portion of the "Works." Second, it allows the observation that *Georgics* 43–203—the ἔργα portion of Vergil's Hesiodic imitation—would have been a more or less continuous reworking of this single Hesiodic passage, had it not been for the interruption of the mythic "Aetiology of *Labor*." Under these circumstances, it becomes more and more interesting to inquire into Vergil's reasons for recasting the first half of Hesiod's poem into this particular form. Certainly the digressive aspect of Vergil's myth, when viewed against its original in terms of structural comparison, becomes even more pronounced. Thus Vergil's treatment of Hesiod's introductory mythic section reinforces our previous conclusions about structural and thematic rationalization of the "Works" and "Days" material in *Georgics* 1. We have seen in several instances that Vergil's pattern of allusion tends to intensify this dualism. Vergil so to speak transfers "Days" material, such as the maxim γυμνὸν σπείρειν, κτλ (*WD* 391–392) from the *opera* section of Hesiod's poem to his own *dies* section, and vice versa. When he had effected this stage of the transformation, which concerns only Hesiod's second ("Works") and third ("Days") sections, it remained for Vergil to deal in some way with the first (mythic) section. The solution he reached was to distribute this material in the form of occasional mythic digressions or asides over the both the "Works" and "Days" sections of *Georgics* 1 and, indeed, over the remaining three books of the poem.

The specific myths to which Vergil refers are by no means exclusively or even primarily Hesiodic.[23] At the same time, however, they normally show a direct thematic relevance to the myths that Hesiod tells in *WD* 11–382. The myth of Pyrrha and Deucalion (*G.* 1.60–63), for example, is first attested in Pindar (*O.* 9.43–57), not Hesiod. But as a creation myth, it clearly belongs in the same class as the stories that Hesiod tells in *Works and Days.* Indeed, Vergil's "Aetiology of *Labor*" may contain a brief allusion to Pindar's version of "Pyrrha and Deucalion." Pindar tells us that "they say indeed that the power of water covered over the dark land, and then that *by the crafts of Zeus* (Ζηνὸς τέχναις 52) the water suddenly receded." In Vergil, "Father [Jupiter] himself wanted the path of agriculture to be difficult, and was the first to move the fields *by craft* (*per artem* 1.121), sharpening mortal minds with anxiety...." The reminiscence is slight; but even if Pindar stands outside of the epic tradition and exerts a correspondingly small influence on the *Georgics* in general, here at least we find useful Pindaric material integrated into Vergil's Hesiodic program.

Later in "Lucky and Unlucky Days"—ostensibly, of course, a technical passage—something similar occurs. As we have seen, this passage uses the mythic *aetia* related by Hesiod in *Works and Days* 765–828 as an excuse to import similar material from other sources—namely from *Theogony* and from *Odyssey* 11. The myths that Vergil relates concern the monsters Coeus, Iapetus, Typhoeus, and the twins Otus and Ephialtes,

23. See Frentz 1967 for full discussion of these passages and their antecedents.

all of them rebels against Zeus/Jupiter's regime. These stories, too, have relevance to "The Myth of Ages"; for if this story is on the one hand a creation myth, it is on the other, as I have already noted, a succession myth; and if the *Georgics* is in any sense about the creation of a new society, it is also concerned with whether the young Caesar, like the young Zeus of the *Theogony*, having assumed his father's place at the head of the state (≈ cosmos) can put an end to the cycle of violent successions, ward off challengers, and securely establish his own regime. Vergil therefore greatly accentuates the succession myth, which is much less pronounced in *Works and Days*, by an allusive variation that involves not only in Hesiod's other masterpiece, but in the Homeric *Odyssey* as well.

To sum up Vergil's way of alluding to the mythic portion of *Works and Days*, we may make the following points. First, *Georgics* 1 imitates *Works and Days* by providing what passes for practical instruction with an explicit mythic setting, and by creating explicit structural, thematic and verbal correspondences between its single most ambitious mythic passage ("The Aetiology of *Labor*") and Hesiod's longest single myth ("The Myth of Ages"). Second, Vergil varies his imitation in several ways. Among them are the use of Hesiodic diction in new or even antithetical senses; the analysis and distribution of Hesiod's "Myth of Ages" over widely separated parts of the *Georgics;* emphasis and suppression of various Hesiodic elements; and the replacement or supplementation of parts of his main source with material from other poems and even other authors.

It is also apparent that Vergil's selective handling of both mythic and technical material from *Works and Days* conforms to the same allusive ideas. On the one hand, Vergil obviously follows Hesiod's general didactic procedure of describing, or at least enumerating, the most important yearly farm operations, along with the best times to perform them; and he also creates an explicit parallelism of structure, theme and expression through his two most obvious Hesiodic showpieces, "The Farmer's *Arma*" and "Lucky and Unlucky Days." On the other hand, the poet varies his imitation by means of the same techniques that he applies to the mythic portion of his Hesiodic material. Not only does Vergil freely alter Hesiod's treatment of myth, an utterly normal and expectable procedure, but he also explicitly contradicts elements of his technical discourse. For example, he borrows Hesiod's evocative description of late summer, but gives it a surprising twist:

> ἦμος δὲ σκόλυμός τ' ἀνθεῖ καὶ ἠχέτα τέττιξ
> δενδρέῳ ἐφεζόμενος λιγυρὴν καταχεύετ' ἀοιδὴν
> πυκνὸν ὑπὸ πτερύγων, θέρεος καματώδεος ὥρῃ,
> **τῆμος πιόταταί τ' αἶγες, καὶ οἶνος ἄριστος,**
> μαχλόταται δὲ γυναῖκες, ἀφαυρότατοι δέ τοι ἄνδρες
> εἰσίν, ἐπεὶ κεφαλὴν καὶ γούνατα Σείριος ἄζει,
> αὐαλέος δέ τε χρὼς ὑπὸ καύματος· ἀλλὰ τότ' ἤδη
> εἴη **πετραίη τε σκιὴ** καὶ βίβλινος οἶνος.
>
> *Works and Days* 582–589

> in primis venerare deos, atque annua magnae
> sacra refer Cereri laetis operatus in herbis
> extremae sub casum hiemis, iam vere sereno.
> **tum pingues agni et tum mollissima vina,**
> tum somni dulces densaeque **in montibus umbrae.**
>
> *Georgics* 1.338–342

Here Vergil has gone out of his way to create a close verbal parallelism, and has followed his normal policy of transferring Hesiod's statements about the schedule of operations found in the "Works" to their (so to speak) proper place among the "Days." Yet we also find him contradicting Hesiod in a completely gratuitous way. Hesiod's lines concern the need to drink and rest in the shade during the "dog days" of late summer; Vergil, bafflingly, assigns this "activity" to early spring, as soon as winter has ended—a time not for rest, as he tells us elsewhere, but to begin the year's work. The point of this disinformation is hard to discern; perhaps its closest parallel is found in "Lucky and Unlucky Days," where Vergil violates—no doubt deliberately—all of Hesiod's and everyone else's advice about what days are best for different jobs. Such passages illustrate the relative importance to Vergil of forging a purely literary relationship as opposed to dispensing accurate instruction.[24]

The same lines also illustrate the distribution of related Hesiodic allusions over widely separated parts of *Georgics* 1. Vergil distributes mythic material borrowed from Hesiod not only throughout Book 1, but in the case of "The Myth of Ages" over Book 2 as well. In such instances, what was originally a single context in *Works and Days* is analyzed into its component parts, which are then incorporated through allusion into a more gradual development of similar themes in the *Georgics*. By the same token, we find "technical" passages of *Works and*

24. Cf. Thomas 1987; for a different view, see Spurr 1986.

Days similarly analyzed though allusion. Some traces of this activity are of minor importance. A phrase such as *rapidive potentia solis* (*G.* 1.92) may well derive from μένος ὀξέος ἠελίοιο (*WD* 414), which occurs in a context that demonstrably influenced Vergil elsewhere (i.e. the passage on woodcutting, the basis for "The Farmer's *Arma*"). But it is difficult to see here evidence for analysis through allusion; the echo is more likely due to Vergil's barely conscious memory of a passage made familiar to him by the effort of intensive imitation, or even to chance. On the other hand, ideas that are found in series in *Works and Days* occur separately in the *Georgics*. In the following passage, Hesiod arranges his list of instructions on readying tools in a very reasonable sequence: (A) Get all your equipment ready well before you will need to start using it; (B) start plowing at the very beginning of Spring; and (C) say a prayer to Zeus and Demeter when you do in fact begin:

φησὶ δ' ἀνὴρ φρένας ἀφνειὸς πήξασθαι ἄμαξαν·
νήπιος, οὐδὲ τὸ οἶδ'· ἑκατὸν δέ τε δούρατ' ἀμάξης,
A τῶν πρόσθεν μελέτην ἐχέμεν οἰκήϊα θέσθαι.
B εὖτ' ἂν δὴ πρώτιστ' ἄροτος θνητοῖσι φανήῃ,
δὴ τότ' ἐφορμηθῆναι, ὁμῶς δμῶές τε καὶ αὐτός,
αὔην καὶ διερὴν ἀρόων ἀρότοιο καθ' ὥρην,
πρωῒ μάλα σπεύδων, ἵνα τοι πλήθωσιν ἄρουραι.
ἔαρι πολεῖν· θέρεος δὲ νεωμένη οὔ σ' ἀπατήσει·
νειὸν δὲ σπείρειν ἔτι κουφίζουσαν ἄρουραν.
νειὸς ἀλεξιάρη παίδων εὐκηλήτειρα.
C εὔχεσθαι δὲ Διὶ χθονίῳ Δημήτερί θ' ἁγνῇ
ἐκτελέα βρίθειν Δημήτερος ἱερὸν ἀκτήν,
ἀρχόμενος τὰ πρῶτ' ἀρότου, ὅτ' ἂν ἄκρον ἐχέτλης
χειρὶ λαβὼν ὅρπηκι βοῶν ἐπὶ νῶτον ἵκηαι
ἔνδρυον ἑλκόντων μεσάβων.

Works and Days 455–469

Vergil imitates each of these points; but he does not associate them. Point A occurs, as we have seen, at *G.* 1.167 in the middle of "The Farmer's *Arma*," where among other things it marks the limit of the Hesiodic passage that is Vergil's source here.[25] Point B has also been discussed: it is found at *G.* 1.43, where it inaugurates the Vergilian "Works."[26] Point C occurs at *G.* 1.339 in a context that features, as we have just seen, other Hesiodic elements:

25. See pp. 72–73.
26. See pp. 136–139.

> in primis venerare deos, atque annua magnae
> sacra refer Cereri laetis operatus in herbis
> extremae sub casum hiemis, iam vere sereno.
> tum pingues agni et tum mollissima vina,
> tum somni dulces densaeque in montibus umbrae.
>
> *Georgics* 1.338–342

Here Vergil repeats Hesiod's injunction to pray to Ceres (≈ Demeter)—omitting Zeus, however—at the very beginning of spring, when winter has barely ended. This emphasis on the season apparently looks to Hesiod's point B; but the corresponding passages in Vergil are widely separated. Instead, Vergil juxtaposes this reference with a quotation from an altogether different section of Hesiod's poem (*WD* 582–589), a passage that concerns an altogether different season, as I have just shown.

What these passages illustrate is the extreme plasticity that Hesiod's exposition had in Vergil's hands. The elements of Hesiod's "technical" argument, as of its mythic underpinning, are willfully manipulated in Vergil's imitation.

Vergil's reworking of Hesiodic source material often intensifies the thematic contrast between "Works" and "Days." I have mentioned this tendency repeatedly, but it bears one further illustration. Here Vergil discusses the tasks that can be performed during inclement weather:

> frigidus agricolam si quando continet imber,
> multa, forent quae mox caelo properanda sereno,
> maturare datur: durum procudit arator
> vomeris obtunsi dentem, cavat arbore lintres,
> aut pecori signum aut numeros impressit acervis.
>
> *Georgics* 1.259–263

The lines look to Hesiod's advice about preparing in winter for plowing season:

> *πὰρ δ' ἴθι χάλκειον θῶκον καὶ ἐπαλέα λέσχην*
> *ὥρῃ χειμερίῃ, ὁπότε κρύος ἀνέρα ἔργων*
> *ἰσχάνει, ἔνθα κ' ἄοκνος ἀνὴρ μέγα οἶκον ὀφέλλοι,*
> *μή σε κακοῦ χειμῶνος ἀμηχανίη καταμάρψῃ*
> *σὺν πενίῃ, λεπτῇ δὲ παχὺν πόδα χειρὶ πιέζῃς.*
>
> *Works and Days* 493–497

These passages in fact illustrate several of the points we have been discussing. Verbal and thematic parallels establish a connection (*si quando* ≈ ὁπότε; *frigidus imber* ≈ κρύος; *agricolam...continet* ≈ ἀνέρα ἔργων | ἰσχάνει). Again, however, Vergil inverts Hesiod's advice, or nearly does so. Where Hesiod warns his listener not to loiter in the warmth of the smithy during the winter's cold, Vergil bids his reader busy himself in the smithy during that season. Hesiod, of course, gives this advice in the "Works" portion of his poem; Vergil, as he does with almost any Hesiodic material involving the *scheduling* of work, "transfers" it to his "Days" imitation.

The passage just examined illustrates still another point, namely, Vergil's reworking of Hesiod's "technical" discourse in the light of material borrowed from other authorities. Specific elements of Vergil's advice on what to do during the winter clearly derive from Hesiod's treatment of this topic; but the passage as a whole shows a much greater affinity with the Roman tradition of agricultural writing. Discussion of what work was possible in bad weather (*G.* 1.259–267) and permissible on holy days (*G.* 1.268–275) goes all the way back to Cato, whom Vergil's account appears to follow directly in some instances.[27] Thus what begins as an imitation of Hesiod actually draws more extensively not only on a different author, but even a different, non-poetic literary tradition.

This observation raises an important point. Some scholars find the metaphrastic aspect of the *Georgics* more important in general than the poem's epic associations. Thus Servius, as I have noted, traces the four-book structure that Vergil employs to Varro's fourfold division of agriculture according to land usage.[28] In the same vein, D. O. Ross, whose work stresses Vergil's debt not only to agricultural writers, but to ancient medicine and physics as well, finds that the argument of Vergil's "Days"—the first half of *Georgics* 1—closely resembles that of Varro in the first half of *DRR* 1.[29] Thus the Vergilian "Works," he argues, "is not Hesiodic in its structure, but rather Varronian...."[30] By the same token, we may recall the way in which Vergil's poem begins. I have already noted the dissimilarity between Vergil's proemium (*G.* 1.1–42) and that of Hesiod (*WD* 1–10).[31] It is no secret that it bears a much stronger likeness

27. See Jahn 1903a.417–418.
28. Cited on p. 28 n. 2.
29. Ross 1987.38–41.
30. Ibid. 38.
31. See p. 136.

to the proemium of, again, Varro's treatise (*DRR* 1.1.4–6).[32] By an argument similar to that which I have used to establish a structural congruity between *Georgics* 1 and *Works and Days*, someone might adduce this introductory Varronian passage to prove that the whole poem should be read as an imitation of *De Re Rustica*. Indeed, at least one scholar has made such a claim.[33] All of these points are well taken; but in my view they need qualification. Specifically, to address Ross's observation, it is true that Vergil and Varro treat of the same essential topics, which are enumerated at *DRR* 1.5.3, each in a different order. The situation is very much like that of the "Weather Signs," which as we have seen handles essentially the same topics as Aratus' "*Diosemiae*" in a revised sequence.[34] But in the *Georgics*, the Varronian material, to say nothing yet of the Aratean, occurs in what is manifestly an overarching contextual allusion to *Works and Days*. To speak of details, Vergil handles Varro's second topic ("necessary equipment") by creating an elaborate allusion to Hesiod—none other than "The Farmer's *Arma*." Because Hesiodic allusion is so obviously important both on the level of the entire book's structure and on the level of individual passages such as this, I find it impossible to agree that the structure of Vergil's "Works" is Varronian *rather than* Hesiodic. But neither am I convinced that it is necessary to make this choice. What I believe a passage like the proem illustrates is another facet of Vergil's Hesiodic program—a tendency to vary imitation of *Works and Days* by assimilating Hesiodic material to (in this case) a Varronian treatment.[35] This variation also takes the form of fairly direct replacement. From the standpoint of Vergil's imitative program, it seems that Varronian material has been substituted for Hesiodic by a process of allusive variation similar to what we have seen in other cases. As for the general allusive program in Book 1, we may say that Vergil has allowed

32. On this point see, above all, Wissowa 1917.

33. Leach 1981.

34. See pp. 79–83.

35. I am aware of the bias that this statement betrays. While the importance of Vergil's Hesiodic program is (I hope) manifest, the influence of Varro, though different perhaps in kind, is hardly less great. It is ultimately pointless to argue that one of these authors preceded the other in shaping Vergil's conception of the *Georgics*. At the same time, I am trying to illuminate Vergil's use of the epic tradition; and this effort inevitably involves an implicit assumption of Hesiod's priority with respect to Varro and other prose authors as a Vergilian source. I must therefore leave the study of Varronian influence, with its attendant biases, to other scholars.

elements of the technical tradition that Varro represents to inform his *aemulatio* of Hesiod.

To conclude, more could perhaps be said, but the nature and extent of Vergil's Hesiod imitation should now be clear. It is in no way a matter of occasional, decorative allusion to *Works and Days*, either for merely "generic" purposes or for any other reason; nor are we to think of the *Georgics* in any strict sense as a Hesiodic poem *in toto*. Instead, we find that Vergil has gone to great lengths to establish structural congruency between *Works and Days* and *Georgics* 1 only; Books 2–4, despite occasional prominent thematic associations with Vergil's imitation of Hesiod, play no part in this imitation of a complete poetic structure. By basing this structural parallelism on three elaborate and extended allusions to specific passages of Hesiod's poem, Vergil in a sense freed himself to exploit other Hesiodic passages in a looser fashion, alluding to them without regard to their original sequence, calling attention to their thematic (rather than structural) relationship to other passages of *Works and Days*, and even to passages in other poets. But to concentrate on these individual elements of allusive variation is to lose sight of Vergil's Hesiodic program, which once we have recognized it can be described quite simply: to rework the structure, major themes, and characteristic forms of expression that he found in *Works and Days* into a new book of poetry, his first book of *Georgics*. This, first and foremost, is the achievement on which Vergil looks back when, having already embarked upon a new allusive program, he speaks so proudly of his Hesiodic song, his *Ascraeum carmen*.

Aratus and Vergil's *Phaenomena*

Soon after "Lucky and Unlucky Days," Vergil's chief source for the remainder of Book 1 becomes the concluding section of Aratus' *Phaenomena*, the "*Diosemiae.*" The change itself is obvious, but a precise understanding of the poem's structure at this point has proved elusive. Many scholars feel that *Georgics* 1 comprises three sections, two Hesiodic sections of "Works" and "Days" followed by an Aratean section of "Weather Signs."[36] Those who subscribe to this view are divided, however,

36. Wilkinson (1969.318), citing the support of Richter (1957.408–409), states that "the weather-signs are distinct from what could conceivably be entitled the 'Days.'" Otis (1964.155) also regards the "Weather Signs" as something added to the bipartite Hesiodic structure of ἔργα and ἡμέραι.

on just where Hesiod's influence leaves off and that of Aratus begins. Some scholars put the border before line 311, others after line 350. These lines define a paragraph on "Vernal and Autumnal Storms"; and the debate over whether this paragraph belongs more properly to the "Days" or to the "Weather Signs" underlines the difficulty of separating Hesiodic from Aratean influence in *Georgics* 1.[37] A few have dodged the issue by labeling this paragraph "transitional." But the general lack of consensus points out the extreme difficulty of viewing Vergil's imitations of Hesiod's "Days" and of Aratus' "*Diosemiae*" as being utterly discrete. Thus a few scholars, very reasonably in my view, simply assume a bipartite structure for *Georgics* 1, taking the "Weather Signs" as being in some sense a part of Vergil's "Days" imitation.

The difficulty of establishing clear divisions, however, is not the only reason for regarding the "Weather Signs" as a part of Vergil's "Days." Prognostication is not alien to *Works and Days:* Hesiod treats of the subject in several connections, chiefly in the "*Nautilia*" (618–694), but also in passing remarks that occur in the "Days" (lines 801 and 828). More important, there exists a continuity of theme from Vergil's "Lucky and Unlucky Days" and "Climatic Zones" paragraphs, which most scholars do regard as part of the "Days" section, all the way through the end of the book—the emphasis throughout, as Erich Burck has stated, being on the relationship between cosmic events and the affairs of humankind, especially as expressed in the georgic world on the one hand and in the political world on the other.[38]

The change of subject matter that occurs toward the end of Book 1 is in itself no reason to assume that Vergil at some point abandons his Hesiodic program. But in examining the sources of the "Days," one discovers quickly that Hesiod is a much less important source of material here than in the "Works." Allusions to specific passages of *Works and Days* continue to occur, but with much less frequency than before. Moreover, when he does allude to Hesiod here, Vergil shows less interest in the Hesiodic "Days" than in the "Works"—and specifically in that portion of the "Works" that supplied Vergil with the bulk of the Hesiodic material that appears in his own "Works." This observation is entirely in keeping with Vergil's policy of transferring thematically "unsuitable" material from the Hesiodic "Works" to more appropriate places in his own "Days." Interestingly, there is little corresponding transferal of original "Days" material to the Vergilian "Works"; nor,

37. Otis 1972.47 et al.
38. Burck 1929.287.

somewhat more surprisingly, does Vergil employ much material from the Hesiodic "Days" in his own "Days." The great exception, of course, is "Lucky and Unlucky Days," the structural anchor of Vergil's "Days" imitation and the only passage of the Hesiodic "Days" that Vergil imitates specifically and at length anywhere in the *Georgics.* This surprising fact is at least partly explained by our discovery that Vergil designed "Lucky and Unlucky Days" as a condensed allusion to the Hesiodic "Days" in its entirety. But this discovery does not explain why Hesiod's direct influence towards the end of *Georgics* 1 falls off to virtually nil.

This decline is readily explained by the corresponding rise in importance of Aratus' "*Diosemiae*" as a major source for Vergil' "Days." To put it in purely quantitative terms, the "*Diosemiae*" provides more material, in fact, than even the Hesiodic "Days" provides. This observation is not incompatible with our previous conclusions about Vergil's program of Hesiodic allusion: the overall *structure* of *Georgics* 1 follows that of *Works and Days.* But in dealing with specific topics, it would seem that Vergil has deliberately substituted Aratean material for Hesiodic in his imitation of Hesiod's "Days." But why did he do so?

It is surprising that the point is never made, but the congruency of structure between *Phaenomena* and *Georgics* 1 is in broad outline similar to that between *Georgics* 1 and *Works and Days.* As in the case of Hesiod's poem, this structural congruency is established by only a few points of obvious contact. To begin at the end, *Phaenomena* concludes with the "*Diosemiae*," which Vergil imitates in his "Weather Signs." Vergil's imitation, of course, involves a fairly extreme internal restructuring of the "*Diosemiae*," and I have already discussed how this treatment of original Aratean material reveals his purpose of creating from it a convincing climax to *Georgics* 1 as a whole.[39] Viewed as parts of two larger contexts, however, both the imitation and its original occupy exactly comparable positions in their respective poems. This structural correspondence looks like a quite deliberate and integral element of Vergil's Aratus imitation; for, as we have seen in the case of Hesiod, extended and elaborate allusion to *Works and Days* at structurally comparable points in *Georgics* 1 signals a more general similarity of form. The "Weather Signs" is just as detailed as any imitation in Book 1, and is more extensive than virtually any other in the entire poem. It only stands to reason that it should be given its full allusive weight—that we

39. See p. 83.

should regard it not simply as a nexus of borrowed themes, phrases, and motifs, but as a carefully planned signal of the structural congruity between *Georgics* 1 and *Phaenomena* as well.

The same is true of the way in which Vergil begins this book. The *Phaenomena*, like *Georgics* 1, plunges directly into its subject without developing a metaphysical background for the exposition, other than the brief remarks on Zeus's providence that occur in the proemium (1–18). Mythology is presented only briefly and tangentially[40]—again, just as in the *Georgics*[41]—until lines 96–136, where Aratus relates in full the *aetion* of the constellation Virgo. This *aetion* amounts to a reworking of Hesiod's "Myth of Ages" (*WD* 109–201) with a view to the requirements of a very different context.[42] Vergil's "Aetiology of *Labor*" is yet another reworking of this material, but one that is clearly indebted both to Hesiod *and* to Aratus—a clear example, in fact, of imitation through *contaminatio*. As often, *contaminatio* is inspired by the basic similarity of two source passages, but reveals itself as a conflation by incorporating different details characteristic of each.[43] Structurally, too, the *Georgics* passage is closer to the *Phaenomena*. In both of the later poems, as in *Works and Days*, the myth of ages is used to develop a metaphysical background for "practical" instruction. But Vergil and Aratus depart from Hesiod, who presents the myth as an essential preliminary to any further instruction; instead, the two imitators present this metaphysical background as a formal digression from "technical" instruction. At the same time, Vergil's "Aetiology" seems to belie its formal status as a digression. The themes that it broaches are as essential to the conception of human life that Vergil develops in the *Georgics* as anything in Hesiod; and in this respect it must be distinguished from Aratus' version, which is in this respect a more truly incidental and less thematically charged digression.

40. Cf. *Ph.* 30–35 on the catasterism of Helice and Cynosura; 71–73, a brief *aetion* of the Corona Borealis.

41. See pp. 143–144.

42. That Aratus imitates Hesiod here was well known in antiquity. Σ *in Ph.* 105 (Maass 1898.357) notes that ὅλην τὴν ἱστορίαν [sc. the former presence of Δίκη among mortals and her subsequent flight from earth] παρ' Ἡσιόδου εἴληφεν. The same point is adduced by the Latin scholium (though not by the Greek) to the seventh- or early eighth-century Latin translation of Aratus *ad loc.* (Maass 1898.201). A similar though more summary judgment is contained in the *Isagoga bis excerpta* (Maass 1898.324).

43. A careful distinction between these two passages, with special attention to their influence on Vergil, is made by Johnston 1980.25–28.

We may point to similar ambiguities of content. While Aratus' digression on Virgo imitates the Hesiodic "Myth of Ages," it would be wrong to infer that it wholly adopts Hesiod's conception of work as a kind of punishment for offenses against the gods: Unlike Hesiod, Aratus makes agriculture a feature of the Golden Age.[44] Whereas Hesiod's golden tribe is fed by the automatic produce of the earth, Aratus' is fed by the agricultural labor of Δίκη. Similarly, though Vergilian *labor* seems in several specific ways like a Jovian curse visited on innocent humankind, it is also represented as an instrument of progress.[45] But Vergil shares with Hesiod rather than with Aratus the primary purpose of using this material to illustrate the need to work. This idea is incidental to or even barely present in Aratus' version. On the other hand, Vergil departs from Hesiod in reversing the order of his account, omitting Prometheus, and ascribing to Jupiter a providential, if depressingly severe, role which is the antithesis of that played by Zeus in *Works and Days*.[46] The motives of Vergil's Jupiter are quite at odds with the those that Hesiod ascribes to Zeus; instead, one suspects here the influence of Aratus' conception of a providential and disarmingly benign Zeus.

Indeed, as we saw in examining Vergil's Hesiodic program, it is necessary to trace motifs found in an obviously imitative passage into other parts of the *Georgics* as well. Moreover, we discovered that, precisely where we previously observed him parting company with Hesiod's "Myth of Ages," Vergil shows greater affinities with Aratus. He follows Aratus rather than Hesiod in identifying the constellation Virgo with *Iustitia* (*G.* 2.473–474), i.e. Δίκη (*Ph.* 96–136), instead of with Αἰδώς or Νέμεσις (*WD* 200–201).[47] It is true that the "Aetiology" proper ignores Aratus' identification of "the life of the farmer...as the mode of life in which Justice can be found,"[48] let alone his assignation of an operative agricultural role to Δίκη (*Ph.* 112–113); but by the same token, we must acknowledge that Vergil accepts the connection between *Iustitia* and the

44. Note, however, that in Aratus Δίκη does the work, not humankind as in Hesiod—a point emphasized by Johnston (1980.28). Edelstein (1967.139 n. 17 and 145 n. 27) sees here evidence that Aratus, unlike Hesiod, was a progressivist, but included this primitivist myth as a traditional generic element: cf. *Phaenomena* 100–101, 768–772.

45. See especially Altevogt 1952; also Taylor 1955, Ryberg 1958, Wilkinson 1963.

46. See the remarks of Otis 1964.157.

47. Cf. *Ecl.* 4.6; Servius *ad G.* 2.537 (Thilo 1887.270); Solmsen 1966.127.

48. Johnston 1980.25.

farmer's life in the finale of Book 2.[49] It is Aratus' conception of Dike's agricultural role in the idealized life of primitive humankind that gives rise to Vergil's celebration of contemporary farming as a type of "golden age" existence in his "Lob des Landlebens,"[50] and to the pronounced dichotomy of tone that so many readers have noticed between this passage and the "Aetiology of *Labor.*" Here then we find something akin to those individual passages of Hesiod and Lucretius to which Vergil refers the reader at several different points in his argument, emphasizing on each occasion distinct aspects of his source. In this case, however, as in some of the other allusive passages that we have examined, two different authors are involved. In any event, Vergil's art is always epitomized by the analytical use of creative allusion to create poetic meaning.

Having recognized in such features as these the Aratean background of Vergil's "Aetiology," one can hardly avoid reading other points as Aratean allusions as well. Vergil's references to astronomy and seafaring, for example (*G.* 1.136–138), are not only topoi of the Golden Age myth, but actually paraphrase the Hesiodic version. On the other hand, they seem to anticipate Aratean motifs explored more fully later in the book, particularly in the "Weather Signs."[51] More generally, the cosmic sympathy on which the notion of prognostication is based—an idea that pervades Aratus' poem—seems to parallel Jupiter's providence throughout Book 1. Further elements could be adduced to illustrate the detail in which Vergil followed Aratus throughout this book, but for my purposes enough has been done to illustrate the general structural and partial thematic congruity of *Georgics* 1 with *Phaenomena.* The problem is now not to trace out in painstaking detail every facet of Vergil's Hesiodic or Aratean program, but rather to make sense of the bare fact that two such extensive and elaborate imitations have been found to coexist in the same five hundred lines of *Georgics* 1.

49. See pp. 146–147.

50. *G.* 2.48–540. I borrow the title from that of the famous essay by Klingner 1931. More recent criticism (Putnam 1979.144; Thomas 1988.1.244–245) has tended to stress more or less exclusively the pessimistic element in these lines. This emphasis has been salutary; but it has also tended, in my view, to obscure the positive qualities that Vergil associates, here and elsewhere, with the farmer's life. For a balanced appreciation of the passage's tensions and ambivalences, see Perkell 1989.111–115.

51. This observation is in line with the scholiast's remark *ad Ph.* 559 that *σφόδρα φροντίζει ὁ Ἄρατος τῶν ναυτιλλομένων καὶ διὰ πολλῶν τεκμηρίων πειρᾶται αὐτοῖς χειμαζομένοις, καθ' ὅσον ἔξεστι, βοηθεῖν. διὸ ἐναντίον ποιεῖ Ἡσιόδῳ· ὁ μὲν γὰρ σφόδρα τῶν γεωργῶν, ὃ δὲ σφόδρα τῶν ναυτιλλομένων ποιεῖται ἐπιμέλειαν* (Maass 1898.449). Vergil, whether or not he was influenced by a similar judgment, varied his imitations of Hesiod with the insertion of nautical material, and those of Aratus with agricultural material.

The Allusive Program of *Georgics* 1

It is apparent that the congruity between *Georgics* 1 and *Phaenomena*, both in structural plan and individual correspondences, is quite as important and open to interpretation as that involving *Works and Days*. When the structural and thematic relevance of both these models is grasped, one at first regards Vergil's allusive program in this book with bemusement as a somewhat cryptic imitative tour de force—even from Vergil's perspective, perhaps, not an unwarranted or unwelcome reaction. And yet, however difficult an accomplishment from the technical point of view, however painstaking the poetic labor it required, the idea of a joint imitation of *Works and Days* and *Phaenomena* is intellectually neither unnatural nor farfetched. Nor, what is more important, is it merely a technical accomplishment, but rather an interpretive statement by an author creatively grappling with the literary history of his language, its people, and their cultural aspirations.

Let us address each of these points, beginning with the formal aspects. It is clear that *Phaenomena* itself, both in its overall structure and in certain specific episodes, imitates *Works and Days*.[52] Explicit imitation of Hesiod is most evident in the proemium[53] and in the Virgo *aetion* discussed above.[54] The relative position of these passages and their Hesiodic models suggests a structural correspondence between *Phaenomena* and *Works and Days*. The way in which Aratus ends his poem, too, seems to recall *Works and Days*. Both the Hesiodic "Days" and the "*Diosemiae*" are departures from what precede them, so much so that the former has been viewed as a spurious addition to the genuine "Works,"[55] the latter as a genuinely Aratean poem, but one that was originally separate from the *Phaenomena*.[56] In each case the opinion is essentially a misunderstanding; but it remains clear that the endings of the two poems veer off in strikingly different directions from their beginnings.

52. See especially Ludwig 1963 and Schwabl 1972.

53. Aratus' proemium is cited as a token of his Hesiodic ζήλωσις by Achilles Tatius (Maass 1898.83).

54. See p. 160. Both the proemium and the passage on the golden tribe are mentioned in the scholiastic tradition as allusions to *Works and Days* (Maass 1898.83).

55. Most scholars today accept the "Days" as authentic. For a summary of arguments against the "Days" and a review of those scholars who have doubted their authenticity see Solmsen 1963 and Bona Quaglia 1973.229–235.

56. The "*Diosemiae*" is named as an Aratean poem separate from the *Phaenomena* by Achilles Tatius (Maass 1898.79) and Σ *in Ph.* 733 (Maass 1898.472–473), and many MSS mark line 733 as the beginning of a new poem (Martin 1956.xxii–xxiii); but the division is now generally regarded as a misunderstanding.

By employing this odd structural and thematic feature, Aratus may well have been following Hesiod. In view of the structural congruity between the beginnings of the poems, it seems likely that Aratus adopted this unusual species of poetic closure to cap an overall program of Hesiodic allusion.

But these formal observations cannot take us beyond the merely preliminary question of how it was *possible* for Vergil to make the first book of his *Georgics* a simultaneous imitation of both *Works and Days* and *Phaenomena*. We must also face the much more important question of *why* Vergil went to the trouble of adopting such a remarkable allusive program. In particular, we must consider whether there is any special significance beyond the poetic craftsman's "fascination with what's difficult" in the decision to combine Hesiod with Aratus.

The obvious answer to this question would be that Vergil, along with the other poets of his age, had learned from Callimachus to revere Aratus as a kind of neo-Hesiod. The *locus classicus* cited in support of this position is Callimachus' famous epigram on *Phaenomena:*

> Ἡσιόδου τό τ' ἄεισμα καὶ ὁ τρόπος· οὐ τὸν ἀοιδῶν
> ἔσχατον, ἀλλ' ὀκνέω μὴ τὸ μελιχρότατον
> τῶν ἐπέων ὁ Σολεὺς ἀπεμάξατο· χαίρετε, λεπταί
> ῥήσιες, Ἀρήτου σύμβολον ἀγρυπνίης.
>
> *Epigram* 27 (Pfeiffer 1953.88)

Although its general meaning is clear, this clever little poem is open to many interpretive questions. What is most important—and, fortunately for us, unambiguous—is that it identifies Aratus as a Hesiodic poet, and that this identification carries the authority of Callimachus, whose masterpiece, the *Aetia*, embodies its author's own views of Hesiodic poetics. The symbolic significance of Hesiod in Callimachean poetry has been studied intensively for some time, and there is no need to rehearse the details here.[57] Likewise the adoption of this literary credo and its translation into Roman terms is now familiar. We have seen in another connection that Catullus' friend Cinna produced a Latin version of this epigram, one that revealed a subtle awareness of Callimachean mannerisms and agreement with Alexandrian literary norms.[58] For the present little more need be said. To anyone conversant with the aesthetic and

57. The main study is Reinsch-Werner 1976.
58. See pp. 46–48.

literary-historical positions of the Alexandrian poets and their Neoteric progeny at Rome, Vergil's decision to create a book of poetry that in its overall structure recalls the common structure of *Phaenomena* and *Works and Days*, and which in the composition of its major episodes recalls several characteristic passages from both those poems—this decision must be read as an affirmation of the Alexandrian/Neoteric position and as a declaration of membership in that tradition.

There is a further point. Earlier in this chapter I begged an important question when I stated that Vergil's "Weather Signs," like the "*Diosemiae*" in Aratus' *Phaenomena*, forms the conclusion of *Georgics*—*if* we ignore Vergil's lines on the death of Julius Caesar (*G.* 1.466–514).[59] The comparison, I think, remains valid; but what about this non-Aratean coda? These lines contain one of the most subtle of all Vergilian allusions.[60] The river Euphrates is named only three times in all of Vergil's poetry, including twice in the *Georgics* (1.509, 4.561). Not only does the name occur at the end of both the first and last books of the poem, but both times it appears in the same *sedes* of the sixth line from the end of the book. In the design of the *Georgics* this bit of wordplay makes good sense. Such symmetry at the end of the first and last books of the poem recalls the treatment of Maecenas' name at the beginning of each book (*G.* 1.2, 2.41, 3.41, 4.2).[61] While the case of his great friend's name may be aptly characterized as a *jeu d'esprit*, that of the Euphrates stands as a pointed and typically Vergilian cross-reference. In Book 1 the peoples of the Euphrates threaten rebellion in response to Caesar's death; in Book 4, Caesar's heir quells this rebellion, dispensing laws to these now compliant nations, and thus makes his way to Olympus. Both the contrast of fortune between the two Caesars and the familial and political connections between them are thematically operative throughout the poem. These passages emphasize both elements because of their prominent, symmetrical placement and because of the index constituted by the repetition of *Euphrates*. When we examine the third occurrence of this word, in the *Aeneid* (8.726), we find a context thematically similar

59. See p. 159.

60. See Scodel–Thomas 1984.

61. Vergil's metrical treatment of his friend's name deserves passing comment. In a normal line with a dactyllic fifth foot, *Maecenas* could conceivably occupy any of seven different positions. Three of these positions would result in a homodyne treatment of the name and in a diaeresis at its end; the other four in a heterodyne treatment with caesura at the end. Vergil chose the latter course, maintaining heterodyne movement in all four occurrences while varying the *sedes* in each of the four books.

to *G.* 4.561; but more important, we find that here too *Euphrates* occurs in the same *sedes* of the sixth line from the end of the book. This treatment of the river's name points to a principle outside the economy of the *Georgics* alone; and those who discovered the pattern have convincingly explained the passages as allusions to Callimachus' *Hymn to Apollo*.[62] The reason: In the sixth line from the end of that hymn we find the phrase Ἀσσυρίου ποταμοῖο, glossed by the scholium *ad loc.* as none other than the Euphrates.

These astonishing allusions are an excellent token of Vergil's respect for Callimachus. We need not assume that an allusion to the end of the *Hymn to Apollo* in these passages requires us to reread all of *Georgics* 1 and 4 as well as *Aeneid* 8 as extended allusions to that poem. In general, Vergil does not refer to Callimachus in this way, as he does to Hesiod, Aratus, and others. Rather, the importance of this passage, at least in *Georgics* 1, is that it points to one of Callimachus' central programmatic statements, in which the poet expresses his preference for poetry on a small scale, poetry that is pure and refined. An allusion to this seminally programmatic passage at the end of *Georgics* 1 makes excellent sense in light of what precedes it. According to my analysis, the allusive program of the book, which incorporates the structure and central thematic elements of both Hesiod's *Works and Days* and Aratus' *Phaenomena*, is based on Callimachus' polemical epigram identifying Aratus as a consummately Hesiodic poet. Scode and Thomasl' identification of *Euphrates* as a reference to another passage in which Callimachus adumbrates his literary views clinches the argument that the basic impulse behind Vergil's Hesiodic/Aratean program is Callimachus.

This, at any rate, is the conclusion that the current orthodoxy would seem to dictate. It contains a good deal of truth. Vergil had boasted of his Callimachean views at the beginning of the *Eclogue* 6, and in the same poem had honored his friend Cornelius Gallus as a literary descendant of Hesiod himself. It makes sense to read the Hesiodic/ Aratean program of *Georgics* 1 in the terms outlined above.

There are, however, elements in the conclusion of the book that cast the allusion in an unusual light. We may begin with the word *Euphrates* itself. The allusion embodied in this word is itself an exquisitely subtle evocation of Callimachus' artistic principles. One would expect nothing less. But why, one may ask, does Vergil in referring his reader to this crucially important literary manifesto employ a word that denotes the

62. Scodel–Thomas 1984.

type of poetry that Apollo and Callimachus deplore? Why also does Vergil make this reference in a passage that abounds in military and political imagery?[63] The assassination of Julius Caesar, provincial revolts, civil war—these are hardly themes for a follower of Callimachus.[64] Is his handling of the allusion to be seen as a sign of his poetic development, his willingness to grapple in the *Georgics* with themes that he had largely avoided in the *Eclogues*?[65] Further, how does the Callimachean reference sort with another, much more obvious allusion in this passage? Vergil caps his despairing meditation on the recent civil wars with a prophetic vignette that relates these terrible events to the farmer's world:

> scilicet et tempus veniet, cum finibus illis
> agricola incurvo terram molitus aratro
> exesa inveniet scabra robigine pila,
> aut gravibus rastris galeas pulsabit inanis
> grandiaque effossis mirabitur ossa sepulcris.
>
> *Georgics* 1.493–497

The impulse for this moving vignette, which compares past grandeur to the degeneracy of later times, comes from Lucretius:

> iamque caput quassans grandis suspirat arator
> crebrius, incassum magnos cecidisse labores,
> et cum tempora temporibus praesentia confert
> praeteritis, laudat fortunas saepe parentis.
>
> *De Rerum Natura* 2.1164–1167

Lucretius' plowman sadly compares the pitiful harvest of his own day with his ancestors' robust crops.[66] In the same way, his Vergilian counterpart marvels at the heroic stature of those who died in the civil wars, men greater than those of his own day. Both authors seem to view the *temporis momen* (*DRN* 2.1169) as one of inevitable decline: *sic omnia fatis | in peius ruere ac retro sublapsa referri* (*G.* 1.199–200), like a boat carried

63. Clauss 1988.309–310.
64. Ibid. 311, 315.
65. Ibid. 318–320.
66. Cf.:

> iamque adeo fracta est aetas effetaque tellus
> vix animalia parva creat quae cuncta creavit
> saecla deditque ferarum ingentia corpora partu.
>
> *De Rerum Natura* 2.1150–1152

downstream (*G.* 1.201–203) or a chariot lurching out of the gate (*G.* 1. 512–514), both beyond the control of helmsman or driver. The thematic congruity of these passages is impressive; similarly, the placement of this allusion at the end of *Georgics* 1 recalls the placement of its model. As in the case of the Callimachean allusion, we need not take this structural correspondence as a signal to reread the entire book as an extended allusion to *De Rerum Natura* 2; but it does round off a series of prominent Lucretian allusions in *Georgics* 1, as we shall see. In recognizing this fact, the critic faces up to an important problem: How are we to understand these allusions to Lucretius in what we have labored to understand as a Hesiodic/Aratean—i.e. Callimachean—program of allusion?

Indeed, it becomes even more difficult to adopt any straightforwardly Callimachean view of Vergil's allusive program in Book 1 after a careful reading of Books 2–4. In the central books of the *Georgics* we shall encounter Lucretius again and again. In the eyes of current scholarship, this poet is related at best obliquely to the main stream Neoteric movement.[67] It is usually thought that Lucretius actually represents better than any other contemporary poet the tradition against which the Neoterics were reacting. Yet moving from Book 1 to Book 2 of the *Georgics*, as the next chapter will demonstrate, Vergil concludes what we may call a very Alexandrian or Neoteric program of allusion, only to begin an equally ambitious and twice as extensive imitation of *De Rerum Natura*.

67. For Lucretius as a poet of *doctrina* see Kenney 1970; on his relation to Callimachus, see Brown 1982.

CHAPTER 5

Lucretius

The Critical Legacy: Ways and Means

LUCRETIUS' INFLUENCE on Vergil has been variously assessed. For some, the famous dictum of W. Y. Sellar remains axiomatic: "The influence, direct and indirect, exercised by Lucretius on the thought, the composition, and the style of the Georgics was perhaps stronger than that ever exercised, before or since, by one great poet on the work of another."[1] For others this view is largely outmoded, and whatever similarities exist between *Georgics* and *De Rerum Natura* are accidents traceable to the two poems' common meter, genre, and so forth.[2] Still others regard Vergil's attitude towards Lucretius as uniformly revisionist, finding a pronounced element of *anti-Lucrèce chez Virgile.*[3] All three positions are usually overstated. Many of the parallelisms between the *Georgics* and *De Rerum Natura* that scholars have adduced are meaningless accidents. Many others, however, can only be deliberate. Of these, a good number use Lucretius' words and images to make a point quite the opposite of anything one might find in *De Rerum Natura.* Others actually intensify Lucretius' argument, restating it in stronger and simpler terms than he himself had used. We have seen enough examples of both procedures to conclude that Vergil does not allude to Lucretius' poem simply to express uniform agreement with or dissent from his philosophy. But the question remains: Exactly what role does Lucretius play in the allusive program of the *Georgics*?

Two apparent answers to this question may be rejected at once.

1. Sellar 1897.199; quoted with approval by Wilkinson 1969.63.
2. See, e.g., Thomas 1988.1.4.
3. Farrington 1963.

First, it is frequently stated that Lucretius was an important generic model for Vergil, that his poem was a mine of typical elements that Vergil found useful in composing his own didactic poem. While this idea seems unobjectionable and almost self-evident, it often takes a form that obscures Lucretius' true influence on the poem. In particular, many have maintained that Lucretius' use of didactic or expository formulae is imitated in the *Georgics*. Understanding of these phrases in the context of Lucretian poetics has improved significantly over the years.[4] By contrast, critical work on Vergil's use of the formulaic style has not advanced appreciably in a century.[5] Sellar pointed to such formulae as *principio, quod superest, his animadversis, nonne vides*, etc. as evidence that "Virgil regarded Lucretius as his technical model."[6] It is in fact true that Lucretius uses a number of words and phrases of this sort over and over again to order his argument, and that Vergil uses several of them as well. In *De Rerum Natura*, however, these repetitions function not merely as a convenient way of marshalling material for a lengthy argument, but as the cornerstone of Lucretius' style. The poet's use of such a technique has vital implications for both his poetics and his method of argumentation and philosophical position. The same is not true of Vergil. Most of the Lucretian formulae that Vergil uses appear in the *Georgics* only once or twice. Their repetition is not a major factor either in Vergil's prosody or in his argument. Taken as a whole, these phrases may indicate in a most general way Vergil's adherence to the rules of the form and genre in which Lucretius preceded him; but to count each occurrence of *nunc age* or *praeterea*, or even to regard Vergil's use of these phrases in general as an allusion to Lucretius in particular, is obviously misleading.

Second, there have been attempts to "measure" Lucretius' influence on the *Georgics* by quasi-statistical means. These efforts derive mainly from the collection of similar passages made early in this century by W. A. Merrill.[7] Merrill assembles a far larger number of verbal parallels between the two poets than can be found elsewhere, and his work has won a measure of real acceptance as the definitive tabulation of relevant

4. The fundamental work was done by Deutsch 1939 and Friedländer 1941. More recent contributions include Minyard 1978, Snyder 1980, and Clay 1983.

5. Sparrow 1931 fails utterly to consider repetition as a true poetic technique. Newton 1953 is concerned less with formulae than with repetitions that occur within a relatively few lines. For the *Aeneid* there is now Moskalew 1982.

6. Sellar 1897.229. Wilkinson 1969.63 echoes this assertion, as do many others.

7. Merrill 1918a.

passages.[8] It is not unusual, for instance, to find standard works on Vergil citing "statistical evidence" for Lucretian influence on the basis of Merrill's compilation.[9] Unlike most commentators, however, Merrill makes no effort to assess the relationship between the two poets. No interpretation is offered, not even an introduction to the tables. There is in fact nothing beyond the title—"Parallels and Coincidences in Lucretius and Vergil"—to suggest what Merrill thought of the wealth of correspondences that he discovered. Even here he does not speak of "influence" or "imitation"; he will go no farther than "parallelism," which could describe either a significant or a meaningless correspondence, before retreating to "coincidence," a term which suggests that some, at least, of these parallels exist thanks only to the law of averages. Finally, even a cursory examination of Merrill's book would reveal that many of his alleged parallels are extremely slight or even imaginary. Others seem simply to miss the point. The reason for these shortcomings is not far to seek. Over the years, Professor Merrill produced two editions of *De Rerum Natura*[10] and a number of monographs devoted to Lucretius, including studies of his relationship to various Latin authors.[11] It is clear that Merrill was most concerned with Lucretius, and that Vergil occupied only the periphery of his interests; and it is a reasonable inference that he hoped to enhance Lucretius' position in the history of Latin literature by demonstrating how extensively he had influenced Vergil. In his zeal to demonstrate this influence, he adduced many meaningless parallels, rendering any statistical argument based on his list utterly worthless.

My point in discussing these two approaches to our problem is simply this: Lucretius has for too long been treated as a special case among Vergil's poetic sources. No one has ever believed that the influence of Theocritus, Hesiod, or Homer on Vergil can be understood purely in

8. Bailey (1947.29) speaks of Merrill's "valuable legacy" as the "'statistician' of Lucretius," although elsewhere (1931.21–39) he seems less appreciative. Merrill's work is accepted, apparently without reservation, as the first among the "annähernd vollständige Tafeln der loci congruentes" by Klepl (1940.5 and n. 1).

9. Bailey (1931.23–24) attempts to indicate the extent of Lucretius' influence on Vergil by estimating the proportion of allowable parallels in Merrill's tabulation. Knight (1932.24–25 and 142 n. 23; 1966.100) approves of the figures that Bailey thus obtains, as does Wilkinson (1969.40); but cf. the counterarguments of Wigodsky (1972.132–33).

10. Merrill 1907 (school edition with commentary [rpt. 1935]) and Merrill 1917 (critical text).

11. E.g. Ennius (Merrill 1918b), Cicero (Merrill 1909, 1921, 1924), and Horace (Merrill 1905) as well as Vergil (Merrill 1918a and 1929).

terms of borrowed formulae or bogus statistics; but, although more useful work has been done, a significant number of studies on Lucretius and the *Georgics* make one or both of these mistakes. Insofar as they do so, they seriously compromise their validity and usefulness. Lucretius' influence on the poem does not differ in kind from that of Hesiod, Aratus, and Homer—or, for that matter, of Theocritus on the *Eclogues* and of Homer and Apollonius on the *Aeneid.* In my opinion, therefore, it is best studied in the same way as that of other poets, i.e. by identifying and analyzing the specific passages where it manifests itself as verbal or structural allusion, and by then attempting to understand these passages as a coherent whole. In other words, we must subject Vergil's program of Lucretian allusion to the same type of analysis that we undertook vis-à-vis Hesiod and Aratus. The procedure has in fact begun with our examination in the Chapter 3 of key passages from *Georgics* 2 and 3. The remainder of our analysis will continue to center on these books; but it will be helpful first to examine some of those passages in *Georgics* 1 where Lucretian influence is apparent.

Georgics 1

Despite the general importance of Hesiod and Aratus in *Georgics* 1, in several important passages their influence is far from obvious; and, not surprisingly, these passages contain what look like allusions to other authors. By far the most prominent of these is Lucretius. We have already seen that an allusion to *De Rerum Natura* occurs unexpectedly at the very end of Book 1. If we scan the commentaries and the specialized studies such as Merrill's, we find them peppered with indications that many turns of phrase in this book derive from Lucretius. Many of these references disappoint us; but a few, even if at first they resist full comprehension, look intriguing. Occasionally, we find several such passages in close proximity, and when as we read and reread the context that unites them, we feel the presence of Lucretius.

It is convenient to begin with a passage that features unmistakable parallels of diction, but more importantly of form, with Lucretius. The Vergilian passage on "Climatic Zones" (*G.* 1.231–258) concerns the division of the earth into habitable and uninhabitable latitudes, a topic of which Lucretius also treats (*DRN* 5.195–234); but the two poets draw opposite conclusions from similar data:

idcirco certis dimensum partibus orbem
per duodena regit mundi sol aureus astra.
quinque tenent caelum zonae: quarum una corusco
semper sole rubens et torrida semper ab igni;
quam circum extremae dextra laevaque trahuntur
caeruleae, glacie concretae atque imbribus atris;
has inter mediamque duae mortalibus aegris
munere concessae divum, et via secta per ambas,
obliquus qua se signorum verteret ordo.
mundus, ut ad Scythiam Riphaeasque arduus arces
consurgit, premitur Libyae devexus in Austros.
hic vertex nobis semper sublimis; at illum
sub pedibus Styx atra videt Manesque profundi.

Georgics 1.231–243

While both poets agree on the low proportion of usable land on the earth's surface, Vergil implies that the existence of an area that humankind can inhabit proves that the universe was rationally designed by a benevolent being. Lucretius regards the much larger proportion of uninhabitable territory as proof that the earth could not have been tailored to human use by a divine providence: *tanta stat praedita culpa* (*DRN* 5.199 = 2.181):

principio quantum caeli tegit impetus ingens,
inde avidam partem montes silvaeque ferarum
possedere, tenent rupes vastaeque paludes
et mare quod late terrarum distinet oras.
inde duas porro prope partis fervidus ardor
assiduusque geli casus mortalibus aufert.
quod superest arvi, tamen id natura sua vi
sentibus obducat ni vis humana resistat
vitai causa valido consueta bidenti
ingemere et terram pressis proscindere aratris.
si non fecundas vertentes vomere glebas
terraique solum subigentes cimus ad ortus,
sponte sua nequeant liquidas exsistere in auras;
et tamen interdum magno quaesita labore
cum iam per terras frondent atque omnia florent,
aut nimiis torret fervoribus aetherius sol
aut subiti perimunt imbres gelidaeque pruinae,
flabraque ventorum violento turbine vexant.

De Rerum Natura 5.200–217

As we have seen in other cases, Vergil's variation on one author often involves simultaneous allusion to others. Such is the case here. Vergil's discussion of "Climatic Zones" looks to a Hellenistic tradition of both science and poetry represented best for us by Eratosthenes of Cyrene:

> *αὐτὴν μέν μιν ἔτετμε μεσήρεα παντὸς Ὀλύμπου*
> *κέντρου ἄπο σφαίρης, διὰ δ' ἄξονος ἠρήρειστο.*
> *πέντε δέ οἱ ζῶναι περιειλάδες ἐσπείρηντο·*
> *αἱ δύο μὲν γλαυκοῖο κελαινότεραι κυάνοιο,*
> *ἡ δὲ μία ψαφαρή τε καὶ ἐκ πυρὸς οἷον ἐρυθρή.*
> *ἡ μὲν ἔην μεσάτη, ἐκέκαυτο δὲ πᾶσα περι⟨πρὸ⟩*
> *τυπτομένη φλογμοῖσιν, ἐπεί ῥά ἑ Μαῖραν ὑπ' αὐτὴν*
> *κεκλιμένην ἀκτῖνες ἀειθερέες πυρόωσιν·*
> *αἱ δὲ δύω ἑκάτερθε πόλοις περιπεπτηυῖαι,*
> *αἰεὶ κρυμαλέαι, αἰεὶ δ' ὕδατι νοτέουσαι·*
> *οὐ μὲν ὕδωρ, ἀλλ' αὐτὸς ἀπ' οὐρανόθεν κρύσταλλος*
> *κεῖτ' αἶάν τ' ἀμπίσχε, περὶ ψῦχος δ' ἐτέτυκτο.*
> *ἀλλὰ τὰ μὲν χερσαῖα* [. . . . .
> . . . .] *ἀνέμβατοι ἀνθρώποισι·*
> *δοιαὶ δ' ἄλλαι ἔασιν ἐναντίαι ἀλλήλῃσι*
> *μεσσηγὺς θέρεός τε καὶ ὑετίου κρυστάλλου,*
> *ἄμφω ἐύκρητοί τε καὶ ὄμπιον ἀλδήσκουσαι*
> *καρπὸν Ἐλευσίνης Δημήτερος· ἐν δέ μιν ἄνδρες*
> *ἀντίποδες ναίουσι.*
>
> Eratosthenes *Hermes* fr. 16
> (Powell 1925.62–63)

Because Vergil accepts the providential assumptions of Hellenistic speculation in preference to Lucretius' argument, it is useful to read this passage not simply as an imitation of a single Lucretian locus, but as an allusive engagement with the whole tradition that it represents. Some have even argued that Vergil's concern here is exclusively with a branch of the tradition that excludes Lucretius.[12] Echoes of other passages from the *De Rerum Natura*, however, clearly establish Lucretius' presence in these lines. Here Vergil's analytical approach to source material is most evident. By echoing several of Lucretius' astronomical passages he is able in effect to make his predecessor testify to the orderliness of celestial mechanics, which is Vergil's chief argument for the existence of a divine intelligence.[13] And yet, his account is not simply

12. Richter 1957.150–152 *ad G.* 1.231ff.; cf. Thomas 1986.195–198.

13. Cf. *G.* 1.234 ≈ *DRN* 5.604 (and 6.968); *G.* 1.249–251 ≈ *DRN* 5.650–659 (and 1.1065–1067); *G.* 1.239 ≈ *DRN* 5.689–695.

more optimistic than that of Lucretius, but more balanced as well; for Vergil is fully able to acknowledge the existence of hellish powers within this dispensation while projecting an attitude of calm and confidence, as Lucretius, in denying the existence of such forces, cannot do:

> nam simul ac ratio tua coepit vociferari
> naturam rerum, divina mente coorta,
> diffugiunt animi terrores, moenia mundi
> discedunt, totum video per inane geri res.
> apparet divum numen sedesque quietae
> quas neque concutiunt venti nec nubila nimbis
> aspergunt neque nix acri concreta pruina
> cana cadens violat semper<que> innubilis aether
> integit, et large diffuso lumine ridet.
> omnia suppeditat porro natura neque ulla
> res animi pacem delibat tempore in ullo.
> at contra nusquam apparent Acherusia templa
> nec tellus obstat quin omnia despiciantur
> sub pedibus quaecumque infra per inane geruntur.
> his ibi me rebus quaedam divina voluptas
> percipit atque horror, quod sic natura tua vi
> tam manifesta patens ex omni parte retecta est.
>
> *De Rerum Natura* 3.14–30

The phrase *sub pedibus* (*G.* 1.243 ≈ *DRN* 3.27) illustrates clearly the subtlety of Vergil's allusive style. In and of itself, the phrase is nothing but the plainest prose. Yet in Lucretius it possesses a terrific thematic significance. The fact that *Acherusia templa* are explicitly said not to appear *sub pedibus*, where one might expect to find them, recalls the imagery of superstition in Book 1: Where *religio* and fear of the gods had once loomed threateningly over a groveling humankind, until challenged and debunked by Epicurus (62–71), his victory has turned the tables (*religio* ***pedibus subiecta*** *vicissim* | *obteritur*; *nos exaequat victoria caelo* 1.78–79).[14] Vergil shows himself sensitive (and, interestingly, sympathetic) to this imagery elsewhere in the *Georgics*.[15] Here, however, he uses the phrase *sub pedibus* to contradict Lucretius quite directly. Vergil's theism stands in marked contrast to Lucretius' essentially atheistic outlook. Neither

14. Similar imagery may be found at *DRN* 3.27 and 5.1139.

15. Cf. 2.490–492: felix qui potuit rerum cognoscere causas
atque metus omnis et inexorabile fatum
subiecit pedibus strepitumque Acherontis avari.

the life of the gods, which is in fact a borrowed poetic fantasy,[16] nor the powers of Acheron find an actual place in Lucretius' world. The vehemence of his assertions and his admission of extreme emotional excitement differ sharply from Vergil's matter-of-fact attitude. If the world of the *Georgics* is made more hazardous by the existence of infernal powers, at least there is some compensation; for the supernatural clemency that exists only in the *intermundia* of the Epicurean cosmos actually exists in the Vergilian temperate zones *mortalibus aegris munere concessae divum*. The *moenia* of Vergil's world need not dissolve to disclose its wonders, so that it is inviting to share the poet's easy confidence in the permanence of god's design.

Vergil rebuts the pessimism of Lucretius' passage by drawing more optimistic conclusions from essentially the same data. Consequently, the "Climatic Zones" passage has been adduced as a typical example of "polemical" allusion to Lucretius.[17] But anyone who accepts this conclusion makes the common mistake of ignoring Vergil's total reaction to a given context. In this case, we are concerned with Vergil's strategy of allusion to one particular Lucretian context, the "Geography." In his "Climatic Zones," Vergil borrows the basic structure and imagery of this passage, but incorporates little of its characteristic diction, changes its tone, and implicitly rejects its essential message, veering instead in the direction of Eratosthenes. Conversely, in an earlier passage, we find Vergil not only quoting Lucretius' "Geography" verbatim, but recalling much of its general outlook on the human condition in a way that bespeaks substantial agreement:

> vidi lecta diu et **mult**o specta**ta labore**
> degenerare *tamen* **ni vis humana** quotannis
> maxima quaeque manu legeret: sic omnia fatis
> in peius ruere ac retro sublapsa referri,...
>
> *Georgics* 1.197–200

> quod superest arvi, *tamen* id natura sua vi
> sentibus obducat **ni vis humana** resistat
> vitai causa valido consueta bidenti
> ingemere et terram pressis proscindere aratris.
> si non fecundas vertentes vomere glebas

16. Lucretius takes details of his vision of the gods' abode from Homer's description of Olympus at *Odyssey* 6.42–46.

17. So Richter 1957.151 *ad loc.*

> terraique solum subigentes cimus ad ortus,
> sponte sua nequeant liquidas exsistere in auras;
> et *tamen* interdum **magno** quaesi**ta labore**
> cum iam per terras frondent atque omnia florent,
> aut nimiis torret fervoribus aetherius sol
> aut subiti perimunt imbres gelidaeque pruinae,
> flabraque ventorum violento turbine vexant.
>
> *De Rerum Natura* 5.206–217

The allusion is unquestionable. The phrases *ni vis humana* and *multo/magno spectata/quaesita labore* are not found elsewhere in Latin epic; in both these passages they occupy the same *sedes.*[18] Vergil places the Lucretian phrases in successive lines and uses the same adversative (*tamen*) that Lucretius repeats in connection with each phrase. The hybrid line thus serves as a *Leitzitat.*

Both the Lucretian and the Vergilian passages share a thematic and rhetorical purpose. Lucretius wants to demonstrate a metaphysical point by means of an agricultural one, while Vergil, as so often in the *Georgics*, suddenly widens his purview from the immediate and mundane concerns of the workaday farmer to the universal stage on which the drama of human life unfolds. Both passages cast this drama in a soberingly pessimistic light. The Lucretian passage is one of the relatively few in *De Rerum Natura* that concerns agriculture. As such, if Lucretius' poem held any interest for Vergil as he composed the *Georgics*, this passage would be sure to draw his attention. Further, we know as we reread *Georgics* 1 that Vergil was intensely interested in this Lucretian passage, which was to be the basis of a different type of allusion later in the same book.

As in the case of the "Climatic Zones," we must consider the entire context in which Vergil embeds this allusion. The paragraph containing it begins with the words

> **possum** multa **tibi** veterum praecepta referre,
> ni refugis tenuisque piget **cognoscere** **c**ura**s.**
>
> *Georgics* 1.176–177

18. The one line that features a similar clausula is [Vergil] *Ciris* 46 (*accipe dona meo* ***multum*** *vigilat**a** **labore***), which occurs in a context rife with Vergilian and Lucretian echoes; see Lyne 1978a.95 *ad Cirim* 1–53. The pseudo-Vergilian line looks also to Cinna fr. 11.1 (Morel–Büchner 1982.116, *carmina Arateis* ***multum*** *invigilat**a** lucernis.*

and so seems to echo an idea expressed by Lucretius at *De Rerum Natura* 1.410–417:

> quod si pigraris paulumve recesseris ab re,
> hoc **tibi** de plano **possum** promittere Memmi:
> usque adeo largos haustus e fontibus magnis
> lingua meo suavis diti de pectore fundet,
> ut verear ne tarda prius per membra senectus
> serpat et in nobis vitai claustra resolvat,
> quam tibi de quavis una re versibus omnis
> argumentorum sit copia missa per auris.

If Vergil is alluding to these lines, then the allusion is rather faint. Verbal correspondence between the two passages is small, and prosody does not seem to be a factor. The idea that the poet/teacher can overwhelm the skeptical or reluctant student with arguments is not exactly a commonplace, but neither can we consider it exclusively Lucretian. One Vergilian detail, however, looks like a pointed variation on Lucretius' treatment of the motif. Where Lucretius' tongue will pour forth from his breast the drafts of inspiration that he has drunk, Vergil has the ability to pass on the precepts of ancient sages. Where Ennian Lucretius stresses the power of individual genius, Callimachean Vergil sings nothing unattested. But this is not all. As if to assure us that he is reacting specifically to Lucretius, Vergil uses the phrase *cognoscere curas.* One is reminded again of the language that Vergil uses at the end of Book 2 (*felix qui potuit rerum* ***cognoscere causas*** 490), but the distinction made between *causas* and *curas* is crucial. Lucretius wishes to dispel *curas* by forcing his reader *rerum cognoscere causas.* Vergil, on the other hand, insists that his reader recognize and face up to the unavoidable *curas* that are the various pitfalls of agriculture. At first these seem to be, as he says, *tenuis curas* (*G.* 1.177)—mice (181), moles (182), toads (183), and the like—but the tone of the passage gradually darkens into Vergil's deepest expression of despair:

> sic omnia fatis
> in peius ruere ac retro sublapsa referri,
> non aliter quam qui adverso vix flumine lembum
> remigiis subigit, si bracchia forte remisit,
> atque illum in praeceps prono rapit alveus amnis.
>
> *Georgics* 1.199–203

What we are dealing with, then, is clearly not just an imitation of a single Lucretian passage as a kind of emulative set piece, but rather a use of Lucretian language to express a characteristically Lucretian idea. This comparison affords an excellent illustration of the ambivalence that pervades Vergil's so-called polemicism. I have mentioned that the parallelism between Lucretius' "Geography" and Vergil's "Climatic Zones" consists in structure rather than in diction. Yet the same words that best express Lucretius' pessimism in the "Geography" are used by Vergil elsewhere as a climax to the mounting pessimism of *Georgics* 1.176–203. This is an important link between the two Vergilian paragraphs, each of which derives in its own way from *De Rerum Natura* 5.195–234. Like the halves of a Lucretian verse used in different parts of the *Georgics*, distinct elements of this Lucretian passage—its structure on the one hand and its diction on the other—have been separated from one another in order to make very different contributions to separate Vergilian paragraphs. The symmetrical structure of the eighty-three lines that contain these two paragraphs reinforces the notion that they are doublets comprising a pair of inverse but complementary reactions to a single source:

G. 1.176–203	The Likelihood of Failure	28 lines
204–230	The Farmer's Calendar	27 lines
231–258	Climatic Zones	28 lines

Vergil's "Climatic Zones" passage borrows its structure from that of Lucretius, but reverses Lucretius' argument against divine providence as a principle of order in the world. "The Likelihood of Failure," on the other hand, uses diction drawn from the same Lucretian passage in developing the idea that the farmer's efforts are all too likely to go unrewarded. Along with the diction, seemingly, goes the pessimistic outlook of the source passage, which is actually intensified in its new context. The two Vergilian paragraphs are in fact opposite sides of the same coin—neither a scornful rebuttal nor a slavish imitation of Lucretius, but rather the serious, reflective, and carefully nuanced reaction that we would expect from this poet.

With this we have made some progress. But if we continue to study allusion to Lucretius without considering Vergil's overall allusive strategy, we shall surely go astray. When we begin to consider the place that

Vergil gave to his double allusion within the architecture of *Georgics* 1, new questions immediately present themselves. We are dealing with a pair of Lucretian episodes embedded in a much larger Hesiodic *and* Aratean context. Vergil's Hesiodic/Aratean program in Book 1 must therefore have something to do with his imitation of Lucretius. In fact, his ambivalent attitude toward Lucretian ideas in "The Likelihood of Failure" and "Climatic Zones" is prefigured in an earlier passage, "The Aetiology of *Labor*" (*G.* 1.118–158). Again this ambivalence can be traced to the poet's imitation of divergent, even conflicting, sources.

Vergil's "Aetiology of *Labor*" and his "Climatic Zones" share a distinctly non-Lucretian element that acts as a thematic framing device. Both paragraphs concern the providential Jovian dispensation of the world.[19] Two passages in *De Rerum Natura* 5, the "Geography" (195–234) and the "Natural History" (783–1457), are specifically directed against the view that the gods had any part either in the creation of the world or in the development of the human race. As with Vergil's "Climatic Zones" and its reply to Lucretius' "Geography," the "Aetiology of *Labor*" in *Georgics* 1 both imitates and rebuts the "Natural History" of *De Rerum Natura* 5. In Vergil's "Aetiology," Jupiter introduces *labor* into the world in order to shake humankind out of its easeful torpor and to spur it towards industry and civilization; and by imputing a civilizing motive to Jupiter, Vergil ironically alludes to Lucretius' characterization of Epicurus in the proemium to Book 5:

pater ipse colendi
haud facilem esse viam voluit, *primus***que per artem**
movit agros, curis acuens mortalia corda
nec torpere gravi passus sua regna veterno.

Georgics 1.121–24

dicendum est, deus ille fuit, deus, inclute Memmi,[20]
qui *princeps* vitae rationem invenit eam quae
nunc apellatur sapientia, qui**que per artem**
fluctibus e tantis vitam tantisque tenebris
in tam tranquillo et tam clara luce locavit.

De Rerum Natura 5.8–12

19. See especially Büchner 1959.249–251, 253–254 (= *RE* 8 A 1271–1273, 1275–1276).
20. Cf. *Ecl.* 5.64, *ipsa sonant arbusta: "deus, deus ille, Menalca!"*

The idea of Jupiter's moving the fields *per artem* (*G.* 1.122) is odd, and has occasioned some disagreement among critics.[21] But a look at the Lucretian passage makes clear that Vergil intended the phrase to invite comparison between the father of gods and men and Epicurus, the founder of Lucretius' faith.[22] Each of them is the first to make a change that fundamentally alters human life (*DRN* 1.62 ≈ *G.* 1.122); Epicurus leads humanity out of darkness (*DRN* 3.1–2), Jupiter out of torpor (*G.* 1.124); both are concerned to provide a method, either a *ratio* (*DRN* 5.9–10) or a *via* (*G.* 1.122). Vergil's implied comparison of Jupiter and Epicurus is not unmotivated. It is Epicurus, Lucretius strenuously insists, who really was a god (*DRN* 5.8), and who made a much more tangible contribution to human happiness than the deities and culture heroes of myth:

> confer enim divina aliorum antiqua reperta.
> namque **Ceres fer**tur fruges Liberque liquoris
> vitigeni laticem **mortali**bus **instituisse;**
> cum tamen his posset sine rebus vita manere,
> ut fama est aliquas etiam nunc vivere gentis.
> at bene non poterat sine puro pectore vivi;
> quo magis hic merito nobis deus esse videtur,
> ex quo nunc etiam per magnas didita gentis
> dulcia permulcent animos solacia vitae.
>
> *De Rerum Natura* 5.13–21

Vergil's reply is brief and to the point:

> prima **Ceres fer**ro **mortali**s vertere terram
> **instituit**, cum iam glandes atque arbuta sacrae
> deficerent silvae et victum Dodona negaret.
>
> *Georgics* 1.147–149

21. Putnam (1979.33 n. 23). Wilkinson (1969.136) translates: "The Father himself did not wish *the way of cultivation* to be easy, and first *caused* the fields to be worked with skill" (cf. Klingner 1967.199). Miles (1980.79) renders: "The *Father of cultivation himself*...was the first to *change* the fields *by design*." (The emphasis in both of these passages is mine.) See also Parry 1972.50–52.

22. Epicurus, too, is *pater* at *DRN* 3.9. Servius Danielis *ad G.* 1.122 (Thilo 1887.161) also glosses *viam* as *rationem*, *methodon* (cf. *DRN* 5.9). The entire line sounds like a reworking of *De Rerum Natura* 5.1154: *haud* ***facilem*** *esse* ***viam*** *voluit, primusque per artem;* cf. *nec* ***facilem*** *est placidam ac pacatam degere* ***vitam***.

In this account, Ceres did in fact teach men to farm once Jupiter's oak stopped providing nourishment. Men had to farm in order to eat; and compared to this necessity, the Epicurean ideal of a *purum pectus* (*DRN* 5.18) seems insignificant. Indeed, Jupiter forced men away from this ideal, *curis acuens mortalia corda* (*G.* 1.123).

Cura reappears later in this paragraph, again in conjunction with the *ars*, as *labor* | *improbus et duris urgens in rebus egestas* (*G.* 1.145–146). Although this passage has been described as Vergil's "religious justification of work" and his "theodicy of labour,"[23] most critics have felt the justification to be incomplete.[24] The *hominumque boumque labores* (118) are simply the straightforward tasks of everyday life on the farm. As such, the word *labor* carries no especially unpleasant connotations. But the *labor improbus* of lines 145–146 has a wider application. *Labor* here is like the *improbus anser* (119) that negates the farmer's work and is a partner of *urgens egestas.* These two forces also appear in one of Lucretius' tirades against the fear of death:

denique avarities et honorum caeca cupido
quae miseros homines cogunt transcendere finis
iuris et interdum socios scelerum atque ministros
A noctes atque dies niti praestante **labore**
ad summas emergere opes, haec vulnera vitae
non minimam partem mortis formidine aluntur.
B **turpis** enim ferme contemptus et acris **egestas**
semota ab dulci vita stabilique videtur
C *et quasi iam leti portas cunctarier ante.*

De Rerum Natura 3.59–67

In the *Aeneid,* Vergil takes up this suggestion, literally stationing *Labor* and *Egestas* (along with *Curae*) before the entrance to the underworld:

23. Wilkinson 1969.134 *et passim;* cf. Wilkinson 1950 and (especially) 1963. It was mainly Wilkinson who lately championed this interpretation of the passage; but he was anticipated by Burck 1929.283, Beutler 1940.420, and Büchner 1959.250 (= *RE* 8 A 1272), among others.

24. See primarily Altevogt 1952. Klingner 1967.201–205 accepts *labor* as an essentially beneficial force, while acknowledging the Epicurean undertones of the passage. La Penna 1962.238–239 concludes that Vergil either did not intend to justify *labor,* or else failed in the attempt, an opinion that is echoed by J. H. Waszink in his response to La Penna's paper (Waszink et al. 1962.258). Paratore 1938.215–218 actually argues that Jupiter is characterized as man's enemy. Within this spectrum fall Ryberg 1958.120–22; Otis 1964.157 and (in response to Wilkinson) 1972.45 and 50; Putnam 1979.32–36.

C *vestibulum ante ipsum primisque in faucibus Orci*
Luctus et ultrices posuere cubilia Curae,
pallentesque habitant Morbi tristisque Senectus,
B et Metus et malesuada Fames ac **turpis Egestas,**
A terribiles visu formae, Letumque **Labo**sque.

Aeneid 6.273–277

This vivid personification, which clearly was written with the passages from *De Rerum Natura* and the *Georgics* both in mind, calls into question the extent to which Jovian *labor* can be justified.[25] Despite his professed belief in Jupiter as the prime mover, Vergil agrees with Lucretius in certain respects concerning subsequent human progress.[26] Here he seems to share in the Lucretian conception of *labor* as a thing to be avoided. In lines 150–154 *labor* is unambiguously a destroyer of crops, the very opposite of what it was at the beginning of the paragraph:

mox et frumentis *labor* additus, ut mala culmos
esset robigo segnisque horreret in arvis
carduus; intereunt segetes, subit aspera silva
lappaeque tribolique, interque nitentia culta
infelix lolium et steriles dominantur avenae.

In alluding to the Epicurean concept of *labor* as a destructive force Vergil seems to have in mind the proemium of *De Rerum Natura* 2:

suave, mari magno turbantibus aequora ventis
e terra magna alterius spectare *laborem;*
non quia vexari quemquamst iucunda voluptas,
sed quibus ipsa malis careas quia cernere suave est.
suave etiam belli certamina magna tueri
per campos instructa tua sine parte pericli.
sed nil dulcius est, bene quam munita tenere
edita doctrina sapientum templa serena,
despicere unde queas alios passimque videre
errare atque viam palentis quaerere vitae,
certare ingenio, contendere nobilitate,
noctes atque dies niti praestante labore
ad summas emergere opes rerumque potiri.

De Rerum Natura 2.1–13

25. For a fuller examination of this parallelism see Michels 1944.135–148.
26. Cf. *DRN* 5.1–54.

The thematic relevance of this passage to Vergil's "Aetiology of *Labor*" is immediately apparent. In concluding this section of *Georgics* 1, Vergil drives home the parallelism with an ironic twist:

> quod nisi et adsiduis herbam insectabere rastris
> et sonitu terrebis avis et ruris opaci
> falce premes umbras votisque vocaveris imbrem,
> heu **magnum alterius** frustra **specta**bis **a**c**er**vu**m**
> concussaque famem in silvis solabere quercu.
>
> *Georgics* 1.155–159

> suave, mari magno turbantibus aequora ventis
> e terra **magnum alterius specta**re l**a**bo**rem**....
>
> *De Rerum Natura* 2.1–2

Line 158 is probably the clearest single-line verbal echo of Lucretius in the entire *Georgics*.[27] The three alterations which Vergil makes in it, as in so many of his imitations, are extremely telling. Most important, as I pointed out in my previous discussion, is the fact that Lucretius' *laborem* has become *acervum;* the focus shifts from means to end. *Labor,* the theme of the Vergilian paragraph, has been exposed as a destructive force, as it is in the *De Rerum Natura;* yet it also produces the food by which man lives.[28] This simple rule of life is enforced by a divine master whom Lucretius prefers, but Vergil is unable, to ignore; rather, Vergil insists, he must be placated with *votis* (157). In the face of grim necessity, the Epicurean ideal of contemplation is in vain (***frustra*** *spectabis* 158); but this line is not the sardonic parody that it is normally taken to be. It is in fact, as it really appears, a genuine cry of despair. Despite his recognition of Jupiter's authority and influence over civilization, Vergil in large measure agrees with Lucretius' conception of *labor.* As Brooks Otis observes, "There is in any event a fatal shadow on the whole picture: man's 'civilization' has a curse on it."[29]

Thus the ambivalent argument of Vergil's "Aetiology" anticipates aspects of both "The Likelihood of Failure" and "Climatic Zones." Lucretius may be wrong to think that the world was not designed by providence, but he is correct that this strange agency has made the

27. But cf. *WD* 395 *πτώσσῃς ἀλλοτρίους οἴκους καὶ μηδὲν ἀνύσσῃς.*

28. It also of course denotes the poet's own activity (e.g. at *DRN* 2.730, 3.419; *G.* 2.39), a clearly non-pejorative use (cf. *DRN* 1.141).

29. Otis 1964.157 n. 1.

world difficult for mortals to inhabit: *tanta stat praedita culpa*. Within this ambivalence we hear the voices of Vergil's three major sources in *Georgics* 1, Hesiod, Aratus, and Lucretius. Both the ambivalence of the passage and the diversity of its source material are analyzed in the paragraphs that follow:

G. 1.118–159	Aetiology of *Labor*
160–175	The Farmer's *Arma*
176–203	The Likelihood of Failure
204–230	The Farmer's Calendar
231–258	Climatic Zones

The more strictly "technical" character of the second and fourth paragraphs sets them apart from the other three, which provide a dialectical commentary on the human condition. By the same token, the two "technical" paragraphs closely imitate the manner and outlook of Hesiod and Aratus respectively, while paragraphs three and five are a bifurcated allusion to Lucretius' "Geography." Finally, as we have seen, these paragraphs analyze the ambivalence of the "Aetiology" into the pessimism of "The Likelihood of Failure," which concentrates on life's difficulties, and the optimism of "Climatic Zones," which focuses on the generosity of god's design. The entire passage displays an interesting psychological continuity in which the three major models represent different narrative points of view. In the ambivalent "Aetiology" (which draws on Hesiod, Aratus, and Lucretius) the narrator moves from relative optimism in his attitude towards divine providence to relative pessimism about the difficulty of the human condition. This shift of mood motivates a discussion of the "weapons" that the farmer needs to win his struggle for survival (Hesiod). This new topic leads naturally into a warning about the enemies that the farmer must face, and about the need to be constant in his struggle (Lucretius). Mentioning the constancy of *labor* "reminds" the narrator that different tasks should be performed at different times, and that these tasks are regulated by a celestial calendar (Aratus); and the certainty and reliability of this mechanism produces another change of mood and an accompanying change of tone from grim resolve to celebration and delight in a world created by a providential god (Lucretius or "anti-Lucretius").

The modulation of tone continues throughout the rest of the book, but there is no need to trace it in detail. It seems clear that Vergil is working generally from Hesiodic pessimism to Aratean optimism. If distrustful superstition (the Hesiodic "Lucky and Unlucky Days" *G.* 1.276–286) rears its head soon after the Hellenistic enlightenment of "Climatic Zones," an Aratean vision of universal order reasserts itself in the "Weather Signs." We began, however, by examining what now turns out to be a reprise of dark Lucretian themes at the very end of the book, as Vergil's plowman meditates on the remains of those who fell to civil war. With this image Vergil returns to the pessimism that he had espoused in lines 176–203 ("The Likelihood of Failure"), but then seemingly renounced in lines 231–58 ("Climatic Zones"). The three passages outline the development, through a process resembling dialectic, of an ambivalent but ultimately rather bleak view of the human condition. They are united in their reliance on Lucretian diction, imagery, and theme. In view of these facts, it is obviously simplistic to seize upon Vergil's reversal of Lucretius' arguments concerning providence in the two poets' respective "Climatic Zones" and "Geography" passages (*G.* 1.231–258 ≈ *DRN* 5.195–234) to prove that Vergil's chief motive in imitating Lucretius is to rebut his predecessor's views on the gods. On the contrary, it is clear that in many ways Vergil has adopted Lucretius' conception of a mechanistic universe in which humankind is forced to do constant battle with an intractable and seemingly hostile environment. If he is more conventionally theistic than Lucretius, his theism manifests itself chiefly his ascribing to the gods an active role in creating the difficult situation that humankind must face—not merely in acknowledging a limited element of providential concern shown by the gods in conceding a small portion of an otherwise uninhabitable world in which to wage against nature a ceaseless war for survival.

It remains true of course that the position that Vergil gradually develops in these passages is not identical with that of Lucretius. I have argued particularly in the case of Hesiod that Vergil's departures from his model should not be taken uniformly as indications of dissent. By the same token, I do not wish to suggest that elements of congruency between *Georgics* 1 and *De Rerum Natura* betoken Vergil's agreement with Lucretius' ideas about the world. Rather, they represent intermediate stages reached in the dialectical exposition that I mentioned above. Friedrich Klingner has spoken of Vergil's "innere Zwiesprache mit Lukrez";[30] but such a formulation is misleading. The dialogue is not between Vergil and Lucretius; it is between all the major sources of

30. Klingner 1967.209.

the *Georgics*, including Lucretius, with Vergil playing the role of tactful moderator.

Georgics 1.118–258 is a continuous meditation on man's place in the universe. Its substance is derived and its tone modulated largely by Vergil's method of *imitatio cum contaminatione*. It is this method of imitation that causes Vergil's argument to shift back and forth between ostensible hope and despair, between seeming acceptance and rejection of the notion of providence, between apparent agreement with or dissent from the views of now one source, now another. Because of its constantly shifting nature, Vergil's exposition is perhaps better characterized as dialectic rather than meditation. Clearly this dialectic pervades the book and involves all of its central themes; clearly, too, these themes are drawn primarily from the work of Hesiod and Aratus, the chief models of *Georgics* 1. Still, Lucretius' role as an interlocutor alongside these ultimately very different poets, and as Vergil's predecessor in finding both a poetic and a Roman voice for these metaphysical concerns, cannot be underestimated. Finally, if Lucretius' importance in *Georgics* 1 is outweighed by that of Hesiod and Aratus, as all must agree it is, one should nevertheless recognize that his presence in that book anticipates the prominence that his language and ideas will assume later in the *Georgics*, especially in the two central books of the poem.

Vergil's *De Rerum Natura*[31]

Our study of allusions to Hesiod and Aratus in *Georgics* 1 permits us to make a few statements of general import about Vergil's treatment of sources. First, allusions to these authors are largely confined to Book 1. That is to say, the first book is an important element not only in the poem's thematic structure and tonal development, but also within what I have called Vergil's program of allusion. Further, these allusions are highly selective. Large sections of both *Works and Days* and *Phaenomena* are basically ignored as Vergil focuses his attention on a few particular passages. This selection of passages represents an interpretation of those poems that reshapes them and the poetic tradition that they define. It is this view of the tradition that Vergil integrates into the verbal, thematic, and imagistic structure of his own poem. This he accomplishes in a few passages of virtuoso *imitatio cum variatione* arranged so as to create patterns of structural correspondence between *Georgics* 1 and both *Works*

31. The title is borrowed from that of Nethercut 1973.

and Days and *Phaenomena*. The result is that we may speak of *Georgics* 1 as an extended imitation of both those poems simultaneously; but we also recognize that it is a new composition, one that gains in power from the ironies it produces by combining its two main sources, and by presenting itself, in effect, as a more perfect version of each.

In the case of Lucretian allusions, the situation is much the same. Despite an important Lucretian presence in Book 1, even a cursory glance shows clearly that the most obvious and extensive allusions to *De Rerum Natura* occur in *Georgics* 2 and 3. When we look to Book 4, however, we find no indication of comparable influence. These facts suggest that the two central books of Vergil's poem together comprise a second structural unit within his allusive program. Again, if the provenance of the allusions in Books 2 and 3 is considered, it turns out that large portions of *De Rerum Natura* are not represented. We also find that most of these allusions occur in the form of elaborate contextual imitations placed at intervals over Books 2 and 3 in such a way as to suggest a structural similarity between those books and their model. Finally, the allusions concern only a few Lucretian passages connected by diction, theme, and image. By means of the numerous types of *variatio* and *contaminatio*, they succeed in creating a general sense of structural and thematic correspondence while contributing to the distinctive development of Vergil's poem. This is just what was observed in the case of Hesiod and Aratus.

We should note a couple of differences between the allusive program of Book 1 and that of Books 2–3. The first of these reflects a difference between the internal structure of *De Rerum Natura* on the one hand, and that of *Works and Days* and *Phaenomena* on the other. Lucretius' poem is more than six times the length of either Hesiod's or Aratus' and, what is more important, it is articulated by books. Vergil need not have made this factor a part of his allusive program, but it is evident that he did so. In *Georgics* 1, as we have seen, Vergil's deployment of allusive contexts suggests a structural correspondence with both *Works and Days* and *Phaenomena* in their entirety. In Books 2–3, the situation is slightly different. In Chapter 3 I discussed two passages that derive from Lucretius, one in *Georgics* 2 ("The Praise of Spring" 315–345), the other in Book 3 ("The Plague of Noricum" 440–566). The main sources of these passages are found in *De Rerum Natura* 5 and 6 respectively. Earlier in this chapter I discussed two passages of *Georgics* 1 ("The Likelihood of Failure" 176–203 and "Climatic Zones" 231–258) as a bifurcated allusion to another passage of *De Rerum Natura* 5. In what follows, we shall see more allusion to passages of these same Lucretian books. On the whole, *De Rerum Natura* 5–6 contribute the most important material to Vergil's Lucretian allusions, particularly those that imitate an extensive episode of Lucretius' poem. While Books 1–2 contribute additional material, allusions to these books tend to be found in the company of allusions to *De Rerum Natura* 5–6 and to incorporate themes common to Books 1–2 and Books 5–6. Allusion to *De Rerum Natura* 3–4 is scarce in the *Georgics* and relatively unimportant.[32] From these facts, it seems reasonable to infer that Vergil's *De Rerum Natura* is essentially confined to the final two books of Lucretius' poem, and that the allusive program of *Georgics* 2–3 is at bottom a structural and thematic allusion not to *De Rerum Natura* as a whole, but to Books 5–6 of that poem. The similarity between this arrangement and that of *Georgics* 1 is that each of the three books corresponds structurally to a discrete poetry book: *Georgics* 1 ≈ *Works and Days/Phaenomena*, *Georgics* 2 ≈ *De Rerum Natura* 5, *Georgics* 3 ≈ *De Rerum Natura* 6.

A second point: In *Georgics* 1, the fact that Vergil simultaneously imitates two related but ultimately very different poems becomes a source of fruitful poetic tension and ambivalence. In Books 2–3, by contrast, Vergil's program of allusion involves only one of two poetry books at a time, each of them the product of the same author and a part of a single poem. But this new strategy does not result in a lessening of the dialectic intensity that we enjoy in Book 1. Vergil's reading of Lucretius is itself deeply ambivalent, and the ambivalences that he finds in *De Rerum Natura* pervade his own poem. We have seen as much in studying allusion to Lucretius in Book 1. We shall find it even more so in Books 2–3.

To begin, let us consider Vergil's deployment of Lucretian allusions within each of the two books.

More than Books 1, 3, and 4, *Georgics* 2 follows an obviously tripartite organization.[33] Three carefully planned "digressions" punctuate the three major sections of Vergil's argument. These passages—lines 136–176 ("*Laudes Italiae*"), 315–345 ("The Praise of Spring"), and 458–540

32. One would expect to find that Lucretius' diatribe against love (*DRN* 4.1058–1287) exerted a powerful influence on *G.* 3.242–283, but specific allusions are extremely difficult to establish.

33. See, e.g., Schmidt 1930.70–73; Richter 1957.85–91; Otis 1964.149–150, 163–169; Wilkinson 1969.85–92.

("Das Lob des Landlebens")[34]—form the thematic core of Book 2. While they deal hardly at all with the announced subject of the book, i.e. arboriculture and viticulture, their purpose is far greater than simply to relieve the tedium of a dry technical discourse. Recent critics are agreed in finding beneath the formal and stylistic variation a substantial and impressive thematic unity. The digressions themselves are united, most obviously, by the theme of praise. A second and, for my purposes, more important bond is their common reliance on related passages of *De Rerum Natura*.

The second of these "digressions" I have already discussed in some detail.[35] It was found that this, the central digression of *Georgics* 2, refers to a number of thematically related passages from *De Rerum Natura*. In these passages Lucretius develops ideas about the earth's creativity; and the development of these ideas culminates in the discussion of the birth of the world, the *novitas mundi*, in *De Rerum Natura* 5. The third as well has been found by many scholars to contain important Lucretian allusions. In it Vergil uses elements of both conventional "Golden Age" lore and Lucretius' image of Epicurean primitivism to construct an idealized vision of the farmer's life.[36]

The importance of Lucretius in both "The Praise of Spring" and "Das Lob des Landlebens" can hardly be disputed. Rather, it is the initial "digression," the first climactic episode of Book 2, that has not been regarded as Lucretian in its inspiration. As we shall soon see, however, the "*Laudes Italiae*" caps a series of arguments constructed on an essentially Lucretian framework.

In Book 3 the relationship between allusion and poetic structure is more direct than in Book 2; for this is the book that Vergil concludes with the most outstanding example of sustained imitation in the entire poem—only the Aratean conclusion of Book 1 is really comparable. I refer, of course, to "The Plague of Noricum" (*G.* 3.440–566), in which Vergil imitates Lucretius' "Plague of Athens" (*DRN* 6.1090–1286). It is also acknowledged that the end of *Georgics* 3 imitates, not only verbally, but structurally and thematically as well, the conclusion of *De Rerum Natura* 6. This pattern closely resembles what we noticed about Vergil's deployment of Hesiodic and Aratean allusions in Book 1 and strongly

34. The titles are conventional and well established. The first paraphrases *G.* 2.138; the third is from the title of Klingner 1931.

35. See pp. 94–104.

36. See in particular Buchheit 1972.45–92.

suggests that Vergil conceived of *Georgics* 3 as an imitation of *De Rerum Natura* 6.

I believe not only that this is correct, but that we can go a bit farther. Because of the material dependence of *Georgics* 2 on *De Rerum Natura* 5 and of *Georgics* 3 on *De Rerum Natura* 6, and because of the similar structural relationship between *Georgics* 2 and 3 on the one hand and *De Rerum Natura* 5 and 6 on the other, it seems reasonable to conclude that Vergil designed the two central books of his poem as an *imitatio cum variatione* of Lucretius' two final books. In what follows I will support this idea by examining other passages of *Georgics* 2 and 3, some of which have not usually been thought to be imitations of Lucretius. Because most instances in which Vergil models his versecraft on that of Lucretius are noted in the commentaries, there is no pressing need to discuss this aspect of *imitatio* per se. Rather, it will reveal much more about the extent and importance of Lucretius' influence on these books to analyze them as contextual imitations from the standpoint of thematics, adducing parallelisms of versecraft where necessary to illustrate the specifically Lucretian background of particular themes.

It is useful to consider *Georgics* 2 as a sustained allusive reworking of two Lucretian themes, variety and creativity. In developing these themes, Lucretius characteristically takes pains to connect them with their opposites—or better, perhaps, their complements. Lucretius the poet takes an evident and contagious delight in observing nature in its various manifestations, which together comprise a rich and variegated tapestry. Lucretius the scientist, however, shares with the reader a feeling of awe that this array is composed of nothing but a vast number of atoms, all but undifferentiated and individually insensible *corpora prima,* moving about in the void. Beneath the external variety of nature the atomist perceives a fundamental sameness. Similarly, the poet delights in exploring with his reader the manifold ways in which these unfeeling elemental bodies combine to produce living, sentient beings; but as a teacher, he never lets his pupil forget the essential identity of birth and death, since the creation of any new thing is made possible only by the destruction of something else. Just as both the poet and the scientist consist in the same mind and sensibility, so do these two realities comprise the paradox of nature.

The theme of variety obviously pervades the opening of *Georgics* 2. Indeed, Erich Burck does not exaggerate in maintaining that the word *varia* in line 9 "is significant not only for the content of the following verses, but for the structure of this entire section of the book."[37] The structure of this first section, i.e. of lines 9–176, resembles a gigantic priamel. A form of this standard exordium topos is used here to arrest the reader's attention with a dazzlingly varied array of images and, finally, to focus that attention on the true object of the poet's scrutiny: Italy. Beginning his discourse with the statement that there are *various* ways of growing trees (9), Vergil distinguishes the ways of *natura* (10–21) from those of *usus* (22–34).[38] Recapitulating this distinction, he discusses the *various* methods of *usus*, noting which ones work best for each species (47–72; cf. 35). The next paragraph continues the idea of variety, but narrows the focus to different grafting techniques (73–82). Vergil has clearly been heading in this direction, as each of the two previous sections (excluding the apostrophe to Maecenas 39–46) concludes with the image of successfully grafted trees as a type of *adynaton*. Persevering in the variety theme, the poet moves from the idea of the various grafting techniques,[39] each useful for different types of tree, to the idea that there are various sub-types even within a species. He begins by naming three different species: *ulmus*, *salix*, and *lotus* (83–84). Adding a fourth, the cypress, the poet specifies a particular variety within the species (*Idaeis cyparissis* 84).[40] This development prepares for the greater diversity of individual types within the olive group (85–86) and in the family of pome fruits (*poma* 87–88).[41] Then follows a twenty-line catalogue of grape varieties (89–108) in which the diversity theme is greatly amplified.[42] The theme also undergoes a subtle change. The grape types in lines 89–108 are identified chiefly by their native regions. This convention, which is in fact anticipated by the epithets found in lines 84 (*Idaeis cyparissis*) and 88 (*Crustumiis Syriisque piris*), prepares the reader

37. Burck 1929.293: "Einen starken Akzent trägt in Vs. 9 das Wort *varia*, das nicht nur für den Inhalt der nächsten Verse, sondern für die Ordnung des ganzen Buchabschnittes bedeutsam ist."

38. On this distinction see Farrell 1983.200–203.

39. Where most ancient authorities recognize three basic methods of grafting (see *Geoponica* 10.75, Columella *De Arboribus* 26.1), Vergil speaks only of two, *insitio* and *oculatio*, thereby echoing, perhaps, the bipartite division between the trees of *natura* and those of *usus*.

40. Theophrastus mentions Cretan cypress (*CP* 2.2.2) and Idaean cypress in particular (3.26). He also mentions lotus and cypress together at 1.5.3 and 5.4.2. Cf. Nicander *Theriaca* 585 Ἰδαίης κυπαρίσσου.

41. The diversity of olive trees is something of a topos in the technical literature: cf. Cato *Agr.* 6; Varro *DRR* 1.24.

42. The diversity of grape types, too, is a topos: cf. Theophrastus *HP* 2.5.7; *CP* 2.4.7.

to shift his attention from vegetation to geography. The list of grape types is thus at the same time a catalogue of different lands, each (as in the case of the various fruits of the lands mentioned in the preceding lines) with its own distinctive quality. A commonplace that includes references to *Libyci aequoris* (105) and *Ionii fluctus* (108) carries this development further by detaching the theme of geographical diversity from that of the diversity found in the plant world.

In a new paragraph (109–135), Vergil begins by concentrating on the land itself: *nec vero* ***terrae*** *ferre omnes omnia possunt* (*G.* 2.109). The reader's attention is, however, focused upon the human factor in this web of agricultural diversity as a veritable parade of nations passes before his eyes.[43] Beginning by noting the kinds of climate and terrain preferred by different plants—the progression here, from trees (*salices* and *alni* 110; *orni* 111; *myrtetis* 112) to the vine (*Bacchus* 113), recapitulates that of the previous paragraph[44]—the poet then returns to the notion of geographical diversity, but in a new connection: *aspice et extremis domitum* ***cultoribus*** *orbem* (114). This transition, too, is subtly and artfully contrived. It begins in the topographical section with the personification of the vine effected by metonymy and paronomasia: ***Bacchus amat*** *colles, Aquilonem et frigora taxi* (113).[45] One hears also, amidst the catalogue of peoples, about the *divisae arboribus patriae* (116). Finally, the names of tribes and nations dominate the reader's attention: *Arabum* and *Gelonos* (115), *India* (116 and 122), and *Sabaeis* (117), *Aethiopum* (120), *Seres* (121), and *Media* (126; cf. *Medi* 134) are all briefly mentioned. Thus the stage is set for the "*Laudes Italiae*" (136–176). The theme of variety, which moves in lines 9–135 from the plant world to the idea of craft to the land itself and to the land's inhabitants, finally is subsumed within the comprehensive variety of singular Italy.[46]

It is at the climax of the "*Laudes Italiae*," of course, that Vergil declares his poem an *Ascraeum carmen;* but, as we saw in Chapter 2, Vergil tends to use such phrases ironically with regard to his own work. The phrase looks back, at this point in Vergil's exposition, to an allusive program that has been completed; a new one is already under way.

43. On the ethnographic background of this entire argument see, in general, Thomas 1982b.1–8 *et passim.*

44. Even the word *Aquilonem* (113) contributes to this effect by recalling *Zephyro* (106) and *Eurus* (107) at the end of the previous paragraph.

45. For *amare* in this sense cf., e.g., Columella *RR* 2.10.3, Pliny *NH* 16.40, and Theophrastus' use of φιλεῖν.

46. Cf. Otis 1964.163–164.

I began this discussion with the remark that the variety theme, with which *Georgics* 2 begins, is a distinctly Lucretian motif. For many, this statement will require justification. Wilkinson, for example, regards the principle of variety as a point of differentiation between the *Georgics* and *De Rerum Natura*.[47] Yet Lucretius does exploit the poetic possibilities of variety. This fact is especially evident in the frequency with which he uses the epithet *varius* in describing the many aspects of nature.[48] That Vergil has in mind here a specifically Lucretian concept of variety is, moreover, guaranteed by his language. Throughout the beginning of *Georgics* 2, the variety theme is consistently glossed by Lucretian reminiscences. Of the first six paragraphs in the book (9–135), five begin by mentioning the variety motif.[49] Four of them also open with obvious Lucretian echoes:[50]

si non ipsa dedit specimen **natura creandi**

De Rerum Natura 5.186

in ***variis*** mundis **varia** ratione **creatis**

De Rerum Natura 5.1345

principio arboribus **varia** est **natura creandis**

Georgics 2.9

inde aliam atque aliam **cultu**ram dulcis agelli
temptabant **fructusque feros m*ansuescere*** terra
cernebant indulgendo blandequ**e colendo**

De Rerum Natura 5.1367–1369

47. Wilkinson 1969.183: "For continuity of argument Virgil's substitute was orchestration, and he relies much more than Lucretius on variety." But Wilkinson also derives Vergil's descriptive style, at least in part, from that of Lucretius (1969.9–13), and this style depends to a significant extent on orchestration and variety.

48. As a simple measure of how often the variety theme occurs in *De Rerum Natura*, Roberts 1968.318 lists 86 passages containing forms of *varius* and *variare*. Moreover, many of these examples occur in formulae (e.g. *variae volucres* 1.589, 2.145, 2.384, 4.1007, 5.801, 5.1078; *varia ratione* 1.341, 5.528, 5.792, and 5.1345), which become memorable through repetition and thus contribute greatly to the reader's sense of variety and its importance to Lucretius. Vergil, by contrast, has these words 59 times in the complete works, 14 times in the *Georgics*; expressed as numbers of occurrences per thousand verses, the figures are *DRN:* 11.6, *Georgics:* 6.4, Vergil [complete works]: 4.6).

49. See *G.* 2.9, 35, 73, 83, 109. The odd paragraph, which begins with line 47, is really a resumption of 35–38, where the argument proper is interrupted by an apostrophe to Maecenas (38–46).

50. There is a possible fifth Lucretian connection in line 73: *nec modus inserere atque oculos imponere simplex;* cf. *G.* 3.482 *nec via mortis erat simplex*, which is inspired in turn by *DRN* 2.918, *et leti vitare vias;* but the connection is distant, and I would not insist on it.

quare agite o proprios generatim discite **cultu**s,
agricolae, **fructusque feros m***ollite* **colendo**

Georgics 2.35–36

praeterea genus huma**num** mutaeque natantes

De Rerum Natura 2.342

praeterea genus haud u**num** nec fortibus ulmis

Georgics 2.83

sed mutarentur, **ferre omnes omnia poss**ent

De Rerum Natura 1.166

nec vero terrae **ferre omnes omnia poss**unt

Georgics 2.109

The progression of allusions suggests a deliberate scheme. The first of them clearly indicates the book's thematic domain; it is most decidedly a book of variety and creativity. More than this, however, the double echo adumbrates the first, immediate offshoot of the variety theme. Variety in this paragraph is represented by two opposing types of cultivation, the methods of *natura* and those of *usus*. The diction of the source line (*DRN* 5.186), moreover, clearly anticipates that of a passage from the "Anthropology" in which Lucretius discusses the origin of agriculture:

at **specimen** sationis et insitionis origo
ipsa fuit rerum primum **natura creatr**ix

De Rerum Natura 5.1361–1362

De Rerum Natura 5.1345, by contrast, comes from Lucretius' account of experiments in early warfare. In combining these two passages, Vergil buttresses the dichotomy in his own account between natural methods of cultivation and the violent ways of *usus*.[51] As if ascribing a similar distinction to his source, Vergil next refers directly to Lucretius' discussion of farming. In echoing Lucretius' *fructusque feros mansuescere* (*DRN* 5.1368), Vergil alters the verb to *mollite* (*G.* 2.36). In view of the imagistic contrast between hard and soft in the preceding paragraph (*G.* 2.9–35), this alteration takes on an added significance, since it is the contrast between "soft" and "hard" methods of growing trees that provides the

51. Theophrastus (*HP* 2.1.1–2.2.5) is the technical source of this distinction. The terms he uses are φυσικωτατός and τεχνή or προαίρεσις.

central motivating impulse behind this entire introductory section on arboriculture.

It is not, however, variety alone, but also its converse that is important in *De Rerum Natura.* Beneath the superficial variety of things lies a fundamental sameness. All things—earth, air, fire, water, animal, mineral, vegetable—all are composed of nothing but atoms and void. Lucretius' observations about variety and sameness thus leave the reader with a vivid sense of paradox, just as when he learns that a single event is at one level merely a reshuffling of atoms, but is perceived as the death of one thing and the birth of another. Vergil shows his awareness of this paradox in his Lucretian glosses on the variety theme. The adverb *generatim* (*G.* 2.35) is the first indication of this awareness.[52] The idea is undeniably present in the third of the series of "mottoes" that mark this entire passage as an imitation of Lucretius. An examination of the relevant contexts, that is, of both the quotation and its original, will clarify the nature of Vergil's allusion:

> **praeterea genus h**aud **unum** nec fortibus ulmis
> nec salici lotoque neque Idaeis cyparissis,
> nec pingues unam in faciem nascuntur olivae,
> orchades et radii et amara pausia baca,
> pomaque et Alcinoi silvae, nec surculus idem
> Crustumiis Syriisque piris gravibusque volemis.
>
> *Georgics* 2.83–88

> **praeterea genus h**uma**num** mutaeque natantes
> squamigerum pecudes et laeta armenta feraeque
> et variae volucres, laetantia quae loca aquarum
> concelebrant circum ripas fontesque lacusque,
> et quae pervulgant nemora avia pervolitantes;
> quorum unum quidvis generatim sumere perge,
> invenies tamen inter se differre figuris.
>
> *De Rerum Natura* 2.342–348

The structure of the two passages is close indeed: Lucretius lists five species—humans, fish, cattle, wild beasts, and birds—spending three lines on the final item. Vergil also mentions five species—elms, willow,

52. Lucretius uses the unusual word eleven times, all but once (*DRN* 4.646) in the same metrical position (before the fifth foot, e.g. *DRN* 1.20, 227, 229, 563, 584, 597; 2.347, 666; 6.1135, a position that agrees exactly with *G.* 2.35, the only occurrence of the word in Vergil.

lotus, cypress, and olive—spending two lines on the final item before moving on into other territory. The theme of both passages is the same, variety within species. While Vergil eliminates Lucretius' explicit conclusion to this effect (*DRN* 2.347–348), he alters the introductory quotation in a way that emphasizes his theme: Lucretius' *genus humanum* becomes *genus haud unum*. The alteration indicates in a restrained fashion an important principle shared by the two poets, the fundamental similarity of all forms of life. Certain traits, such as the existence of individual characteristics within kinds, apply to all species, both animal (as in Lucretius) and plant (as in Vergil). The same principle is further illustrated by a related parallel. In Lucretius the passage cited above illustrates a basic point of Epicurean physics, the variety of forms among atoms (*DRN* 2.333–341).[53] Although in the *Georgics* the atomic theory behind this principle is lacking, the principle itself still holds. Various trees grow in various ways (*G.* 2.9–35), and respond to different methods of cultivation (47–72), and even to variations on a single method, such as the different grafting techniques (73–82). This is because the variety of trees extends even to subspecies: there are not just elm, willow, lotus, and cypress, but different kinds of each (83–108). And one important way of distinguishing types is according to where they grow (109–135). Thus, alongside the variety theme, the idea of intergeneric similarity runs through the entire *Georgics* passage, extending from plants to the land and, finally, in the climactic "*Laudes Italiae*" (136–176), to humankind itself.

This series of Lucretian "mottoes" marks an abrupt change of focus in *Georgics* 2. Not only the Hesiodic/Aratean program of Book 1, but even Vergil's approach in that book to *De Rerum Natura* has been thrust into the background. No longer are Lucretius' words made to lament the *temporis momen* towards the worse, or twisted into a paean to divine providence. Instead, they introduce and develop the theme of variety, a principle source of delight and wonderment in the most upbeat book of the *Georgics*. Not only in these opening paragraphs, but throughout this book—particularly in the effusive "Praise of Spring" and "Das Lob des

53. This idea goes back to the very first proof in the poem, which begins with the proposition that *nullam rem e nilo gigni divinitus umquam* (*DRN* 1.150); see pp. 98–99. The implication of this passage, however, is not only that different species tend to preserve their types, but that the *foedera naturai* involve affinities between different elements in the scheme of things. Certain things are born at certain times and certain places, and they grow in certain ways according to their atomic structure.

Landlebens"—allusion to Lucretius attends the celebration of the most productive and pleasurable aspects of the farmer's work.

As always, however, Vergil eventually reminds us that this story has two sides. The series of Lucretian "mottoes" ends with an allusion that celebrates the underlying similarity of different species. We have seen several instances in which Vergil imitates a passage twice, each time exploiting a different facet of it and thus bringing out complexities in his source. Two passages of Book 1, "The Likelihood of Failure" and "Climatic Zones," both allusions to Lucretius, provide a clear illustration of this technique, which is also in evidence here. Continuing where the passage quoted above leaves off, Lucretius illustrates the principle of individuation within all species with a moving vignette of a cow searching in vain for her calf, which has been used as a *victima* of human religion (*DRN* 2.352–366). Vergil's imitation of this passage occurs in a different context, the "Plague" finale of *Georgics* 3 (515–530); but if the circumstances are different, the emotional effect is quite similar. In Chapter 3 I discussed this allusion in the context of Vergil's "Plague" imitation. There I argued that while the Lucretian passage has nothing to do with the "Plague" per se, nor even with disease, its function in this context is to strengthen the analogy between the human and the animal world in Vergil's account. The basic motif of Lucretius' human plague is fortified with additions from *De Rerum Natura*, additions that point to both the human (epilepsy *DRN* 3.487–498 ≈ *G.* 3.511–517) and the animal (mother/plowmate *DRN* 2.352–366 ≈ *G.* 3.515–530) sphere. When we consider the Lucretian passage in its totality, we see that the imitation concerns not just the immediate context of Vergil's "Plague," but the entire thematic structure of *Georgics* 2–3. Vergil alludes to the first line of the Lucretian paragraph in the last of the "mottoes" that open Book 2. There he suggestively alters Lucretius' *praeterea genus humanum* (*DRN* 2.342) to *praeterea genus haud unum* (*G.* 2.83) as an illustration that certain traits, such as differentiation among individuals or individual types, apply to all species.[54] By dividing this passage in two, incorporating its beginning and its general principle into the series of Lucretian "mottoes" that opens *Georgics* 2 while reserving Lucretius' illustration of the principle (the mother cow searching for its calf, which has been sacrificed) for the plague imitation of *Georgics* 3 (the grief of the plow ox over its fallen yokemate), Vergil follows what has become a

54. The theme receives further development in Book 3 (e.g. in the story of Hero and Leander: *amor omnibus idem* 244) and Book 4 (the bees).

familiar procedure. This particular example is only one of many that suggest an unusually close relationship between *Georgics* 2 and 3. As usual where a divided source passage is concerned—as in Vergil's double allusion to Lucretius' "Geography," for instance—the relationship involves contrast. Here a unified Lucretian argument is analyzed into two parts, one that stresses the almost limitless variety of individuals within species, another that concerns the similarities that exist even between different species. The former component comes as a dispassionate statement of one of the general principles that underlies the *rerum natura*, the latter as a "proof" of this principle, taking the form of a particular illustration that is pregnant with emotion. After analysis, Vergil integrates the general principle into his own argument concerning the principles of arboriculture, an argument that lays the groundwork for the effusive passages on the productivity of Italy, of spring, and of the farmer's life. By the same token, the particular illustration, resonant with the themes of death and perverse religiosity, is united with Vergil's chief reworking of those Lucretian themes in the conclusion of Book 3, "The Plague of Noricum."

A similar contrast presents itself in another double allusion. In "Das Lob des Landlebens," as I have mentioned, Vergil's borrows Lucretius' contrast between a virtuous rustic's existence and that of the luxurious urbanite (*G.* 2.461–474, 493–540). In "The Plague of Noricum," the life of the mourning plow ox is depicted in very similar terms:

> non umbrae altorum nemorum, non mollia possunt
> prata movere animum, non qui per saxa volutus
> purior electro campum petit amnis; at ima
> solvuntur latera, atque oculos stupor urget inertis
> ad terramque fluit devexo pondere cervix.
> quid labor aut benefacta iuvant? quid vomere terras
> invertisse gravis? atqui non Massica Bacchi
> munera, non illis epulae nocuere repostae:
> frondibus et victu pascuntur simplicis herbae
> pocula sunt fontes liquidi atque exercita cursu
> flumina, nec somnos abrumpit cura salubris.
>
> *Georgics* 3.520–530

This characterization strengthens the comparison between the human and animal spheres mentioned above; the plow ox leads the same existence as does the virtuous farmer of "Das Lob des Landlebens." Here, however, the simple rewards of the rustic life are nullified by the grief attendant to death, which disrupts the orderly pattern of the georgic world, much the same, indeed, as it topples humankind from the pinnacle of civilization in *De Rerum Natura* 6. The Epicurean ideal provides a model for the the georgic ideal, each unsullied by the perversions of civilized luxury; and yet both are undone by the grisly and inevitable horror of death.

I began this discussion of *Georgics* 2–3 by noting that in Book 2, the bulk of the Lucretian material that Vergil borrows comes from *De Rerum Natura* 5; in Book 3, imitation of Lucretius consists mainly of the extensive "Plague of Noricum," which depends heavily on *De Rerum Natura* 6. We have now seen that *Georgics* 2 and 3 are simultaneously linked and contrasted by shared motifs, and that the most important of these motifs frequently involve allusion to *De Rerum Natura.* It is also clear that the main principle that Vergil used to organize Lucretian allusions across the two books is a thematic polarity between birth and death. Finally we must consider the thematic structure and tonal movement of *Georgics* 2–3 within the poem as a whole and in comparison with that of *De Rerum Natura* 5–6 as well.

In regarding *De Rerum Natura* 5–6 as a unified pair, Vergil clearly takes his cue from Lucretius himself. Book 5 is an anomaly in that it lacks the element with which the other five books of Lucretius' poem conclude, an image of decay.[55] Indeed, Cyril Bailey goes so far as to suggest that the book is unfinished, and that Lucretius obviously intended to provide it with an ending in keeping with those of the other books.[56] This is an unnecessary supposition, however, and is furthermore not in keeping with the evidence that the existing ending presents. The *cacumen*, the pinnacle of civilization, of which Lucretius speaks at the book's end (*DRN* 5.1457), is clearly Athens, which he salutes as the birthplace of Epicureanism in the proemium to Book 6. This element

55. Book 1 ends with an image of the destruction of the world (1083–1113); Book 2 with the plowman who is unable to produce crops as big as those of his ancestors (1150–1174); Book 3 with the idea that no period of life, no matter how long, can detract from the eternity of death that awaits us (1076–1094); Book 4 with the image of water gradually wearing away stone (1278–1287); and Book 6 with the "Plague of Athens" (1090–1286). On the ending of Book 5, see the following note.

56. Bailey 1947.1548: "the book ends abruptly and without the elaborate conclusions of the others"; but the reason for this "omission" is surely that given by Minadeo (1969.47), who notes that "its intrusion would have fatally disturbed the blending of that book's close with the Proem of VI." Minadeo's analysis of the tonal contrast between the proemia and epilogues of all six books of the poem is especially complete and satisfying: see his pp. 33–50.

of continuity is reinforced by a thematic balance between the two books. *De Rerum Natura* 5 begins with a warning that the world, as a composite entity—i.e. one which is born—is mortal (235–415). This assertion is based on an elementary physical principle established much earlier in the poem (*DRN* 1.1083–1113). The remainder of the book concerns the formation of the world (416–563), the orderly nature of celestial phenomena (564–771), and finally the origins of life and the development of civilization (772–1457). Book 6, on the other hand, deals generally with disturbing and destructive phenomena, such as electrical storms (96–422), earthquakes (535–607), volcanoes (639–702), and plagues (1090–1286), including the "Plague of Athens" (1138–1286). The contrast between the two books is thus between development and relative stability on the one hand and cataclysmic violence and destruction on the other, both in the world at large and in human civilization in particular. The pivotal nature of the juncture between the two books is intensified by the placement of the word *cacumen*, the final word in Book 5; for, as Lucretius makes clear early in the poem, all things grow until they reach a natural limit—*donec alescendi summum tetigere cacumen* (*DRN* 2.1130)—after which they must inevitably decline (*DRN* 2.1122–1143). The very clear echo of *De Rerum Natura* 2.1130 that is heard in the concluding line of Book 5 (*artibus ad summum donec venere cacumen* 1457) signals the change in Lucretius' focus between Books 5 and 6 from the growth to the dissolution of the physical world and of civilization. The tonal movement of *De Rerum Natura*, both as a whole and in its individual parts, from great exuberance to considerable gloom, simply reflects the poet's view of the natural order of things. The miracle of birth and growth is balanced by gradual senescence and, finally, death.

That Vergil appreciated and imitated this contrast is apparent in the thematic structure of *Georgics* 2–3. As with *De Rerum Natura* 5–6, the end of Book 2 is closely bound together with the beginning of 3.[57] At the center of Book 2 is a cluster of allusions that refer to a passage from *De Rerum Natura* 5, but include an important element drawn from earlier books. The *hieros gamos* motif, that universal myth of the world's birth, is taken from Lucretius' argument that the birth of one entity is made possible by, in fact necessitates, the death of another. As I have shown, Vergil exploits and develops Lucretius' treatment of the birth theme but rigorously excludes any mention of death from his imitation.

57. See especially Buchheit 1972.45–159.

Georgics 2 considers the idea of growth and development both in the yearly operation of the farm and, implicitly, in the development of the Roman state. Death, however, becomes the chief subject of the finale to *Georgics* 3, just as it is in the sixth book of *De Rerum Natura*. Vergil, of course, pays more attention to the idea that death is a personal tragedy suffered by lonely individuals (and does so, ironically, by importing an allusion to Lucretius' grieving cow into the context of the "Plague"); but both passages treat of death on a massive scale as a primarily social disaster that affects the political and the georgic worlds alike.

Lucretius' account of social development in Book 5 is couched in general terms; but he makes it clear immediately in Book 6 that he considers not Rome, but Athens at the height of its powers as the *cacumen* of human civilization. It is Athens that he praises in the proemium of that book, and Athens whose dissolution he depicts at the book's conclusion. Vergil reverses this scheme. Where Lucretius does not identify his paragon of civilization in Book 5, Vergil in his imitation leaves no doubt: *sic...rerum facta est pulcherrima Roma* (*G.* 2.533–534); but where Lucretius concludes Book 6 by describing a historical occurrence, Vergil gives his plague an essentially imaginary setting. Neither Rome nor Athens is the pinnacle from which civilization falls, but rather the countryside of an unknown place in Noricum.[58] And yet Athens assumes a prominence elsewhere in the book that recalls its role in *De Rerum Natura* 6. Fairly early in *Georgics* 3, Vergil curiously emphasizes the role of the Greeks in developing the science of animal husbandry. The reader is reminded of the excellence of Epirotic and Mycenaean horses (*G.* 3.121) or of those from Epidaurus (44); of the potential glory of an Olympic (19, 49, 180, 202) or a Nemean (19) victory; even of pertinent Greek legends or words (90, 148). Athenian inventiveness in particular is mentioned in lines 113–117:

> **primus** Ericthonius currus et quattuor **ausus**
> iungere equos rapidusque rotis in**sistere victor.**
> frena Pelethronii Lapithae gyrosque dedere
> impositi dorso, atque equitem docuere sub armis
> insultare solo et gressus glomerare superbos.

58. Noricum, as Richter 1957 *ad G.* 3.474ff. observes, denotes an area corresponding roughly to modern Austria; but Vergil is deliberately vague about the exact location of his plague, even to the point of giving the landlocked area a coastline (see p. 85). On the supposed historicity of the plague, see Harrison 1979.4–6.

The *protos heuretes* motif recalls Lucretius' praise of Athens itself in the proemium to Book 6:

> **prim**ae frugiparos fetus mortalibus aegris
> dididerunt quondam praeclaro nomine Athenae
> et recreaverunt vitam legesque rogarunt,
> et primae dederunt solacia dulcia vitae,
> cum genuere virum tali cum corde repertum,
> omnia veridico qui quondam ex ore profudit;
> cuius et extincti propter divina reperta
> divulgata vetus iam ad caelum gloria fertur.
>
> *De Rerum Natura* 6.1–8

as well as his initial description of Epicurus' achievement:

> **primu**m Graius homo mortalis tollere contra
> est oculos **ausus primus**que ob**sistere** contra
> . . . .
> ergo vivida vis animi *pervicit*, et extra
> processit longe flammantia moenia mundi
> atque omne immensum peragravit mente animoque,
> unde refert nobis **victor** quid possit oriri,
> quid nequeat, finita potestas denique cuique
> quanam sit ratione atque alte terminus haerens.
> quare religio pedibus subiecta vicissim
> obteritur, nos exaequat victoria caelo.
>
> *De Rerum Natura* 1.66–79

Ericthonius, one of the early kings of Athens, is also the inventor of the chariot, a symbol of the civilization that Athens gave to the world.[59] At the beginning of Book 6, Lucretius' Athens gives the world a different governing principle, the *ratio* of Epicureanism;[60] but at the end of that

59. For the Lapiths as an emblem of civilization, cf. *G.* 2.455–457, where they are contrasted with the bestial, drunken Centaurs. This passage helps to explain the role in which the Lapiths are cast here: Richter (1957.277–278) notes that the Centaurs are normally credited with inventing the art of equitation. The bestiality of the Centaurs in *Georgics* 2, however, would be inconsistent with the restraining, civilizing *aetion* of Book 3. For the chariot as an emblem of civilization in Vergil, see Wilhelm 1982.

60. Although he was not, strictly speaking, an Athenian by birth (he was from Samos, but his father, Neocles, retained his Athenian citizenship as a cleruch; see Diogenes Laertius 10.1–27) Epicurus is clearly identified with Athens in the proemium to *De Rerum Natura* 6.

book, the same city, the pinnacle of human civilization, is destroyed by plague, a victim of the seeming irrationality of nature. The same is true of Vergil's "Plague," where the reader is ironically reminded of the Lucretian original and of the Greek culture heroes who had made that civilization possible:

> praeterea iam nec mutari pabula refert,
> quaesitaeque nocent artes; cessere magistri,
> Phillyrides Chiron Amythaoniusque Melampus.
> saevit et in lucem Stygiis emissa tenebris
> pallida Tisiphone Morbos agit ante Metumque,
> inque dies avidum surgens caput altius effert.
>
> *Georgics* 3.548–553

> nec requies erat ulla mali: defessa iacebant
> corpora. mussabat tacito medicina timore,
> quippe patentia cum totiens ardentia morbis
> lumina versarent oculorum expertia somno.
>
> *De Rerum Natura* 6.1178–1181

Not anonymous *medicina*, as in Lucretius, but the very culture heroes who personify the art of healing are depicted, much as Ericthonius earlier in *Georgics* 3 and Epicurus in *De Rerum Natura* 6 had personified the civilizing power of Athenian culture. Vergil makes explicit in his "Plague" what is a chilling, but merely implicit possibility in *De Rerum Natura*, namely that the terrifying power of death can overcome even the soundest and most perfect rational system. This, I think, is the point of Vergil's vivid personification not only of Chiron and Melampus, but of Styx, Tisiphone, Sickness, and Fear, whose victory reverses that of Epicurus over *religio*.

In *Georgics* 2–3, then, Vergil reproduces the rise and subsequent collapse of civilizing forces that is the basic structural and thematic motif of *De Rerum Natura* 6. Given the tonal movement and overall structure of the two poems, such an imitative scheme speaks volumes about about Vergil's Lucretian program. The tone of *De Rerum Natura* moves generally from joy to gloom within single books and over the course of the poem as a whole. The tonal pattern of the *Georgics* is slightly more complicated, moving from joy to gloom in Books 1 and 3, and from apprehension to relative optimism in Books 2 and 4. Most recent work, of course, following a pattern established in connection with the rest of Vergil's *oeuvre*, has tended to hold that the poem's subtext is basically and consistently pessimistic; but whatever Vergil's view of fate or of the nascent Augustan regime may have been, however qualified his approval of any given thing may normally be, one must try to avoid the mistake of attending only to the subtext. Overt statements count as well.[61] In this connection, I find little to quibble with in Otis' analysis of the "*variatio* of tone and mood" between "the relative gaiety and lightness of Books II and IV" on the one hand, and "the sombre and heavy character of I and III."[62] Within this framework, a book-length imitation of a Lucretian book faithful to the tonal progression of the original would really be feasible only in *Georgics* 1 or 3, while an imitation involving most consecutive books of *De Rerum Natura* would not allow tonal congruency between one of the source books and its derivative. But, thanks to the structural peculiarity that binds *De Rerum Natura* 5 to Book 6, it was possible to reproduce, as it were, this pair of books without tonal distortion in a correspondingly balanced pair of books in the *Georgics*. Such a reproduction could take place only in the two central books of the poem, or the crucial tonal progression from relative hopefulness to despair would be lost. This is exactly what is found in *Georgics* 2–3. In fact, congruency is found not only in abstract patterns such as these, but in theme and even, to a large extent, in subject matter as well.

We see in this imitation, which must surely have been part of the poet's earliest plans for the *Georgics*, the familiar traces of Vergil's allusive dialectic. By basing the tonal movement of *Georgics* 2–3 on that of *De Rerum Natura* 5–6, and by actually intensifying the thematic and tonal opposition that he found in Lucretius, Vergil uses his source in much the same way as he uses Hesiod and Aratus in Book 1. As before, the source text is reformed to suit new purposes. It is common to speak of Lucretius' "influence" on the *Georgics*, as in the dictum of Sellar with which I opened this chapter. Without denying Lucretius his place in Vergil's imagination, however, I suggest that we would more properly celebrate Vergil's remaking of Lucretius, not in a polemical sense, but in the spirit of respectful emulation, of declaring at the same time Lucretius'

61. For a forthright attempt to address the dichotomy between single-mindedly "optimistic" and "pessimistic" readings, see Perkell 1989.

62. Otis 1964.151. Although Otis goes on to stress what he feels is an even more fundamental distinction between "an inanimate [sc. in Books 1–2] and an animate [Books 3–4] subject matter," (p. 153), it is the alternating tone of the even and odd numbered books that I want to consider here.

place in the tradition of didactic epos, and the newly won position of the *Georgics* as the culmination of that tradition. This sort of tendentious reinterpretation of the poetic past is exactly what we observed in the case of Hesiod and Aratus. The addition of Lucretius indicates on the one hand the consistency of Vergil's allusive program, and on the other a new, non-Callimachean component of his literary-historical goals.

I have suggested that the similarity of Vergil's "Plague of Noricum" to Lucretius' "Plague of Athens"—like that between Vergil's "Weather Signs" and Aratus' "*Diosemiae*"—indicates a structural relationship with what precedes each of those passages. But the "Plague" resembles the "Weather Signs" in another way: it signals the end of a stage in Vergil's allusive program. Although casual echoes of Lucretian passages that are thematically important to *Georgics* 1–3 can be found in *Georgics* 4, the final book of the poem contains no new contexts of allusion based chiefly on Lucretius. This situation parallels the case of Hesiod and Aratus, the primary models of *Georgics* 1, who exerted practically no material influence on *Georgics* 2–4. Even more than in the earlier case, however, the conclusion of Vergil's Lucretian program sounds a note of particular finality. We have been given no clue, as in Book 1, of what might follow. Many surprises are in store. After the frightening and depressing spectacle of the plague we revel in the mock-heroics of the bees; and after such a challenging encounter with Lucretius, it is with amazement that we contemplate Vergil's fresh and unexpected reading of the poet who would most engage his imagination over the remainder of his life.

CHAPTER 6

Homer

The "Homeric Problem" of the *Georgics*

In Chapter 3 we gained some preliminary insight into Vergil's program of Homeric allusion in the *Georgics*. Here we shall confront the same issue in the light of knowledge gained from studying the poet's allusions to Hesiod, Aratus, and Lucretius. At the same time, we must face the much more formidable problem of understanding the Homeric program as an element in the allusive strategy of the *Georgics* as a whole.

This has been our goal all along; but throughout Books 1–3, the task is not so difficult. Their obvious differences notwithstanding, Hesiod, Aratus and Lucretius fit without great discomfort into one or two broadly unified poetic categories. Despite what some scholars say, ancient critics recognize as "didactic poetry" a non-narrative branch of the epic tradition that treats not of kings and battles, either legendary or historical, but of knowledge, whether practical and technical or theoretical and philosophical.[1] The tradition begins with Hesiod, continues, *pace* Aristotle, with Parmenides, Empedocles, and other pre-Socratic philosophical poets, undergoes a typically Hellenistic transformation in poets like Aratus, finds expression among the Romans in the work of Lucretius, and is in a sense summed up in the *Georgics* itself. In choosing his models for Books 1–3, then, Vergil was clearly at pains to define the tradition to which he wished to be heir. We see this impulse behind

1. See p. 61 n. 2. Servius *praefatio in G.* (Thilo 1887.129.9–12) notes that because the *Georgics* is didactic it therefore must be addressed to someone, like *Works and Days* and *De Rerum Natura*. One may accordingly question whether Servius and his sources would have included the poems of Empedocles and Aratus under this rubric.

the selection of many secondary models as well. Among those whose influence we can still in some measure evaluate for ourselves I include Eratosthenes of Cyrene and Varro of Atax. The former, best known as an outstanding scientific scholar and as head of the Alexandrian library, also composed an epic poem called *Hermes* on the structure of the cosmos; and this poem informs Vergil's discussion of "Climatic Zones" in Book 1.[2] The latter, not to be confused with Vergil's other Varro, the more famous senator and polymath, is known as the author of several translations or close imitations of Hellenistic poetry.[3] The Vergil scholia quote a sizeable passage of his *Ephemeris*, which follows very closely a passage of Aratus' "*Diosemiae*," in order to show how much Vergil borrowed from Varro in his own imitation of Aratus.[4] Among those poets whose influence is attested, but cannot be adequately illustrated, we know that Vergil did imitate Nicander of Colophon, according to Quintilian (*IO* 10.1.56), and that his *Georgica* gave Vergil the title of his own poem, according to Servius.[5]

Clearly, then, Vergil's interest in the expository branch of the epic tradition is pronounced. But viewed in these terms, his decision to make the concluding episode of the *Georgics*, the "Aristaeus," a sophisticated allusion to both the *Iliad* and the *Odyssey* is a complete departure from his practice in the rest of the poem. In fact, we can measure the differences in many ways. Diction: the "Aristaeus" is full of words that are used nowhere else in the poem while at the same time being almost devoid of specifically agricultural language. Mythology: nowhere else in

2. Fragments in Powell 1925.58–63, Lloyd-Jones and Parsons 1983.183–186, 424–425, frr. 397–398, 922). On the subject of the poem see Hiller 1872.64. On Vergil's imitation see [Probus] *ad G.* 1.233 (Hagen 1902.364), Richter 1957.150–152 *ad loc.*, and pp. 172–176 of the present work. Although Vergil alludes to the *Hermes* in detail here, it was not an important model for the *Georgics* overall. Parsons (1974.6; cf. Lloyd-Jones and Parsons 1983.184) has shown that Eratosthenes' poem was about 1600 lines long, but the passage on the five zones seems to have been the only one that interested Vergil.

3. Fragments in Morel–Büchner 1982.121–129. He produced an *Argonautae* in four books, evidently modeled closely on Apollonius of Rhodes (frr. 1–10); a *Chorographia* (frr. 11–20), of which one fragment (fr. 13) deals with the five zones theme; an *Ephemeris* (frr. 21–22; the title is a matter of dispute), on which see the following note; and a *Bellum Sequanicum* in at least two books (frr. 23–24).

4. Servius Danielis *ad G.* 1.375 (Thilo 1887.205 = fr. 21 Morel–Büchner 1982.127); *Brevis exp. ad G.* 1.397 (Hagen 1902.265 = fr. 22 Morel–Büchner 1982.127). The relevant passage of Aratus is *Phaenomena* 938–957.

5. Vergil adapted a few passages of Nicander's *Theriaca* (21–27, 51–56, 179–180, 359–371) in Book 3 (for the details see Thomas 1988.2.119–123 *et passim*), and Servius detects

the *Georgics* do we encounter such a fully developed treatment of myth.[6] The same can be said with regard to narrative. Only in a very few passages does Vergil organize his didactic discourse in a form approaching that of a conventional narrative. The most prominent of these is "The Plague of Noricum," which obviously does not make use of the full technical panoply of narrative as does "Aristaeus." Finally, the characters who appear in this story are the only ones in the poem who are not alluded to merely in passing, and the only ones actually permitted to speak. Of course, we can explain these differences in any number of ways; but ultimately, all explanations take us back to one fact. The "Aristaeus" differs from the rest of the *Georgics* in precisely the same ways that its model, Homer, differs from Hesiod, Aratus, and Lucretius. We can therefore make sense of Homer's presence in the *Georgics* only in the broadest terms, in that all of Vergil's major poetic sources are epic poets; but it is abundantly clear that Homer belongs to a different branch of the epic tradition than Hesiod, Aratus, Lucretius, and the rest.

Before addressing the challenge of the "Aristaeus" itself, let us review the evidence for Homeric allusion in the earlier books of the *Georgics*. In Book 1 we found that allusion to Lucretius takes its place in a predominantly Hesiodic and Aratean program, and that this fact helps us to understand the place of all three authors in Vergil's poem-long strategy of allusion. We may reasonably hope for similar assistance from the same quarter in the case of Homer.

---

a passing reference at *Georgics* 2.215, Servius Danielis at 3.391 (Thilo 1887.239 and 307–308; cf. Macrobius *Sat.* 5.22.10). No Vergilian references to the surviving fragments of the *Georgica* have been identified: see Gow–Scholfield 1953.209. These fragments, twenty-four in all (frr. 68–91 Schneider 1856.79–122), comprising some 152 lines, deal chiefly with garden crops, including flowers (fr. 74, the longest at 72 lines)—a topic that Vergil explicitly "excludes" (*G.* 4.116–148; cf. Columella *RR* 10 *proemium* 3)—although trees (frr. 69, 75, 76, 80), wine (frr. 86 and 91), oxherds and mule teams (fr. 90), pigeon-breeding (fr. 73), and even seafood (fr. 83, probably in a context that concerns seasoning, a recurring theme in the surviving fragments) are mentioned as well. Wilamowitz (1924.1.85) thought the surviving fragments representative of the entire poem, while Kroll (*RE* 17.255) argued that they actually reflect the interests of Athenaeus, our source for all but two of the fragments (90 and 91).

6. This is not to deny the general importance of mythology throughout the poem, on which see Frentz 1967, who for some reason, however, does not deal at all with "Aristaeus" (see p. 4).

Georgics 1–3

Homeric influence on the first three books of *Georgics* is subtle and unobtrusive, but what is there is of great importance. The major commentaries and a few specialized studies have uncovered a number of references to the *Iliad* and *Odyssey* in these books, but have differed on their significance.[7] The most recent of the specialized studies is that of Knauer,[8] who brings to bear on the problem the expertise gained from his earlier study of the *Aeneid* and Homer. Knauer's research on the *Georgics* represents an extreme in that he strenuously downplays the importance of the alleged Homeric allusions that have been found in Books 1–3 as compared with Book 4, particularly the "Aristaeus." To summarize his argument: (1) Characteristic elements of heroic epos are scarcer in the early books than in Book 4. For example, Books 1–3 contain just seven similes, while Book 4 alone has ten.[9] Furthermore, few of the similes found in the early books show any material dependence on particular Homeric similes.[10] (2) Many of those Homeric elements that do occur in the early books are "of a kind which could, by their being typically Homeric formulae, be easily isolated and endowed with a history of their own. Thus they might be integrated into the 'Georgics.' The very fact of their being 'typical' makes it difficult at the outset to decide on the character of a possible direct 'quotation' by Vergil."[11] (3) In the relatively few passages in Books 1–3 that seem to be modeled on a specific Homeric exemplar, it is possible to show that Vergil in fact has either a different model or else no particular model in mind. In these cases, the imitated elements are not typical or formulaic within

7. Erdmann (1912.6) numbers Homer as one of the poem's most important stylistic models along with Hesiod, Ennius, and Lucretius. Büchner, in his discussion of sources (1959.305–309 [= *RE* 8 A 1327–1331]) does not even mention Homer in connection with the *Georgics*. Wilkinson (1969.68) classes Homer as chief among those literary sources that Vergil "uses casually in the course of the poem, but which cannot be said to have contributed to the original conception." Apropos of the "Aristaeus," he notes that "Homer may have provided the framework; but the detailed treatment is in the Hellenistic manner. The ancestor of this kind of poetry is Callimachus" (p. 216). Thomas (1988.1.5), who in general also stresses the Callimacheanism of the poem, states that "Homer was the poet with whom Virgil was most familiar, in this poem as the *Aeneid*," and cites many Homeric parallels in his commentary.

8. 1981b.890–918.

9. Knauer 1981b.893.

10. Knauer 1981b.893–5.

11. Knauer 1981b.908.

the Homeric poems, but through frequent imitation by other poets "had been severed from their original contexts and had become freely 'available.'"[12]

Knauer's survey brings out some important points about Vergil's use of Homer in the *Georgics*, and warns critics against making easy assumptions. On the other hand, in some cases his skepticism seems excessive, and a different interpretation seems possible. One is in fact encouraged to see things differently by Vergil's first adaptation of a Homeric simile in the *Georgics*. Early in Book 1 our attention turns to the complementary tasks of irrigation and drainage. The former topic is introduced by way of a remarkable Homeric allusion:[13]

> quid dicam, iacto qui semine comminus arva
> insequitur, cumulosque ruit male pinguis harenae
> deinde satis fluvium inducit rivosque sequentis,
> et, cum exustus ager morientibus aestuat herbis,
> ecce supercilio clivosi tramitis undam
> elicit? illa cadens raucum per levia murmur
> saxa ciet, scatebrisque arentia temperat arva.
>
> *Georgics* 1.104–110

> φεῦγ᾽, ὁ δ᾽ ὄπισθε ῥέων ἕπετο μεγάλῳ ὀρυμαγδῷ.
> ὡς δ᾽ ὅτ᾽ ἀνὴρ ὀχετηγὸς ἀπὸ κρήνης μελανύδρου
> ἂμ φυτὰ καὶ κήπους ὕδατι ῥόον ἡγεμονεύῃ
> χερσὶ μάκελλαν ἔχων, ἀμάρης ἐξ ἔχματα βάλλων·
> τοῦ μέν τε προρέοντος ὑπὸ ψηφῖδες ἅπασαι
> ὀχλεῦνται· τὸ δέ τ᾽ ὦκα κατειβόμενον κελαρύζει
> χώρῳ ἔνι προαλεῖ, φθάνει δέ τε καὶ τὸν ἄγοντα·
> ὣς αἰεὶ Ἀχιλῆα κιχήσατο κῦμα ῥόοιο
> καὶ λαιψηρὸν ἐόντα· θεοὶ δέ τε φέρτεροι ἀνδρῶν.
>
> *Iliad* 21.256–264

In his perceptive discussion of this allusion, Ross notes that "It is Virgil's practice in the *Georgics*...to turn literary topics, abstractions, similes, metaphors, and the like into realities...."[14] Here indeed Homeric simile

12. Knauer 1981b.906.

13. Knauer does not mention this passage, first identified as an allusion to Homer by Scaliger 1561.244a.A–B and subsequently discussed by Jahn 1903a.403–404; cf. Miles 1980.77, Thomas 1986.178–179, and especially Ross 1987.48–54, who finds anticipation of the Homeric reference in lines 102–103.

14. Ross 1987.51.

becomes Vergilian reality, and we must consider carefully just what this transformation implies. In the first place, we think of theme. Whatever his reason for alluding to Homer at this point, Vergil clearly chose to quote a passage thematically consistent with his own exposition. This is, perhaps, only natural; but every such choice is not only a selection, but a rejection as well. In general terms, the primary world of Homeric epos and of the *Iliad* in particular is that of men at war. For the poet of the *Georgics*, thematically apposite material was to be found not in this primary world, but in the secondary world of the similes. Here, Vergil has borrowed the vehicle of the simile, an illustration of one of the farmer's mundane tasks. What he has not borrowed, the tenor of the simile, is one of the *Iliad*'s most transcendent moments, Achilles' hubristic battle against the river god Xanthus. But the tenor is not altogether lost. Rather, through allusion, something extremely significant takes place: Homer's secondary world, the agricultural world of the simile, becomes the primary world of the *Georgics*. Conversely, Homer's primary world, the world of men at war, recedes into metaphor. It is in this passage that Vergil begins to invest the farmer's tasks with soldier's trappings, a process that will continue throughout the poem.[15] Here we learn that he must "train the land" and "command the fields" (***exercet**que frequens tellurem atque **imperat** arvis* 99).[16] Thus Achilles' singular struggle with the river Xanthus finds its place in the recurring struggles of the georgic world.

This transformation of Homeric material has an important literary-historical dimension as well. By converting Homer's simile into a straightforward agricultural *exemplum*, Vergil has also converted a generic element typical of heroic epos into something that fits more comfortably into the generic vocabulary of the *Georgics*. Simile is in many ways the Homeric figure *par excellence*. Here Vergil borrows the material of a Homeric simile, but strips it of its characteristic form: it appears no longer as a simile, but as a straightforward "technical" *exemplum*. This process too, like that of making poetic figures and conceits into agricultural realities, can be observed throughout the *Georgics*. Such a treatment of Homer is not, however, typical of this poem or this poet

15. On this topic in general see Betensky 1979.

16. Vergil's diction here is carefully chosen. Untrained land, like a green recruit, is merely "soil" (*tellus*), not yet ready to respond to the farmer's orders; after basic training, it becomes proper "farmland" (*arva*), capable of doing his bidding.

alone. We meet the same procedure in the Hellenistic poets, particularly Theocritus.

In his study of the anterior material from which Theocritus constructed his pastoral world, David Halperin writes that

> the anti-heroic material in Homer is particularly important to Theocritus because it furnishes him with an authoritative poetic precedent *within* the genre in which he elected to compose; it represents a traditional means of accomplishing his innovative purpose—to remake the genre of *epos* into a suitable vehicle for "modern" (Alexandrian) aesthetic ideals and themes. In this sense, Theocritus' inversion of epic subjects can be related to the composite nature of the comparisons in Homer's similes, for Homer is notoriously fond of juxtaposing the homely activities of humble people to the glorious struggle of the heroes.[17]

This observation fits exactly the Homeric allusion discussed above. Vergil, the student of Theocritus, uses Homeric material to establish the literary pedigree of his didactic essay. Such an attitude towards and treatment of Homer sorts well with what we have learned about the allusive program of *Georgics* 1. As Knauer puts it, "A typical element of the Homeric epic is the simile. In Hesiod similes are rare. In Georgics 1–3 Vergil follows his model Hesiod."[18] Moreover, the fact that this is Vergil's first allusion to Homer in the poem gives it a programmatic import. By alluding to a Homeric passage in non-Homeric form, Vergil in effect molds Homeric epos to his own didactic requirements—requirements shaped by Hellenistic poets on the alternative epic model provided by Hesiod—making it, in Vergil's own pregnant phrase, a *deductum carmen*.

This interpretation suggests that Homer may well be an important factor in the *Georgics* even before Book 4. A survey of the other Homeric material in Book 1 confirms the impression. In what follows I shall examine this material, paying particular attention to how Vergil integrated Homeric allusions into passages more obviously modeled on his didactic sources.

17. Halperin 1983.228; cf. Hommel 1958.743, Vischer 1965.32–33, Ott 1972.149.

18. Knauer 1981b.893. Knauer is referring here to the general paucity of similes in *Georgics* 1–3 as compared with Book 4 (cf. p. 210), but his syllogism applies particularly well to this argument. Note, however, that Knauer adopts the traditional view of Hesiod as *the* model of the *Georgics*.

Homer and Hesiod

In some cases the commentators encourage us to choose between two possible sources or models for a given passage. A case in point is the following quotation *cum variatione* of a line from *Works and Days:*

> **Pleïadas, Hyadas,** claramque Lycaonis Arcton
>
> *Georgics* 1.138
>
> **Πληϊάδες** θ' **Ὑάδες** τε τό τε σθένος Ὠρίωνος
>
> *Works and Days* 615

Such a close Hesiodic reminiscence is only to be expected in a book so extensively indebted to *Works and Days.* Neither are we surprised that, when we compare the immediate contexts of the two passages, we find them to be quite different. Hesiod of course mentions the setting of the Pleiades, of the Hyades, and of Orion as a signal to plow (*WD* 612–617). Vergil transfers the line to another passage modeled on Hesiod, his "Aetiology of *Labor*".

> tunc alnos primum fluvii sensere cavatas;
> navita tum stellis numeros et nomina fecit
> **Pleïadas, Hyadas,** claramque Lycaonis Arcton;
> tum laqueis captare feras et fallere visco
> inventum et magnos canibus circumdare saltus.
>
> *Georgics* 1.136–140

This transferal agrees entirely with Vergil's normal principles of *variatio.* The theme of navigating by the stars functions in the "Aetiology of *Labor*"—i.e. in Vergil's imitation of the Hesiodic "Myth of Ages"—as a token of human degeneracy. Again, the theme of identifying the constellations reminds us of the second major source of the "Aetiology," Aratus' *Phaenomena.*

The situation is complicated, however, by the fact that certain details of the Vergilian passage point unmistakably to Homer as well. The single line in surviving Greek poetry that the Vergilian passage most nearly resembles, as every commentator notes, occurs in the *Iliad:*

> *ἐν μὲν γαῖαν ἔτευξ', ἐν δ' οὐρανόν, ἐν δὲ θάλασσαν,*
> *ἠέλιόν τ' ἀκάμαντα σελήνην τε πλήθουσαν,*

> ἐν δὲ τὰ τείρεα πάντα, τά τ' οὐρανὸς ἐστεφάνωται,
> **Πληϊάδας** θ' **Ὑάδας** τε τό τε σθένος Ὠρίωνος
> **Ἄρκτον** θ', ἣν καὶ Ἄμαξαν ἐπίκλησιν καλέουσιν,
> ἥ τ' αὐτοῦ στρέφεται καί τ' Ὠρίωνα δοκεύει,
> οἴη δ' ἄμμορός ἐστι λοετρῶν Ὠκεανοῖο.
>
> *Iliad* 18.483–489

The Homeric line is nearly identical with Vergil's up to the penthemimeres, naming a pair of constellations, the Pleiades and the Hyades.[19] Homer follows with Orion, Vergil with Ursa Major (which he calls by its Greek name, Arctus); and when we examine the two contexts, we find that the mention of this constellation also forms a part of the Homeric quotation.

Insofar as *Works and Days* 615 and *Iliad* 18.486 are virtually identical, the Vergilian line can be considered a conflation of the two. If one is intent on identifying a single source for such a quotation, Hesiod would clearly seem to be a more relevant choice in a passage like this, which is obviously otherwise indebted to him. Perhaps for this very reason Vergil has provided sufficient detail to ensure that his reference to Homer will be unmistakable. In the first place, the two constellations named in *Georgics* 1.138 are in the accusative, as also in *Iliad* 18.486, while in *Works and Days* 615 they are in the nominative. In the second place, it is obvious that Vergil reproduces only the first half of his model lines, up to the penthemimeres. For the second half, he ignores the Hesiodic passage entirely and for Orion substitutes Arctus, which occurs at the beginning of the next line in Homer only. Thus *Iliad* 18.486 and 487 are "combined" into *Georgics* 1.138. Further, when Vergil speaks of the *naming* of the constellations by navigators (*nomina fecit* | 136)—a detail that I previously interpreted as a glance towards Aratus[20]—it now seems probable that he is also alluding to the the clausula of the second Homeric line (ἐπίκλησιν καλέουσιν | *Iliad* 18.486).

19. Cf. *Odyssey* 5.270–275:

> αὐτὰρ ὁ πηδαλίῳ ἰθύνετο τεχνηέντως
> ἥμενος· οὐδέ οἱ ὕπνος ἐπὶ βλεφάροισιν ἔπιπτε
> **Πληϊάδας** ἐσορῶντι καὶ ὀψὲ δύοντα Βοώτην
> **Ἄρκτον** θ', ἣν καὶ Ἄμαξαν ἐπίκλησιν καλέουσιν,
> ἥ τ' αὐτοῦ στρέφεται καί τ' Ὠρίωνα δοκεύει,
> οἴη δ' ἄμμορός ἐστι λοετρῶν Ὠκεανοῖο.

20. See p. 214.

In the light of these formal similarities, one might equally argue that the Homeric passage, not the Hesiodic, was uppermost in Vergil's mind as he composed this line. But this judgment is too facile as well. While the Homeric passage shows greater similarity of verbal detail with Vergil's imitation, Hesiod's conception of seafaring as a symbol of humanity's decline from an original, more blessed state is the basis for Vergil's mentioning navigation at all, and the allusive context as a whole looks mainly to his "Myth of Ages." The point is that Vergil appears to have deliberately created this ambiguity as a means of establishing common ground between Homeric and Hesiodic epos. The passage no doubt contains an element of bravado, a wish to demonstrate the thoroughgoing familiarity with specific texts that makes it possible to identify and to capitalize on such similarities as these. At the same time, this particular example points quite suggestively to an important aspect of Vergil's allusive program. Such an allusion possesses an integrative function, linking poetic traditions that might normally be regarded as discrete.[21] By itself, of course, such a limited instance of this phenomenon will not mean very much. But as we have seen, Vergilian allusion functions in similar ways at the level of versecraft and with respect to larger discursive structures. The interpretation advanced here has in fact received corroboration in advance from another, more extensive combination of Hesiodic and Homeric material. In Chapter 3, we discovered that Vergil's discourse on "Lucky and Unlucky Days," an elaborate and involved imitation of Hesiod's "Days," incorporates significant Homeric allusion as well, to the myth of the Aloidae; and we concluded that allusion of this sort becomes a way of finding common ground between distinct poems and even distinct authors.[22] What is now becoming clear is how much of Vergil's task involved the incorporation of Homeric material into the Hesiodic world that he had established in *Georgics* 1. As we proceed through the poem, we shall find that many parallel examples involving all of Vergil's chief models will bear out our interpretation of this passage.

21. It is of course possible that Vergil was not being completely inventive here. To the extent that Homer was regarded as the source of all poetry, Vergil may well have regarded such textual similarities as evidence of Hesiod's borrowings from the *Iliad* and *Odyssey*. The allusion would then be read as an acknowledgement that Homer was the source of the didactic as well as the heroic tradition. On this topic see, in general, Neitzel 1975.

For a possible allusion to Callimachus in this passage, see Thomas 1986.1.194; cf. 1988.91 *ad loc.*

22. See pp. 123–127.

Homer and Aratus

In Chapter 4, we saw how Vergil adopts the Callimachean view of Aratus as a "Hesiodic" poet. Neither Callimachus nor Vergil is especially tendentious in expressing this opinion, nor do they seem far from what Aratus himself might have said. Nevertheless, we should be wary of ignoring the historical realities of ancient criticism. In claiming Aratus as a poet in the mold of Hesiod, Callimachus was entering the lists, as he so loved to do, of a literary controversy. It was not simply a question of whether Aratus wrote Hesiodic poetry; rather, it was the choice of Hesiod as a model *in preference to Homer* that counted. The Aratus scholia clearly attest the lively debate over the poet's true model, Hesiod or Homer, in passages such as this:

> (4) ζηλωτὴς δὲ ἐγένετο τοῦ Ὁμηρικοῦ χαρακτῆρος κατὰ τὴν τῶν ἐπεῶν σύνθεσιν. ἔνιοι δὲ αὐτὸν λέγουσιν Ἡσιόδου μιμητήν γεγονέναι· καθάπερ γὰρ ὁ Ἡσίοδος Ἔργων καὶ Ἡμερῶν ἀρχόμενος ὑμνῶν ἀπὸ Διὸς ἤρξατο, λέγων "Μοῦσαι Πιερίηθεν ἀοιδῇσι κλείουσαι," οὕτω καὶ ὁ Ἄρατος τῆς ποιήσεως ἀρχόμενος ἔφη "ἐκ Διὸς ἀρχώμεσθα," τά τε περὶ τοῦ χρυσοῦ γένους ὁμοίως τῷ Ἡσιόδῳ. (5) Βόηθος δὲ ὁ Σιδώνιος ἐν τῷ αὖ περὶ αὐτοῦ φησιν οὐχ Ἡσιόδου ἀλλ' Ὁμήρου ζηλωτὴν γεγονέναι· τὸ γὰρ πλάσμα τῆς ποιήσεως μεῖζον ἢ κατὰ Ἡσίοδον.
>
> *Isagoga bis excerpta* 2.4–5
> (Maass 1898.324)

It seems clear as we read through *Georgics* 1 that Vergil chose his models with this literary *disputatio* very much in mind. For instance, when we notice that Books 2–3 of the poem allude primarily to Lucretius, and that Book 4 and the poem concludes with an extraordinarily ambitious reworking of Homer, we may wonder about the literary-historical implications of this progression. Finally, when we find Vergil working to establish contact between Hesiod and Homer in his "Aetiology of *Labor*" and "Lucky and Unlucky Days," we may be curious as to whether he took seriously the non-Callimachean tradition about Aratus and regarded his Hellenistic model as an imitator not only of Hesiod, but of Homer as well.

We have seen a faint indication that this may be so in the Homeric quotation that is embedded in the "Aetiology of *Labor*."[23] If we now

23. See pp. 214–216.

turn to the prelude affixed to Vergil's imitation of the "*Diosemiae*," the paragraph on "Vernal and Autumnal Storms" (*G.* 1.311–350), we will be able to watch the process continue. The great storm that Vergil describes at lines 311–334 calls forth the admonition to watch the monthly motions of the heavenly bodies:

> hoc metuens caeli mensis et sidera serva,
> frigida Saturni sese quo stella receptet,
> quos ignis caelo Cyllenius erret in orbis.
>
> *Georgics* 1.335–337

The subsequent lines (338–350) concern the importance of religious observance; but the opening of the "Weather Signs" proper looks back to the storm itself, relegating the charming description of festivals held in honor of Ceres and Bacchus to the status of a digression:

> Atque haec ut certis possemus discere signis,
> aestusque pluviasque et agentis frigora ventos,
> ipse pater statuit quid menstrua luna moneret,
> quo signo caderent Austri, quid saepe videntes
> agricolae propius stabulis armenta tenerent.
>
> *Georgics* 1.351–355

Although the ensuing "Weather Signs" itself, of course, is to a large extent a translation of Aratus, scholars have long understood that it begins with a Homeric flourish in the great storm:

> A Quid tempestates **autumni** et sidera dicam?
> atque, **ubi** iam breviorque dies et mollior aestas,
> quae vigilanda viris? vel cum ruit imbriferum ver,
> spicea iam campis cum messis inhorruit et cum
> frumenta in viridi stipula lactentia turgent? 315
> saepe ego, cum flavis messorem induceret arvis
> agricola et fragili iam stringeret hordea culmo,
> omnia ventorum concurrere proelia vidi,
> quae gravidam late segetem ab radicibus imis
> B sublimem expulsam eruerent: ita **turbine nigro** 320
> ferret hiems culmumque levem stipulasque volantis.
> saepe etiam immensum caelo venit agmen aquarum
> et foedam glomerant tempestatem imbribus atris
> collectae ex alto nubes; ruit arduus aether

C et pluvia ingenti sata laeta **boumque labores** 325
C, D **diluit;** implentur fossae et cava **flumina crescunt**
cum sonitu fervetque fretis spirantibus aequor.
E **ipse pater** media nimborum in nocte corusca
fulmina molitur dextra, quo maxima motu
terra tremit, fugere ferae et mortalia corda 330
per gentis humilis stravit pavor; ille flagranti
aut Atho aut Rhodopen aut alta Ceraunia telo
F deicit; ingeminant Austri et **densissimus imber;**
nunc nemora ingenti vento, nunc litora plangunt.

Georgics 1.311–334

B ὡς δ᾽ ὑπὸ **λαίλαπι** πᾶσα **κελαινὴ** βέβριθε χθὼν
A, F **ἤματ᾽ ὀπωρινῷ, ὅτε λαβρότατον** χέει **ὕδωρ** **385**
E **Ζεύς,** ὅτε δή ῥ᾽ ἄνδρεσσι κοτεσσάμενος χαλεπήνῃ,
οἳ βίῃ εἰν ἀγορῇ σκολιὰς κρίνωσι θέμιστας,
ἐκ δὲ δίκην ἐλάσωσι, θεῶν ὄπιν οὐκ ἀλέγοντες·
D τῶν δέ τε πάντες μὲν **ποταμοὶ πλήθουσι** ῥέοντες,
πολλὰς δὲ κλιτῦς τότ᾽ ἀποτμήγουσι χαράδραι, 390
ἐς δ᾽ ἅλα πορφυρέην μεγάλα στενάχουσι ῥέουσαι[24]
C ἐξ ὀρέων ἐπικάρ, **μινύθει δέ τε ἔργ᾽ ἀνθρώπων·**
ὣς ἵπποι Τρῳαὶ μεγάλα στενάχοντο θέουσαι.

Iliad 16.384–393

In this case, too, commentators have taken little account of the Homeric context. The passage occurs in the "Aristeia of Patroclus" at the point when the Danaans begin to rout the Trojan attackers and to repel them headlong from the ships. The relevant lines, as in the first Homeric quotation of the poem, are found in a simile. The Trojans raise a vast, disorganized cry of fear as when a cloud from Olympus moves into the sky in place of clear air, when Zeus brings storm (*Iliad* 16.364–367). Hector himself is in desperate retreat, hard pressed by Patroclus, and the simile is repeated with emphasis in the form quoted above. The Homeric storm is a cosmic reflection of the disorder with which the Trojans flee. Significantly, it is a storm that Zeus sends in anger at men who use violence to pass crooked decrees in assembly, who banish Justice and care not for the gods' regard (386–388). These details, which Vergil himself omits, are important. On the one hand, they highlight the

24. Vergil imitates this line in a later passage (*Georgics* 4.372–373), as noted by Conington–Nettleship 1898.380 *ad G.* 4.373.

seemingly arbitrary nature of Vergil's storm, which strikes humiliating fear into all humanity (330–331), not into a small group of transgressors, without any explicit reason.[25] On the other hand, the departure of Justice, an Aratean symbol of the end of the Golden Age (*Ph.* 134), becomes a significant image for Vergil as well (*G.* 2.473–474); and the final "storm" of the "Weather Signs," the murder of Caesar and the subsequent civil wars, shares two motifs with Homer's storm. First, the assassination too is a cosmic reflection of disorder in human affairs. Second, it is also characterized as a punishment for unjust ways—specifically, the injustice of Rome's Trojan ancestors: *satis iam pridem sanguine nostro* | *Laomedonteae luimus periuria Troiae* (*G.* 1.501–502).[26]

The extensive Homeric background of this paragraph casts significant light both on Vergil's argument and on his reading of the *Iliad.* Through his familiar habit of borrowing the similes that in Homer illustrate the heroic deeds of men at war, he imparts a martial character to his agricultural discourse. As in the case of Achilles and the River Xanthus, the martial element is preserved in Vergil's diction: thus in the passage quoted above, the winds meet in battle (*omnia ventorum concurrere proelia vidi* 318); the rain advances like an army on the march (*saepe etiam immensum caelo venit agmen aquarum* 322); Jupiter's thunderbolt is a weapon (*telo* 322).[27] Moreover, Vergil's choice of Iliadic material attests to his selective approach to Homer. Where before he had found in that martial epic *par excellence* a convenient vignette of irrigation, in Homer's

25. This motif imparts a highly Lucretian color to the passage. Cf. *De Rerum Natura* 1.62–79, Epicurus' triumph over superstition: when humankind lay prostrate on the ground (*mortalia corda...humilis stravit pavor, G.* 1.330–331), Epicurus was the first to raise *mortalis...oculos* (*DRN* 1.66–67) against the specter of *Religio*, unchecked by *fama deum, fulmina*, or *minitanti* | *murmure...caelum* (*DRN* 1.68–69; cf. *ille flagranti* | *aut Atho aut Rhodope aut alta Ceraunia telo* | *deicit, G.* 1.331–333). Cf. Lucretius' preface to his explanation of thunder and lightning (*DRN* 6.50–95) and the remarks of Cole 1979.27.

26. This passage is recalled, restored to simile form, in Aeneas' vision of the destruction of Troy:

> in segetem veluti cum flamma furentibus Austris
> incidit, aut rapidus montano flumine torrens
> sternit agros, sternit **sata laeta boumque labores**
> praecipitisque trahit silvas; stupet inscius alto
> accipiens sonitum saxi de vertice pastor.
>
> *Aeneid* 2.304–308

Cf. *Georgics* 1.325, and for further discussion see Briggs 1980.17–19.

27. Jahn (1903.423) and Cole (1979.15) cite Lucretian passages in which storms are described as battles. See also Hardie 1986.180–187, 200–209.

storm simile he finds an equally apposite description of the storms that threaten to undo the farmer's work. Thus not only does he borrow something of Homer's heroic, martial character to lend dignity and weight to his own more humble theme, but he in effect turns the *Iliad* into a source for a poem on agriculture.

This Homeric introduction to the "Weather Signs" creates an important link between the *Iliad* and the *Phaenomena*. What is more, it creates an expectation that there are further similarities between these two models. This expectation is confirmed when we inspect the provenance of Vergil's signs of rain (*G.* 1.373–392). The bulk of this passage comes directly from Aratus, with the usual compression and rearrangement.[28] Not all of the signs that Vergil mentions, however, are found in Aratus' discussion of rain signs (*Ph.* 933–987). Indeed, the first sign that Vergil gives—cranes flying before an approaching storm—is not from those lines of Aratus that are his main model here, but rather are "transposed" from a different section of the "*Diosemiae*":

> numquam imprudentibus imber
> obfuit: aut illum surgentem vallibus imis
> aëriae fugere grues, aut bucula caelum
> suspiciens patulis captavit naribus auras,
> aut arguta lacus circumvolitavit hirundo
> et veterem in limo ranae cecinere querelam.
>
> *Georgics* 1.373–378

> *οὐδ' ὑψοῦ γεράνων μακραὶ στίχες αὐτὰ κέλευθα*
> *τείνονται, στροφάδες δὲ παλιμπετὲς ἀπονέονται.*
>
> *Phaenomena* 1031–1032

We may ask why Vergil has performed this rearrangement. The behavior of cranes is commonly attested as an indicator of rain,[29] but this in itself will not have induced Vergil to mention it so prominently. It is not unlikely that he was materially influenced by some other source, since his version differs so much from the *Phaenomena*, not only by virtue of its placement, but materially as well. Aratus states that cranes fly in a

28. On the structure of Vergil and his model see Beede 1936. The fullest critical discussion of Vergil's adaptation is Cole 1979.62–92. See also Perrotta 1924; Jermyn 1951. 34–37.

29. E.g. Aristotle *HA* 614b and Theophrastus *De Signis* 52, both of whom, however, state that when cranes in flight notice an approaching storm, they return to earth and lie low until the weather improves.

whirling gyre at the approach of rain. Elsewhere he notes that a long and orderly flight formation indicates fair weather:

> *καὶ δ' ἄν που γέρανοι μαλακῆς προπάροιθε γαλήνης*
> *ἀσφαλέως τανυσαίεν ἕνα δρόμον ἤλιθα πᾶσαι,*
> *οὐδὲ παλιρρόθιοί κεν ὑπεύδιοι φορέοιντο.*
>
> *Phaenomena* 1010–1012

Cole, who remarks upon this discrepancy, notes that Hesiod, too, connects cranes with rainy weather, and that he does so in a passage that immediately follows his instructions on gathering wood for farm equipment, including the plow.[30] She suggests that, since Hesiod's plow passage was obviously important to Vergil, his nearby treatment of cranes as a sign of rainy weather "inspired Vergil to include them as his first sign here."[31] Perhaps; but Hesiod focuses on the cry of the crane, not on its flight, and is more interested in this sign as the beginning of plowing season than as a sign of rain per se. There is, moreover, nothing in the language of either Aratus or Hesiod to connect them closely with Vergil here.

Once again, however, Vergil's passage shows a close resemblance to a passage from the *Iliad.* In fact there are two passages, both similes, that are important here. The first occurs at the beginning of *Iliad* 3, where the Trojans advance with a shout like that of cranes:

> *ἠΰτε περ κλαγγὴ γεράνων πέλει οὐρανόθι πρό,*
> *αἵ τ' ἐπεὶ οὖν χειμῶνα φύγον καὶ ἀθέσφατον ὄμβρον,*
> *κλαγγῇ ταί γε πέτονται ἐπ' Ὠκεανοῖο ῥοάων*
> *ἀνδράσι Πυγμαίοισι φόνον καὶ κῆρα φέρουσαι·*
> *ἠέριαι δ' ἄρα ταί γε κακὴν ἔριδα προφέρονται.*
>
> *Iliad* 3.3–7

Homer's cranes, like Vergil's, flee from a storm and unspeakable rain (*χειμῶνα φύγον καὶ ἀθέσφατον ὄμβρον* 4 ≈ *illum* (sc. the *imber* of the previous line)...*fugere* 374–375).[32] More telling is the fact that they are in both poets *ἠέριαι* (7) ≈ *aëriae* (375).[33] The Greek word is one of varying

30. Cole 1979.64, citing *WD* 448–451.
31. Ibid. 65.
32. Cole 1979.65.
33. Conington–Nettleship 1898.210 *ad G.* 1.375.

meanings, from "early-morning" (< ἦρι) to "misty," "airy," or "airborne" (all < ἀήρ).[34] Apollonius uses ἠέριος in several different senses,[35] and the Alexandrian poets in general loved such words, which occasioned the display of great erudition in poetic debates about the words' true meanings. This motive probably has something to do with Vergil's use of *aërius* here. By using its Latin equivalent, Vergil enters the Hellenistic tradition of lexical imitation of Homer. Moreover, he guarantees contextual reference by his treatment of the word. First, he uses it in the same metrical *sedes* as Homer.[36] Second, with respect to the Aratean tradition, he transfers the epithet from one sign to another. Immediately after the cranes Vergil mentions the heifer's behavior as a sign of rain: *aut bucula caelum | suspiciens patulis captavit naribus auras* (*G.* 1.375–376). The breezes that the heifer sniffs are unmodified. In the original, Aratus' βόες sniff from the air (ἀπ' αἰθέρος ὠσφρήσαντο *Ph.* 955). Cicero rendered this as *umiferum duxere ex aëre sucum* (*Aratea* 955). Varro of Atax managed to improve on Cicero, writing *naribus aërium patulis decerpsit odorem.* Now Vergil knew well and obviously admired Varro's translation, to judge from the excerpt that the Vergilian scholia preserve.[37] If we consider our *Georgics* passage in light of this tradition alone, it is clear that Vergil has simply transferred the Varronian epithet *aërius* from the breezes sniffed by the heifer to the cranes. But by doing so, he effects a reference to Homer's cranes as well.

For some, Varro's translation of Aratus will be a sufficient and more likely source of Vergil's *aërias* than the history of learned inquiry into Homeric diction. But there is more to the Homeric reference than this. The passage to which Vergil alludes here is, as I have noted, the simile describing the Trojans as they advance against the Achaeans at the beginning of the *Iliad.* The simile sharply distinguishes between the two peoples, for while the Trojans advance noisily—this being the point of the crane simile—the Achaeans advance in silence, "breathing valor" (*Iliad* 3.8). There is an irony in this distinction, however; for not long before this passage, the Achaean host too is compared to a flock of cranes:

34. Both derivations are recognized by Frisk 1960.624 s.v.

35. See Livrea 1973.350 *ad Arg.* 4.1239 (with further references), and cf. his remarks on pp. 277 *ad* 954 and 437 *ad* 1577.

36. Cole 1979.65.

37. Fr. 22 (Morel–Büchner 1982.127), from his *Ephemeris* (the title given by *Brevis exp. ad G.* 1.397 [Hagen 1902.265]).

τῶν δ᾽, ὥς τ᾽ ὀρνίθων πετεηνῶν ἔθνεα πολλά,
χηνῶν ἢ γεράνων ἢ κύκνων δουλιχοδείρων,
᾽Ασίῳ ἐν λειμῶνι, Καϋστρίου ἀμφὶ ῥέεθρα,
ἔνθα καὶ ἔνθα ποτῶνται ἀγαλλόμενα πτερύγεσσι,
κλαγγηδὸν προκαθιζόντων, σμαραγεῖ δέ τε λειμών,
ὣς τῶν ἔθνεα πολλὰ νεῶν ἄπο καὶ κλισιάων
ἐς πεδίον προχέοντο Σκαμάνδριον....

Iliad 2.459–465

Vergil was evidently mindful of this irony, because soon after his reference to the other crane simile, he alludes to this one as well.[38] In listing the signs of rain, he begins with cranes, then moves on to the heifer, the swallow, frogs, the ant, the rainbow, and crows (375–382). Then more birds:

iam variae pelagi volucres et quae Asia circum
dulcibus in stagnis rimantur prata Caystri—
certatim largos umeris infundere rores,
nunc caput obiectare fretis, nunc currere in undas
et studio incassum videas gestire lavandi.

Georgics 1.383–387

This sequence of references amounts to a penetrating commentary on Homer. In the second of the two similes, to which Vergil alludes first, the screaming Trojans are likened to cranes, but differentiated from the silent Achaeans. But then Vergil reminds us of the earlier simile, in which the Achaeans too are compared to cranes.

The allusion to both these crane similes seems to pose a question: Are these bitter enemies really so different? The irony latent in his source must have appealed to Vergil, who was later to portray the feeble King Latinus gazing in wonder at two great warriors born at opposite ends of the earth preparing to come together in battle (*Aeneid* 12.707–709). By the time Aeneas and Turnus meet in single combat, the Italians in particular have made much of the supposed differences between themselves, the hardy westerners, and the effeminate orientals.[39] Latinus, without perhaps disparaging Aeneas personally, must have such differences in mind as he looks on the two heroes. But to the omniscient narrator, Aeneas and Turnus do not look so different at all; they are as similar as a pair of bulls fighting over control of a herd (12.715–724).

38. As is noted by Servius *ad G.* 1.383 (Thilo 1887.206–207).
39. Most overtly perhaps in the speech of Numanus Remulus (9.599–620).

The idea that enemies come inevitably to resemble one another occurs over and over again in Vergil;[40] but it is especially interesting to find it here in the "Weather Signs," serving as a kind of gloss on his program of Homeric allusion.

A final remark: These two Homeric similes are linked by the simple fact that they both involve cranes; but they bear at least as strong a structural relationship. The earlier simile is one of several that Homer uses to launch the "Catalogue of Ships" in *Iliad* 2. The latter, at the beginning of Book 3, closes out the catalogue of Trojan forces. In effect, they bracket the long passage in which Homer enumerates the forces that fought at Troy.[41] This will not be the last time that Vergil alludes to the great catalogue; nor is it the first time that he has bestowed a glimmer of heroic grandeur upon his humble georgic discourse through the integrative medium of the epic simile. In such a case, the device becomes much more than ornamental. Indeed, what are integrated here are more than just the parallel worlds of Heroic poetry, the primary world of the narrative and the illustrative world of the simile. Allusion here integrates the world of heroic epos with the didactic world delineated by the *Georgics;* and furthermore, this particular allusion shows the reader both worlds in a state of severe disruption. The approach of storm is implicitly likened to the approach of war, the equivalent terms in the respective worlds of Aratus and Homer.

Homer and Lucretius

From these discussions it seems clear that Vergil not only alludes to Homer in *Georgics* 1, but that he worked hard to incorporate Homeric material into predominantly Hesiodic and Aratean contexts. It is equally obvious that his plan was not so much to imitate Homer by incorporating lots of similes into his poem, but almost to disguise his use of specific Homeric similes by depriving them of their characteristic form and presenting them as straightforward narrative. We shall notice a related procedure as we examine Homeric elements in Lucretian contexts.

Knauer dismisses most of the alleged Homeric parallels in *Georgics* 1–3 on the grounds that they involve either Homeric formulae or else

40. One thinks especially of the "*Laudes Italiae,*" in which Italy and its inhabitants come more and more to resemble the orient with which the poet initially contrasts them: see Putnam 1975.172–176.

41. Cole 1979.82.

non-formulaic expressions and motifs that had by Vergil's time become clichés detached from their original context, and thus could not constitute meaningful allusions to specific passages of Homer. In this category Knauer places Vergil's use of the clausula *mortalibus aegris* | (1.237):

> This verse-ending has, of course, been compared with the Homeric δειλοῖσι βροτοῖσιν, which occurs 6 times in the 'Iliad' and 'Odyssey.' Already Ennius had rendered βροτός by *mortalis*,[42] but only in Lucretius can we confidently establish a Homeric assonance (6.1):
>
> *primae frugiparos fetus mortalibus aegris*
> *dididerunt quondam...Athenae*
>
> It is from this model that Vergil took over his line-ending in the 'Georgics,' not from Homer. Later we meet with it three times in the 'Aeneid'—the ending has become a Vergilian formula.[43]

The general principle that formulae and clichés are dicey material on which to hang an allusion is sound, but not absolute; the idea that the source of a given phrase must be a single passage of a single author is, as we have seen repeatedly, even more arguable. In any event, we must, as always, consider the alleged allusion in context. The passage in question is Vergil's "Climatic Zones":

idcirco certis dimensum partibus orbem
per duodena regit mundi sol aureus astra.
quinque tenent caelum zonae: quarum una corusco
semper sole rubens et torrida semper ab igni;
quam circum extremae dextra laevaque trahuntur 235
caeruleae, glacie concretae atque imbribus atris;
has inter mediamque duae **mortalibus aegris**
munere concessae divum, et via secta per ambas,
obliquus qua se signorum verteret ordo.
mundus, ut ad Scythiam Riphaeasque arduus arces 240
consurgit, premitur Libyae devexus in Austros.
hic vertix nobis semper sublimis; at illum
sub pedibus Styx atra videt Manesque profundi.
maximus hic flexu sinuoso elabitur Anguis
circum perque duas in morem fluminis Arctos, 245

42. The word occurs five times in the *Annales* (Skutsch 1985.820 s.v.), in Book 3 of the *Saturae* (fr. 6 Vahlen 1903.205), in the *Epicharmus* (fr. 7.5 Vahlen 1903.222), and in the tragedy *Achilles* (fr. 8 Jocelyn 1967.70). It had already been used by Naevius in the *Bellum Punicum* (fr. 6.1 Morel–Büchner 1982.23).

43. Knauer 1981b.906.

Arctos Oceani metuentis aequore tingi.
illic, ut perhibent, aut **intempesta** silet **nox**
semper et **obtenta** densentur **nocte** tenebrae;
aut redit a nobis Aurora diemque reducit,
nosque ubi primus equis Oriens adflavit anhelis 250
illic sera rubens accendit lumina Vesper.
hinc tempestates dubio praediscere caelo
possumus, hinc messisque diem tempusque serendi,
et quando infidum remis impellere marmor
conveniat, quando armatas deducere classis, 255
aut tempestivam silvis evertere pinum;
nec frustra signorum obitus speculamur et ortus
temporibusque parem diversis quattuor annum.

Georgics 1.231–258

The Lucretian background of this passage, which I discussed in the previous chapter, would seem to confirm Knauer's assertion that Lucretius is the immediate source of the clausula *mortalibus aegris*, which he in turn either found in a now lost passage of Ennius, or else modeled directly on the Homeric formula. A similar line of development can be postulated for the phrase *nox intempesta*. Ennius evidently modeled the phrase on similar Homeric noun-epithet combinations. Lucretius then imitated the Ennian formula, perhaps even unaware of its Homeric provenance. This is a plausible reconstruction of the route by which such phrases might have presented themselves to Vergil. We are, however, dealing with a poet obsessed with the literary past, and who expresses that interest in his allusions. If we consider his use of these phrases more carefully, the effort turns out to be well worthwhile.

At least one Ennian occurrence of *nox intempesta* is instructive in that it appears to have a specific model in the *Iliad:*

bellum aequis manibus **nox intempesta** diremit

Ennius *Annales* 160
(Skutsch 1985.85)

Ζεὺς δ' ἐπὶ ***νύκτ' ὀλοὴν*** *τάνυσε κρατερῇ ὑσμίνῃ*

Iliad 16.567

Here Ennius' *nox intempesta* ("originally 'timeless night,' i.e. the period of complete darkness when there is no obvious indication of time")[44] is

44. Skutsch 1985.193 *ad Ann.* 33.

not a particularly close translation of ὀλοός ("destructive, deadly").[45] Indeed, it is particularly inappropriate here for the simple reason that a battle can hardly be postponed by "the dead of night." What such an example shows, I think, is that *intempestus* serves Ennius as a "perpetual epithet" of *nox*.[46] This is a case of allusive variation in which Ennius substitutes his own formula for one of Homer's, imitating the Homeric style in general as well as the details of a particular passage.

At any rate, the Ennian formula somehow passed on to Lucretius and from Lucretius to Vergil, who uses it in a passage, quoted above, of marked allusion to *De Rerum Natura* 5. But is Vergil really thinking only of Lucretius? The phrase occurs only once in *De Rerum Natura* in a passage that refreshes the formula by means of *variatio:*

> atque **intempesta** cedebant **nocte** paventes
>
> *De Rerum Natura* 5.986

While Ennius is no doubt Lucretius' source for the phrase, the imitator shows independence in his prosody. With Vergil things are different. *Intempestus* is not Vergil's usual epithet for *nox;*[47] in fact, he so uses it only thrice.[48] Two of these three passages are in the *Aeneid;* the other is the *Georgics* passage under discussion. None of the *Aeneid* passages has anything to do with Lucretius. One of them, however, is clearly based on Ennius:

> et *lunam* in nimbo **nox intempesta tene**bat
>
> *Aeneid* 3.587
>
> quom *superum lumen* **nox intempesta tene**ret
>
> Ennius *Annales* 33
> (Skutsch 1985.73)

45. LSJ s.v.

46. Ennian nights are also "starry" (*hinc nox processit stellis ardentibus apta* 348 [Skutsch 1985.100], *nox quando mediis signis praecincta volabit* 414 [Skutsch 1985.107]; cf. Homeric ἀστερόεις, normally an epithet of οὐρανός; Homer never uses it to modify νύξ) and "clear" (*in nocte serena* 387 [Skutsch 1985.104 and 552 *ad loc.*, citing Cicero's use of the same clausula at *Aratea* 104 to render Aratus *Ph.* 323 καθαρῇ ἐνὶ νυκτί). Otherwise, *nox* is unmodified in the *Annales*.

47. Actually, there is no such thing. Vergil uses twenty different epithets that denote perpetual attributes of night, the most common of which are *atra* (six times), *sera*, and *umida* (five each). Four occur three times each, four twice, and nine once.

48. The epithet is used in only one other passage (*Aeneid* 10.184), where it modifies the town of Graviscae; for the meaning in this context see Servius *ad loc.* (Thilo 1884.955).

The other (*Aen.* 12.846) has the phrase in the same *sedes*. Similarly, at *Georgics* 1.247 *intempesta* occupies its normal Ennian *sedes*, while *nox* has been "moved" to final position, resulting in a hyper-Ennian line. A final monosyllable, so frequent in the surviving fragments of the *Annales*, is rare enough in the *Aeneid* and extremely rare in the *Georgics*.[49] In the *Aeneid*, this clausula often occurs in a context redolent of Ennius.[50] It therefore seems likely that at *Georgics* 1.247 Vergil partially altered the prosody of *nox intempesta* here to underline its Ennian character.[51]

The argument here is essentially parallel to what we observed in the case of the "Aetiology of *Labor.*" In that basically Hesiodic context is a line that closely resembles one of Hesiod, but even more closely resembles one of Homer. Here we have a basically Lucretian context, the "Climatic Zones," in which is found a phrase used by Lucretius, but in a form that is much closer to an occurrence of the same phrase in Ennius. In respect of the "Aetiology" I suggested that Vergil employs double allusion to identify a Hesiodic borrowing from Homer. In this passage, the inference that he uses a particular form of the phrase *nox intempesta* to identify Ennius as the source of his immediate model, Lucretius, is even more readily made.

What then of Ennius' source? The argument that we are testing is that, if we can identify a more immediate source than Homer for such epic formulae as *mortalibus aegris* and *nox intempesta*, then we cannot suppose that Vergil is alluding to Homer when he uses these phrases. But what do we then make of a line such as this:

> ἀλλ' ἐπὶ νὺξ ὀλοὴ τέταται δειλοῖσι βροτοῖσιν
>
> *Odyssey* 11.19

It is obvious that δειλοῖσι βροτοῖσιν is the model of Vergil's *mortalibus aegris*, and we have seen that νὺξ ὀλοή lies behind Ennius' *nox intempesta*.

49. Eight percent in the *Annales* as compared with 0.7% in the *Aeneid*, according to Skutsch 1985.49; the figure for the *Georgics* is about 0.2% (five examples in 2,187 lines). This prosody is more common in Lucretius (2.6% according to Bailey 1947.1.115) than in Vergil, but not so common as in Ennius, with whom the mannerism is normally identified.

50. E.g. the formula *divum pater atque hominum rex*| (*Aen.* 1.65, 2.648, 10.2, 743, *Ann.* 203 [Skutsch 1985.88]; cf. *Aen.* 3.375, 12.851, *Ann.* 591 [Skutsch 1986.126]). For other sources and effects of this rhythm, see Norden 1957.440–441; Austin 1955.61–62 *ad Aen.* 4.132.

51. This point is made by Thomas 1986.180–181, who adds that the iambus *silet* preceding *nox* contributes to the Ennian rhythm of the line.

When both of these Latin phrases occur in a single Vergilian passage *and* both of their Greek analogues in a single line of Homer, does it not follow that Vergil may have the Homeric verse in mind? In fact, Vergil's diction here resembles Homer's rather closely. The repetition *nox* | ...*nocte* (247–248) is a signal; for in the second element *obtenta*...*nocte* we find an allusion to the Homeric ἐπί νὺξ...τέταται. The very same diction occurs in the passage with which we began, Ζεὺς δ' **ἐπὶ νύκτ' ὀλοὴν τάνυσε** κρατερῇ ὑσμίνῃ (*Iliad* 16.567), the passage from which Ennius had first "translated" νὺξ ὀλοή as *nox intempesta*.

It is of course thoroughly characteristic of the Alexandrian poets to allude to Homer in this way, with chalcenteric attention to lexical detail, and often without concern for the original context. Here, however, we can observe Vergil's usual sensitivity to contextual propriety as well. If we consider the passage in which this Homeric line occurs, we find it apposite to Vergil's theme:

> ἡ [sc. νηῦς] δ' ἐς πείραθ' ἵκανε βαθυρρόου 'Ωκεανοῖο
> ἔνθα δὲ Κιμμερίων ἀνδρῶν δῆμός τε πόλις τε,
> ἠέρι καὶ νεφέλῃ κεκαλυμμένοι· οὐδέ ποτ' αὐτοὺς
> 'Ηέλιος φαέθων καταδέρκεται ἀκτίνεσσιν,
> οὔθ' ὁπότ' ἂν στείχῃσι πρὸς οὔρανον ἀστερόεντα,
> οὔθ' ὅτ' ἂν ἂψ ἐπὶ γαῖαν ἀπ' ουρανόθεν προτράπηται,
> ἀλλ' ἐπὶ νὺξ ὀλοὴ τέταται δειλοῖσι βροτοῖσιν.
>
> *Odyssey* 11.13–19

The Cimmerians, who inhabit a land on which the sun never shines, are a close Homeric equivalent of those who dwell at the antipodes of an Eratosthenic globe. It is in their country, of course, that Odysseus' "*Nekyia*" takes place; not by accident, therefore, does Vergil note of the two poles of his globe:

> hic vertix nobis semper sublimis; at illum
> sub pedibus Styx atra videt Manesque profundi.
>
> *Georgics* 1.242–243

Homer's Cimmerians live in proximity to the dead ψυχαί in perpetual gloom; the inhabitants of Vergil's southern hemisphere live below the Styx and the Manes of the deceased, either in gloom or in a night and day that alternate with those of the northern hemisphere. Specific terms have been altered to reflect a world view that embraces both the traditional beliefs of Roman religion and the calculations of Hellenistic science, but the allusion remains unmistakable.

With this discovery, the purpose of Vergil's allusion becomes clear. On the one hand, he is in a sense tracing the history of Greek and Latin epic through specific instances of typical language, from Homer to Ennius to Lucretius; on the other, having already used similar means to bridge the gap between Homer, Hesiod, and Aratus, he is now drawing Homer into another portion of his allusive program, using allusion to establish common ground between Homer, Lucretius, and Eratosthenes.

Other elements in this passage contribute to the same end. For instance, the lines

> maximus hic flexu sinuoso elabitur Anguis
> circum perque duas in morem fluminis Arctos,
> **Arctos Oceani metuentis aequore tingi.**
>
> *Georgics* 1.244–246

conclude with an obvious allusion to Homer:[52]

> *ἐν μὲν γαῖαν ἔτευξ᾽, ἐν δ᾽ οὐρανόν, ἐν δὲ θάλασσαν,*
> *ἠέλιόν τ᾽ ἀκάμαντα σελήνην τε πλήθουσαν,*
> *ἐν δὲ τὰ τείρεα πάντα τά τ᾽ οὐρανὸς ἐστεφάνωται,*
> *Πληϊάδας θ᾽ Ὑάδας τε τό τε σθένος Ὠρίωνος*
> **Ἄρκτον** *θ᾽, ἣν καὶ Ἄμαξαν ἐπίκλησιν καλέουσιν,*
> *ἥ τ᾽ αὐτοῦ στρέφεται καί τ᾽ Ὠρίωνα δοκεύει,*
> **οἴη δ᾽ ἄμμορός ἐστι λοετρῶν Ὠκεανοῖο.**[53]
>
> *Iliad* 18.483–489

Again repetition serves a referential function; for, just a hundred lines earlier, Vergil had alluded to this same passage in a way that drew attention to the constellation Arctus in particular:

52. Acknowledged by Conington–Nettleship 1898.195 *ad* 246, Jahn 1903a.416, Thomas 1986.197.

53. Cf. *Odyssey* 5.275 (quoted on p. 215 n. 19). The parallelism is noted by Cerda 1647.226.2e–f *ad* 16 (= *G.* 1.246); Jahn 1903a.416.

> tunc alnos primum fluvii sensere cavatas;
> navita tum stellis numeros et nomina fecit
> **Pleïadas, Hyadas,** claramque Lycaonis **Arcton;**
> tum laqueis captare feras et fallere visco
> inventum et magnos canibus circumdare saltus.
>
> *Georgics* 1.136–140

We have seen this sort of allusive cross-reference in the case of Hesiod and Lucretius. We now find an example that calls our attention to the presence of Homeric allusion in predominantly Hesiodic and Lucretian contexts.

In fact, Homeric allusions of this kind are rather abundant in *Georgics* 1–3. Two more examples will suffice. In Book 2, Vergil trots out another, obvious homerism:

> non ego cuncta meis amplecti versibus opto
> non, mihi si linguae centum sint, oraque centum,
> ferrea vox....
>
> *Georgics* 2.42–44

The history of this motif has been thoroughly charted.[54] Homer, of course, is the originator; it later appears in Ennius, Hostius, and, on Servius' authority, Lucretius.[55] There are, of course, limits to our understanding. We know nothing of the context in which Lucretius might have used this motif;[56] nor are we much better off with Hostius[57] or

54. Weinreich 1916–1919.172–173, Courcelle 1955, Pascucci 1959, Skutsch 1985. 627–629; with reference to our passage, Knauer 1981b.905–906.

55. The passages in question are *Iliad* 2.488–493, *Annales* 469–470 (Skutsch 1985. 114), Hostius fr. 3 (Morel–Büchner 1982.45), Lucretius fr. 1 (Martin 1963.281; attributed to Lucretius by Servius *ad G.* 2.42 [Thilo 1887.221] ≈ *Aen.* 6.625 [Thilo 1884.88]).

56. Lachmann inserted the fragment after *DRN* 6.839 (arguments: 1850.2.398–399). It has more usually been associated with the invocation of Calliope at 6.92–95 (see Boeck 1958.243–246), but a more recent argument (Giancotti 1976) places it in the lacuna that precedes 2.165. Some scholars question the attribution, citing the frequent scribal confusion of *lucretius* and *lucilius;* see Lucilius fr. dub. 7 Mueller 1872.159, 287 = 1364–1365 Marx 1904.93 and 1905.433–434=1383–1384 Krenkel 1970.726–727; cf. Regel 1907.81–82, Wigodsky 1972.98–99, Zetzel 1977.41 n. 8.

57. We know virtually nothing about the poem or its author. Macrobius (*Sat.* 6.3.6) states that Hostius borrowed the motif from *Iliad* 2.489 and used it in the second book of his *Bellum Histricum,* and it is a reasonable, but unprovable, inference that he followed Homer by outlining the origin of the conflict in Book 1 and mustering the forces in Book

even Ennius.[58] In both these cases, however, we can at least be sure that we are dealing with a historical epic clad in the conventional finery established on the Homeric model by Ennius. Homer, of course, employs the motif to open the famous "Catalogue of Ships," a passage to which Vergil alludes elsewhere in the *Georgics*, as we have seen. In the *Aeneid* it appears as a closure motif—not for Vergil's imitation of the Homeric "Catalogue," however, nor in the narrator's persona, but to enable the Cumaean Sibyl to terminate her description of the torments suffered by those in Tartarus (6.625–627). Obviously, therefore, this conceit is primarily concerned with scope: the speaker would need many voices, tongues, or what have you to encompass the sheer numbers in question. But it is equally clear that the motif was felt to be at home in serious, august, even rather frightening contexts: the onset of the Greek host, or the punishment of the damned. These are topics that would require bronze or iron equipment, and lots of it, to be described properly.

With this in mind, we turn to *Georgics* 2.42–44. Here Vergil strains to outdo his predecessors: Homer had spoken of ten tongues, ten mouths, an unbreakable voice, and a heart of bronze. Ennius, too, would have needed ten mouths and tongues, but would also have wanted his *cor* and *pectus* plated with iron. Hostius raised the ante to a hundred, but said nothing that we know of about armor. Lucretius, we may infer, also needed a hundred mouths, but specified, according to Servius, that they be of bronze. Vergil, then, shows some restraint by not asking for a thousand tongues of adamant; but he takes the highest number yet available, Hostius' and Lucretius' hundred, and combines it with the stronger metal, Ennius' iron. This is the mighty vocal apparatus he would need if he were to catalogue completely the known methods of—tree cultivation!

This passage must surely be one of the most broadly humorous in the *Georgics*. Indeed, it is hard to imagine that Vergil didn't smile when he borrowed these lines intact to play a much more conventionally heroic role in the *Aeneid*. Even if we agree that the motif had become detached from any specific context, there is no indication that it ever appeared in such an incongruous guise as this. But, when we regard it in

---

2. Bardon (1952.1.179) points out that there were probably at least three books, or testimonia such as Servius *ad Aen.* 12.121 (= fr. 1 Morel–Büchner 1982.44) would have spoken of a *liber prior* rather than a *liber primus*.

58. Skutsch (1985.627–628) includes it among the *sedis incertae fragmenta* and discusses its possible context.

light of all the other Homeric allusions that we have discussed, a familiar pattern appears. The implied comparison between the Homeric and Vergilian catalogues is merely a more humorous example of what we found in the allusion to the Xanthus simile: a means of measuring the gap between the heroic and the georgic world, between Homeric and Vergilian epos. At the same time, both passages do in fact bring Homer into the georgic world, into Vergilian epos. Indeed, this allusion does so more frankly than any yet discussed. Finally, the tradition here outlined—Homer, Ennius, and Lucretius—is also familiar, and is again made to enrich a poem that presents itself as the product of another tradition, as an *Ascraeum carmen:* a learned imitation of Hesiod in the Alexandrian and Neoteric manner.

Our final example from the first three books of *Georgics* concerns the chariot race of Book 3:

> **non**ne vides, cu**m** **praecipiti certamine campum**
> **corripuere, ruunt**que effusi carcere currus,
> **cum spes adrectae iuvenum, exsultantiaque haurit**
> **corda pavor pulsans?** illi instant verbere torto
> et proni **dant lora,** volat vi fervidus axis;
> iamque humiles iamque elati sublime videntur
> aëra per vacuum ferri atque adsurgere in auras.
> **nec mora nec requies;** at fulvae nimbus harenae
> tollitur, umescunt spumis flatuque sequentum:
> tantus amor laudum, tantae est victoria curae.
>
> *Georgics* 3.103–112

This passage has long been felt to imitate the chariot race episode in Homer's "Funeral Games of Patroclus."[59] Certainly it contributed to the more ambitious imitation of the same Homeric passage, the boat race episode in Vergil's "Funeral Games of Anchises":

> considunt transtris, intentaque bracchia remis;
> intenti exspectant signum, **exsultantiaque haurit**
> **corda pavor pulsans** laudumque arrecta cupido
> inde ubi clara dedit sonitum tuba, finibus omnes,
> haud mora, prosiluere suis; ferit aethera clamor
> nauticus, adductis spumant freta versa lacertis.

59. The parallelisms between *Georgics* 3.108–109 ≈ *Iliad* 23.368–369 and 3.111 ≈ 23.380–381 are noted by Macrobius *Sat.* 5.13.7 and 5.13.3 respectively. Richter 1957.277 adds *Iliad* 23.500–508.

infindunt pariter sulcos, totumque dehiscit
convulsum remis rostrisque tridentibus aequor.
non ta**m** **praecipit**es biiugo **certamine campum**
corripuere ruunt effusi carcere currus,
nec sic immissis aurigae un**danti**a **lora**
concussere iugis pronique in verbera pendent.
tum plausu fremituque virum studiisque faventum
consonat omne nemus, vocemque inclusa volutant
litora, pulsati colles clamore resultant.

Aeneid 5.136–150

This second version borrows language from the *Georgics* passage and, what is more, explicitly compares the boat race to the chariot races described in the *Georgics* and the *Iliad.* These facts would seem to guarantee that Vergil was alluding to Homer in the earlier version as well as the later. Knauer, however, sees it differently. "That Vergil in the 'Georgics' was thinking of the 'Iliad' cannot, in the light of other cases discussed so far, be supported by the thorough imitation of the ἆθλα in the games in honour of the deceased Anchises...."[60] His reasoning, again, is that both Vergilian passages incorporate a wealth of Lucretian and Ennian diction:

nonne vides etiam patefactis tempore puncto
carceribus non posse tamen prorumpere equorum
vim cupidam tam de subito quam mens avet ipsa?

De Rerum Natura 2.263–265

nonne vides...

Georgics 3.103

quom a **carcer**e **fusi**
currus cum sonitu magno permittere certant

Annales 463–464
(Skutsch 1985.113)

expectant veluti consul cum mittere signum
volt, omnes avidi spectant ad **carcer**is oras
quam mox emittat pictos e faucibus **currus**

Annales 79–81
(Skutsch 1985.77)

praecipiti certamine campum
corripuere, ruuntque ef**fusi** carcere **currus**

Georgics 3.103–104

60. Knauer 1981b.909.

hic **exsultat** enim **pavor** ac metus...

De Rerum Natura 3.141

cum spes adrectae iuvenum, **exsultant**iaque haurit
corda **pavor** pulsans?

Georgics 3.105–106

nec mora nec requies interdatur ulla fluendi

De Rerum Natura 4.227 = 6.931

nec mora nec requies...

Georgics 3.110

Knauer holds that the *Georgics* passage was assembled out of various Ennian and Lucretian treatments of the same general theme, and that Vergil only later had the clever idea of incorporating the result into his boat race episode, itself an imitation of a Homeric horse race. He further notes that "the ἄθλα of the 'Iliad' are in themselves typical epic building material anyway,"[61] meaning that even if Homer must be regarded as the ultimate source of the Ennian and Lucretian—and therefore the Vergilian—contests, horse race episodes on the Homeric model had become so common that no specific reference was possible outside the elaborate structural correspondences present in *Aeneid* 5. Of course, nowhere in the *Georgics* outside the "Aristaeus" do we find structural imitation of Homer on any significant scale. We have, however, repeatedly found it possible to establish the presence of allusion through a detailed comparison of versecraft; and this procedure is sufficient to confirm the relevance of Homer here as well. Though his ostensible subject is the spirit of a champion racehorse, Vergil follows Homer in focusing on the emotions of the driver (*cum spes adrectae iuvenum, exsultantiaque haurit | corda pavor pulsans?* 105–106 ≈ τοὶ δ' ἐλατῆρες | ἕστασαν ἐν δίφροισι, πάτασσε δὲ θυμὸς ἑκάστου | νίκης ἱεμένων *Il.* 23.369–371). This focus of course lends weight to one of the key themes of *Georgics* 3, the essential similarity between animal and human. The details of the race too are the same. As the teams cover the course (*cum praecipiti certamine campum | corripuere* 103–104 ≈ οἳ δ' ὦκα διέπρησσον πεδίοιο *Il.* 23.364), and the drivers go repeatedly to the whip (*illi instant verbere torto | et proni dant lora* 106–107 ≈ οἱ δ' ἅμα πάντες ἐφ' ἵπποιιν μάστιγας ἄειραν, | πέπληγόν θ' ἱμᾶσιν *Il.* 23.362–363), both poets show us a

61. Knauer 1981b.909.

cloud of dust mixed with the hot breath of laboring steeds (*at fulvae nimbus harenae* | *tollitur, umescunt spumis flatuque sequentum* 110–111 ≈ ὑπὸ δὲ στέρνοισι κονίη | ἵστατ᾽ ἀειρομένη ὥς τε νέφος ἠὲ θύελλα | χαῖται δ᾽ ἐρρώοντο μετὰ πνοιῇς ἀνέμοιο *Il.* 23.365–367), the chariots rising and falling as they speed over the course (*iamque humiles iamque elati sublime videntur* | *aëra per vacuum ferri atque adsurgere in auras* 108–109 ≈ ἅρματα δ᾽ ἄλλοτε μὲν χθονὶ πίλνατο πουλυβοτείρῃ, | ἄλλοτε δ᾽ ἀΐξασκε μετήορα *Il.* 23.368–369). The fact that descriptions of such scenes were commonplace in ancient poetry cannot obscure the typically Vergilian procedure of identifying the *Urtext* and integrating it into his own discourse.

The question that remains is one of interpretation. As in the previous example, which derives ultimately from the "Catalogue of Ships," we find here a motif that enjoyed an extensive life of its own between the times of Homer and Vergil.[62] Again in this case Vergil quotes Ennius and Lucretius as representatives of the tradition. This is the inverse of the situation that we found with respect to Hesiod and Aratus. When borrowing passages from those authors, Vergil quietly and almost surreptitiously weaves into them imitations of compatible Homeric passages. The technique is comparable to those employed by Hellenistic poets when alluding to Homer. Here he draws much more frankly on a fund of heroic language and motifs developed by Latin poets in open imitation of Homer.[63] While the artistry of Vergil's allusive style and the importance of the didactic tradition are equally evident in both cases, the two types of imitation clearly move in opposite directions. On the one hand we have Hesiod and Aratus, who represent a distinct alternative to the grandiose heroism of Homeric epos, whose didactic mode is guaranteed as epic by its similarities with Homer, but defined on its own terms by its equally important differences. On the other there is Lucretius, who conceived of didactic epos on a grand scale, borrowing the epic style of Homer and Ennius while rejecting their eschatology.[64] The allusive structure of the *Georgics* suggests that we see in this diverse treatment of sources not a dichotomy,

62. Perhaps the best known imitation of Homer's horse race is Sophocles *Electra* 709–760, parts of which may have influenced Vergil; see Thomas 1988.2.59 *ad* 110–111, who also (*ad* 111) places Callimachus' *Victoria Berenices* from *Aetia* 3 (see Parsons 1977.9 *ad* 16ff., Lloyd-Jones and Parsons 1983.100–117, frr. 254–269) in this tradition.

63. This point is made by Erdmann 1912.15.

64. See *De Rerum Natura* 1.102–135. I will have more to say on this topic in Chapter 7.

but a progression. Book 1 is dominated by an ambitious Hesiodic and Aratean program, a program into which Homer is and can be integrated only in the most surprising ways. Books 2 and 3 derive from Lucretius, and here an open spirit of Homeric imitation is much more recognizable. Book 4, of course, is primarily concerned with Homer himself. It is difficult to avoid the conclusion that Vergil's arrangement of sources is deliberate, and that its purpose is to integrate Homer into the didactic branch of the epic tradition; or, to put it another way, to show how even that branch of the tradition derives, ultimately, from Homer. Such a program would of course have the ulterior purpose of presenting the Vergil's own poem as a triumphant summary of the entire tradition from its beginning down to the poet's own day. Only a thorough analysis of Book 4 can prove or disprove this interpretation, which still leaves several important questions unanswered; but it seems at this point the most natural working hypothesis, and, as we shall see, further evidence will bear it out.

In conclusion, three points. First, I hope it is now obvious that the evidence indicates a definite Homeric presence in *Georgics* 1–3. The allusions are not as extensive or important, perhaps, as the Hesiodic, Aratean, and Lucretian imitations that we have discussed; but this is no doubt part of the point. The treatment of Homeric material in Books 1–3 is learned, delicate, and tactful, as the ancient heroic epos is assimilated to the purposes of modern didactic epos. Second, the transformation is effected by pointed reference to passages in which Homer's poetry in some measure resembles that of Vergil's didactic predecessors, or in which these predecessors themselves made use of Homeric material, drawing Homer into the generic ambit of didactic epos. Third, occasional allusion to Homer in the earlier books, by establishing a Homeric presence in the *Georgics*, prepares the reader for the poem's much more ambitiously Homeric conclusion. It remains to be seen what Homeric elements inform the didactic portion of Book 4, the account of apiculture, and how appropriate a conclusion to this didactic poem the heroic story of Aristaeus actually is.

Georgics 4

Book 4 falls clearly into two parts, the "technical" discourse on bees and the aetiological epyllion that is the imaginative conclusion of both book and poem. These parts are unified, however, by their reliance on

Homer as a poetic model. Each of the sections approaches Homer in a different way; yet each is recognizably a part of the allusive program that I have traced in Books 1–3.

The Bees (Georgics *4.8–314)*

With his account of apiculture in Book 4, Vergil's meditation on human and, more specifically, Roman civilization through the lives of animals becomes more explicit than ever. In Book 1, human *labor* is opposed to nature, locked in an unremitting battle for survival; but the same forces that indicate disruption of the natural world through storm also attend the destruction of the political world through civil war. In Book 2, humankind is viewed largely as a product of the land: various peoples differ from one another as do the climates and produce of different regions, with Italy and Rome representing, if only ironically, the ideal. The animals of Book 3 suffer the same afflictions of love and disease as do their human counterparts, and in the pandemic fury and suffering that these afflictions cause, the reader begins to sense that the georgic world resembles his own by possessing a social dimension as well. With the bees of Book 4, this line of development culminates in a lengthy account of the social fabric of the hive and of a way of life that includes harvesting, grazing, civil war, and many of civilization's discontents.[65]

The didactic element in Vergil's account of the bee state draws on several prose traditions extrinsic to our line of inquiry.[66] Among didactic poems, it may have owed something to Nicander's lost *Melissurgica*.[67] But in describing the bee state, Vergil again and again imparts a heroic color to his verse, and in doing so he regularly harks back to Homer. The essential characteristics of this heroic—or better, perhaps, mock-heroic[68]—account of the hive are those that we have encountered in the

65. These parallels between the societies of bees and men are ironically reinforced by the interlude of the Corycian gardener (*G.* 4.116–148); see Perkell 1981.

66. The apicultural lore is drawn primarily from Varro; see Jahn 1904. For the description of the bee state according to the conventions of ancient ethnography see Dahlmann 1954 and Thomas 1982b.

67. This is however pure speculation. See Gow–Scholfield 1953.215.

68. I do not hold with those who see the bee discourse as a lighthearted or humorous *jeu d'esprit*, like the *Batrachomyomachia;* nevertheless, I do not see how we can deny that Vergil was writing with the conventions of such poetry in mind. Most obvious is the application of conventionally heroic (i.e. Homeric) diction to the deeds of these tiny beasts in both poems; although Vergil stops well short of the bathos that pervades the typical scenes

early books of *Georgics:* delicacy, *doctrina*, inversion, and reduction. Yet the generally homerizing quality of the entire episode, and particularly of the passages describing warfare among the bees, is allowed to show through much more clearly than before.

A clear idea of how Vergil's Homeric program develops can be had by considering his growing use of similes. Knauer has remarked on the large number of similes in this part of the poem: six, as compared with seven in Books 1–3 and only four in the much more frankly Homeric "Aristaeus."[69] This fact alone indicates that Vergil has adopted a much more conventionally heroic style in this episode. But the manner in which these similes are deployed is still more telling. Although the poet has, as Knauer observes, generally avoided taking over Homeric similes in their entirety,[70] he on occasion imitates with care the style and purpose of Homeric similes. Hermann Fränkel has clearly demonstrated a consistency in the relationships between tenor and vehicle in different classes of Homer's similes,[71] and his observations can be applied to several of Vergil's similes here as well. The first and last similes of the bee discourse are cases in point. Lines 67–87 discuss the phenomenon of swarming, and their technical source is Varro (*DRR* 3.16.29–31), who actually inspires some of Vergil's imagery here; but Vergil departs from Varro in characterizing the swarm as a battle (*pugnam* 67). No commentator ever fails to note the political and martial language of the passage: *discordia* (68) has arisen between two kings,[72] and has resulted in war (*bello* 69); the bees produce a *Martius...canor* (70) not unlike the sound of trumpets (71),[73] they sharpen their weapons and don their armor (74), gather around the general's headquarters (*praetoria* 75), and call the enemy (*hostem* 76) to battle. The conflict takes place in spring—i.e. when the season for campaigning begins again—on a battlefield (*campos* 77). As they burst from the gates and join battle, the war cry sounds in the heavens (*aethere in alto* | *fit sonitus* 78–79). Only after all this battle

---

of the *Batrachomyomachia* (e.g. the heroic speeches of Physignathus and Psicharpax and the other combatants, or the *concilium deorum* of 168–198). If he knew the *Batrachomyomachia* as a Homeric poem, then we must consider even this fact in evaluating his Homeric program.

69. Knauer 1981.895.
70. Knauer 1981.900.
71. Fränkel 1921.16–25.
72. Cf. Varro *DRR* 3.16.18 on *seditiones.*
73. Cf. Varro *DRR* 3.16.9 *ut imitatione tubae.*

imagery does Vergil's language return to a level appropriate to the technical subject: the bees tangle themselves up in a big ball and fall to the ground (79–80). It is this last point that becomes the tenor of the first simile in Book 4:

> magnum mixtae glomerantur in orbem
> praecipitesque cadunt; non densior aëre grando,
> nec de concussa tantum pluit ilice glandis.
>
> *Georgics* 4.79–81

The final simile of Vergil's bee discourse illustrates the manner in which new bees burst forth from the carcass of a bullock:

> interea teneris tepefactus in ossibus umor
> aestuat, et visenda modis animalia miris,
> trunca pedum primo, mox et stridentia pennis,
> miscentur, tenuemque magis magis aëra carpunt,
> donec ut aestivis effusus nubibus imber
> erupere, aut ut nervo pulsante sagittae,
> prima leves ineunt si quando proelia Parthi.
>
> *Georgics* 4.308–314

Homeric warriors, of course, are not compared to hailstones, acorns, raindrops, or arrows, but the martial character of such weather similes is quite consistent. Vergil is clearly using these similes in a manner similar to that of Homer, just as he was to do later in the *Aeneid*.[74] We may in a more general way compare these passages with the first Homeric allusion of Book 1. There Vergil uses Achilles' battle with the river Xanthus to describe the irrigation of fields. The allusion involves a specific Homeric passage, which is stripped of its characteristic form and is inverted with respect to its illustrative function. Here Vergil alludes to no single Homeric passage in particular, but rather fashions a simile of Homeric appearance out of generically heroic language and material, and returns to the Homeric practice of describing warfare in terms of nature. This procedure encourages a reading of Vergil's bee lore as an essay in the conventions of heroic epos in the Homeric mold.

Other Vergilian similes seem rather inconsistent with the tone of heroic epos, and therefore un-Homeric; but appearances can deceive. In

74. Snow, rain, or hail in Homer usually illustrate a thick volley of weapons (e.g. *Iliad* 12.156–158, 278–286); cf. *Aen.* 9.668–671.

describing what he considers the inferior species, Vergil versifies a passage of Varro on signs of sickness in bees:

> minus valentium signa, si sunt pilosae et horridae, ut pulverulentae, nisi opificii eas urget tempus.
>
> *DRR* 3.16.20

> namque aliae turpis horrent ceu pulvere ab alto
> cum venit et sicco terram spuit ore viator
> aridus.
>
> *Georgics* 4.96–98

Varro's casual comparison may have inspired Vergil's more developed simile. Some have felt that Callimachus may also be on Vergil's mind:[75]

> *μηδ' ὄκ' ἀφ' αὐαλέων στομάτων πτύωμες ἄπαστοι*
>
> *Hymn* 6.6

Neither of these sources resembles Homer either in tone or subject. The latter, however, is Homeric in a peculiarly Alexandrian way. Callimachus' *πτύωμες* is a learned allusion to the language of Homeric epos—a *hapax legomenon.*[76] This fact was not lost on Vergil: the word he uses to translate *πτύωμες*, *spuit*, is cognate with it, begins at the same point in its line and, most important, is itself a Vergilian *unicum*. It is of course impossible to prove that Vergil, in composing the *Aeneid*, deliberately avoided the word so as to maintain its uniqueness in his poetry; but it looks very much as though this were the case. The episode in which the Homeric *hapax* occurs is the boxing match of *Iliad* 23, where Epeius has just dealt Euryalus a knockout blow:

> *φίλοι δ' ἀμφέσταν ἑταῖροι,*
> *οἵ μιν ἄγον δι' ἀγῶνος ἐφελκομένοισι ποδέσσιν*
> *αἷμα παχὺ* **πτύοντα,** *κάρη βάλλονθ' ἑτέρωσε·*
> *κὰδ δ' ἀλλοφρονέοντα μετὰ σφίσιν εἷσαν ἄγοντες,*
> *αὐτοὶ δ' οἰχόμενοι κόμισαν δέπας ἀμφικύπελλον.*
>
> *Iliad* 23.695–699

75. First advanced by Richter (1957.342); cautiously accepted by Knauer (1981.900) and Thomas (1988.164).

76. See Hopkinson 1984.84 *ad loc.* The verb *πτύω* is not uncommon outside of Homer; but it is clear that Greek scholars were interested not only in Homeric rarities that occurred nowhere else, but also in more common words used by Homer only once, and even in words used frequently in one of the Homeric epics, but only once in the other. See, in general, Kumpf 1974.1–19.

Vergil of course imitates this passage quite closely in the boxing match of *Aeneid* 5:

> ast illum fidi aequales genua aegra trahentem
> iactantemque utroque caput crassumque cruorem
> ore **eiectantem** mixtosque in sanguine dentes
> ducunt ad navis; galeamque ensemque vocati
> accipiunt, palmam Entello taurumque relinquunt.
>
> *Aeneid* 5.468–472

Here *eiectantem* renders the corresponding Greek participle πτύοντα. Both words end at the penthemimeres. Vergil may have had other reasons for translating the Homeric *hapax* with a word that occurs five times in the *Aeneid*.[77] It seems to me likely that one reason was that he had already translated the word qua *hapax* in the *Georgics*, using its Latin cognate *spuo*, which he wished to preserve as an *unicum* in his own poetry.[78]

A similar allusive subtlety appears in the next example. The famous simile of the Cyclopes in their forge (4.170–175) is not borrowed directly from Homer, but is Homeric in form and purpose. The inspiration for this vignette of the toiling Cyclopes must be a similar scene in Callimachus' *Hymn to Artemis* in which the goddess visits the Cyclopean forge to commission arms. She finds them hard at work:

> αὖθι δὲ Κύκλωπας μετεκίαθε· τοὺς μὲν ἔτετμε
> νήσῳ ἐνὶ Λιπάρῃ (Λιπάρη νέον, ἀλλὰ τότ᾽ ἔσκεν
> οὔνομά οἱ Μελιγουνίς) ἐπ᾽ ἄκμοσιν Ἡφαίστοιο
> ἑσταότας περὶ μύδρον· ἐπείγετο γὰρ μέγα ἔργον·
> ἱππείην τετύκοντο Ποσειδάωνι ποτίστρην.

77. The tone of the word *spuo* probably disqualifies it for the higher genres, except in special circumstances. L&S and *OLD* s.v. cite occurrences in Plautus, Varro *Men.*, Propertius, and Martial, and in the medical tradition Lucretius *DRN* 6.1188 (translating Hippocrates *Progn.* 2.145 πτύελον), Celsus 2.8.24, 8.9.1c, Pliny *NH* 17.9.6.52, 28.4.7.35, 28.36. It is not found in Cicero, but occurs in Petronius (131.4), Seneca (*Const. Sap.* 1.3), Solinus (1.74), and the Vulgate Bible (*Num.* 12.14).

78. For a similar instance of lexical imitation see Wills 1987. A general study of this phenomenon in Hellenistic poetry is needed. At present, valuable material can be found in commentaries and other specialized studies. On *hapax legomena* in general see Kumpf 1974. On their use by Apollonius see Kassegger 1967, Giangrande 1967, Garson 1972; on Callimachus von Jan 1893, Hopkinson 1984.200 s.v "Homer, *hapax* and *dis legomena*," Bulloch 1985.28; on Theocritus Chryssafis 1981.287 s.v. "Word-choice." In some cases, of course, the lexical borrowing signals a contextual allusion; see, e.g., Bulloch 1985.29.

αἱ νύμφαι δ' ἔδδεισαν, ὅπως ἴδον αἰνὰ πέλωρα
πρηόσιν Ὀσσαίοισιν ἐοικότα (πᾶσι δ' ὑπ' ὀφρύν
φάεα μουνόγληνα σάκει ἴσα τετραβοείῳ
δεινὸν ὑπογλαύσσοντα) καὶ ὁππότε δοῦπον ἄκουσαν
ἄκμονος ἠχήσαντος ἐπὶ μέγα πουλύ τ' ἄημα
φυσάων αὐτῶν τε βαρὺν στόνον· αὖε γαρ Αἴτνη,
αὖε δὲ Τρινακρίη Σικανῶν ἕδος, αὖε δὲ γείτων
Ἰταλίη, μεγάλην δὲ βοὴν ἐπὶ Κύρνος ἀΰτει,
εὖθ' οἵγε ῥαιστῆρας ἀειράμενοι ὑπὲρ ὤμων
ἢ χαλκὸν ζείοντα καμινόθεν ἠὲ σίδηρον
ἀμβολαδὶς τετύποντες ἐπὶ μέγα μυχθίσσειαν.

Callimachus *Hymn* 3.46–61

Within this Callimachean context, however, lurks a direct allusion to Homer. When Vergil's Cyclopes dip the screeching bronze in the basin, the action recalls a Homeric simile. Odysseus and his men blind the Cyclops Polyphemus by plunging a sharpened, heated stake into his eye. When they do so, the eye sizzles upon the stake, making a sound like that of a hot axe head dipped in a basin for tempering:

πάντα δέ οἱ βλέφαρ' ἀμφὶ καὶ ὀφρύας εὖσεν ἀϋτμὴ
γλήνης καιομένης· σφαραγεῦντο δέ οἱ πυρὶ ῥίζαι.
ὡς δ' ὅτ' ἀνὴρ χαλκεὺς πέλεκυν μέγαν ἠὲ σκέπαρνον
εἰν ὕδατι ψυχρῷ βάπτῃ μεγάλα ἰάχοντα
φαρμάσσων· *τὸ γὰρ αὖτε σιδήρου γε κράτος ἐστίν·*
ὣς τοῦ σίζ' ὀφθαλμὸς ἐλαϊνέῳ περὶ μοχλῷ.

Odyssey 9.389–394

Vergil carefully alludes to the element that Callimachus borrows from Homer (*alii stridentia tingunt* | *aera lacu* 172–173). By doing so, he discloses the literary history of the motif. In Homer the blacksmith's activity is part of a simile that illustrates the blinding of Polyphemus—the typical pattern, in which the homely and familiar glosses the outlandish. In Callimachus, the Cyclopes perform the very activities of the Homeric simile, and thus themselves allude to the blinding of their brother in the *Odyssey*. (We may compare the allusive strategy of Theocritus' eleventh *Idyll*.) Vergil then seizes upon the detail that binds these passages together. By combining the Homeric and Callimachean passages, Vergil not only furthers his Homeric program, but does so in the context of the learned, subtle poetic technique that he observes throughout the *Georgics*. The effect is one of both raising the poem's tone to a level approaching that of Homer's heroic epos and, conversely, bringing Homer down to the level of Vergil's own didactic epos, which has been painstakingly developed by imitation of Hesiod, Aratus, and Lucretius.

Against the background of the last two examples, the next takes on a different complexion than it might otherwise have. When Vergil informs us that bees stabilize themselves for flying in windy conditions by carrying a pebble (*G.* 4.194–196) he is following Aristotle (*HA* 626b 24). The idea that they thus act like boats taking on ballast against high seas (195) appears to be Vergil's own invention. We may wonder about the poetic implications of this comparison, and about the relevance of these implications to Vergil's own endeavors.

Nautical imagery is not infrequent in the *Georgics*, often appearing in a way that makes the sailor's life seem parallel to that of the farmer.[79] It also appears in the first and last books of the poem as a metaphor for poetic composition.[80] Its most prominent role however is in Vergil's invitation to Maecenas in Book 2 to accompany him on his poetic voyage:

> tuque ades inceptumque una decurre laborem,
> o decus, o famae merito pars maxima nostrae,
> Maecenas, pelagoque volans da vela patenti.
> non ego cuncta meis amplecti versibus opto,
> non, mihi si linguae centum sint oraque centum,
> ferrea vox. ades et primi lege litoris oram;
> in manibus terrae. non hic te carmine ficto
> atque per ambages et longa exorsa tenebo.
>
> *Georgics* 2.39–46

79. This is especially true in Book 1, where for example both sailing and agriculture are Jovian innovations at 118–159. See Thomas 1988.1.76–77 *ad G.* 1.50, and cf. 1.197–203, 371–373. Likewise Italy is praised both for its agricultural products and its maritime situation (2.158–164); the plague of Noricum attacks the creatures of land and sea, despite the fact that the historical Noricum had no coastline (3.541–545); and Aristaeus, the farmer-hero of Book 4 who is included in Vergil's invocation of rustic divinities (1.14–15), is the child of a river nymph. This parallelism is at some variance with both the Roman agricultural tradition (Cato *Agr.* proem. 1, 3) and the outlook of Hesiod (*WD* 618–694). Aratus' *Phaenomena*, however, is written with seafaring very much in mind, and this fact may have influenced Vergil.

80. *da facilem cursum* 1.40; *vela traham et terris festinem advertere proram* 4.117.

This passage, like the more obvious "Marble Temple" proemium of Book 3, carefully contrasts the poem that we are reading with a different, hypothetical one, whether on this or some other topic. I have already discussed the Homeric disclaimer of lines 42–44, here put to a purpose that is altogether new.[81] We may further note that the motif here occupies a cardinal position between two very different poetic alternatives. At first, Maecenas is encouraged to see through to the end the work that the poet has begun—or, to preserve the metaphor of *decurre* (39), to bring it into port.[82] This metaphor sounds the nautical theme, but relates to the writing of poetry only in the most general way: Vergil means to finish what he has started.[83] He then invites Maecenas to fly along spreading his sails before the open sea (41). This familiar image—a ship far from land, skimming along with sails unfurled—typically connotes poetry of a grand sort with respect to scope, style, and theme.[84] But then, having declared the grandeur of his program, Vergil retreats: *non ego cuncta*, etc. (42–44). By deflating the Homeric topos Vergil makes his disavowal of heroic designs all the more effective. He then revises his nautical metaphor: join me and skirt the edge of the coast, keeping land nearby (44–45). This new image connotes a very different sort of poetry than was glimpsed in the previous one. The shallow waters of a coastal voyage connote poetry that is small in scale, unpretentious in style, humble in theme.[85]

The program adumbrated in this carefully crafted passage is entirely in line with Vergil's treatment of heroic sources in the earlier books of the *Georgics*. Further, the self-reflexive literary import of nautical imagery in the poem is not confined to this passage. Therefore, when we find

81. See pp. 232–234.

82. *OLD* s.v. 4a.

83. Along with exordia a common application of the nautical topos (see n. 80; Ovid *Ars* 1.771–772, 3.747–748, *Rem.* 70, 811–812, *Fasti* 1.4, 4.18). Eventually it became primarily an exordium topos (Curtius 1973.129–130).

84. Cf. Pindar *P.* 2.62–63, 4.3, *N.* 3.26–27, 4.69–72, 5.50–51, 6.28–29; Aeschylus *Choephori* 821–822; Horace *Carm.* 4.15.1–4 (echoing Vergil *Ecl.* 6.3–5); Propertius 3.3.22–24, 3.9.3–4, 3.9.35–36; Ovid *Met.* 15.176–177, *Tristia* 2.329–330, 547–562. On this imagery in general see Becker 1937.71–72; Wimmel 1960.222–233; Cairns 1972.235 nn. 29–30; Curtius 1973.128–129, where related passages in prose authors (e.g. Cicero *Tusc.* 4.5.9, Quintilian *IO* 12 *proemium*, Pliny *Epist.* 8.4.5) are mentioned; Cody 1976.82–89; Bonnafé 1987.92–96.

85. On the ambivalences of this crucial programmatic passage see Thomas 1985. 69–70. Pindar (*P.* 10.51–52, 11.39–40) uses similar imagery to check a flight of fancy and bring himself back to the subject at hand. Cf. in particular the Horatian and Propertian passages cited in the previous note.

a simile in which bees—themselves a common symbol of poetry—are expressly compared with ships,[86] we may well expect to find a veiled programmatic statement.

What is the nature of this statement? To answer this question, we must first know what sort of poetry bees connote. Often, of course, bees are nothing more than a straightforward metaphor for poets in general.[87] Pindar once invokes bee imagery—in the company, as it happens, of nautical imagery—to check the impetus of his imagination and to provide closure to an ode:

> *κώπαν σχάσον, ταχὺ δ᾽ ἄγκυραν ἔπρεισον χθονὶ*
> *πρῴραθε, χοιράδος ἄλκαρ πέτρας.*
> *ἐγκωμίων γὰρ ἄωτος ὕμνων*
> *ἐπ᾽ ἄλλοτ᾽ ἄλλον ὥτε μέλισσα θύνει λόγον.*
>
> *Pythian* 10.51–54

Its poetic voyage completed, Pindar's song of praise is compared to a bee browsing in a meadow, lighting now here, now there. There is more to sing, in other words, but the singer prefers a selective style to all-inclusive bombast. It is to express this position that he invokes the bee simile. Likewise Callimachus marks the difference between the grand style championed by Envy and the slender style favored by Apollo:

> "'*Ἀσσυρίου ποταμοῖο μέγας ῥόος, ἀλλὰ τὰ πολλὰ*
> *λύματα γῆς καὶ πολλὸν ἐφ᾽ ὕδατι συρφετὸν ἕλκει·*
> *Δηοῖ δ᾽ οὐκ ἀπὸ παντὸς ὕδωρ φορέουσι μέλισσαι,*
> *ἀλλ᾽ ἥτις καθαρή τε καὶ ἀχράαντος ἀνέρπει*
> *πίδακος ἐξ ἱερῆς ὀλίγη λιβὰς ἄκρον ἄωτον.*"
>
> *Hymn* 2.108–112

In this famous passage, the Euphrates is made the symbol of poetry conceived and written on a grand scale, while the droplets offered by the bees represent a more careful, miniaturist aesthetic. This is in general the type of poetry that bees connote. The emphasis in such passages is on the sweetness of the verse, not its power or scope.[88]

86. Pindar pairs the two images at *Pythian* 10.51–54, discussed immediately below.

87. Pindar *O.* 10.97–99, *I.* 5.53–54; Bacchylides 10.10; Aristophanes *Aves* 748–750 (cf. 907), *Eccl.* 974; Plato *Ion* 534a–b; Hermesianax fr. 7.57 (Powell 1925.99); Christodorus *AP* 2.69 and 110 ; Varro *DRR* 3.16.7; Lucretius *DRN* 3.10; Horace *Epist.* 1.3.21, 1.19.44–45. See Pucci 1977.19–21; Waszink 1974.

88. See Pucci 1977.27–29, Scheinberg 1979.16–28. Leonidas' (or Meleager's?) epitaph for Erinna (*AP* 7.13 [Gow–Page 1965.138, 2563–2566), which is rife with echoes of

Returning to Vergil, what do we find? His bees carry pebbles to stabilize themselves on windy days, just as boats take on ballast against high seas. His word for "boats," *cumbae* (195), means "a small boat" or "skiff"[89] as opposed to a "ship" (*navis*). It is the word used by Propertius (3.3.22) to suggest that the higher genres are beyond his small powers. But what does it mean in this context for such a boat to take on ballast? Surely that the boat has ventured rather farther than usual out to sea, where it is tossed by waves (*fluctu iactante* 195). Occurring as it does in this book of the *Georgics*, in which Vergil has left behind allusion to the epos of Hesiod, Aratus, and Lucretius in favor of Homer, and in this remarkable series of similes, which are experiments in the variety of ways of incorporating Homeric epos into a didactic program, can this image of bees and boats taking on ballast connote anything other than the Homeric program of this very book, and the poet's determination to appropriate to his didactic poem the elements of heroic epos?[90]

With our final example we encounter an elaborate simile borrowed directly from Homer. The passage is a description of what happens when sickness attacks the hive. The bees make a sound that is deeper than usual, a sound like that of wind in the trees, waves against the shore, or fire in a furnace:

> tum sonus auditur gravior, tractimque susurrant,
> frigidus **ut** quondam silvis immurmurat Auster,
> **ut** mare sollicitum stridit refluentibus undis,
> aestuat **ut** clausis rapidus fornacibus ignis.
>
> *Georgics* 4.260–263

The triple form of the simile together with the elemental nature of the vehicles mark this as an unusually elevated comparison, and the reader's awareness of the subtext only strengthens this impression. Homer uses these same terms to connote the sound made by warriors marching into battle:

Callimachus' second *Hymn* and of *Aetia* 1 fr. 1.16–17 (Pfeiffer 1949.4), belongs in this tradition. Cf. Horace *Carm.* 4.2.25–32.

89. *OLD* s.v.

90. On the double nature of the bees as a symbol of the Neoteric ideal and as Vergil's first true essay in the heroic style, see Wimmel 1960.175 n. 3.

ἐκλύσθη δὲ θάλασσα ποτὶ κλισίας τε νέας τε
᾿Αργείων· οἱ δὲ ξύνισαν μεγάλῳ ἀλαλητῷ.
οὔτε *θαλάσσης κῦμα τόσον βοάᾳ ποτὶ χέρσον,*
ποντόθεν ὀρνύμενον πνοιῇ Βορέω ἀλεγεινῇ·
οὔτε *πυρὸς τόσσος γε ποτὶ βρόμος αἰθομένοιο*
οὔρεος ἐν βήσσῃς, ὅτε τ᾿ ὤρετο καιέμεν ὕλην·
οὔτ᾿ *ἄνεμος τόσσον γε περὶ δρυσὶν ὑψικόμοισι*
ἠπύει, ὅς τε μάλιστα μέγα βρέμεται χαλεπαίνων,
ὅσση ἄρα Τρώων και ᾿Αχαιῶν ἔπλετο φωνὴ
δεινὸν ἀϋσάντων, ὅτ᾿ ἐπ᾿ ἀλλήλοισιν ὄρουσαν.

Iliad 14.392–401

This allusion has been recognized at least since the time of Servius, who emphasizes the triple form of the simile and the characteristically Vergilian compression, saying nothing of the equally characteristic pun *ut*/*οὔτε*.[91] Also important, however, is the Homeric context from which this simile is borrowed. It is the moment when, after Hera has beguiled Zeus with sex and sleep upon Mt. Ida ("The Deception of Zeus"), the Trojan and Achaean forces fall upon one another. The tenor of the simile is the cry itself, the noise that the enemy forces make as they advance to do battle (400–401). We may remember in passing the two similes from *Iliad* 2 and 3 that Vergil imitates in Book 1. In the latter simile the Trojans advance as noisily as cranes in flight before a storm, while the Achaeans march in disciplined silence. In the former, however, the Achaeans are also likened to a flock of cranes, not in the sound that they make, but in their numbers. I have argued that Vergil's decision to borrow from these two similes in the space of a few lines of the *Georgics* is an ironic commentary on the fundamental similarity of the opposing armies, despite Homer's observation to the contrary. Here he imitates another simile describing these forces as they enter battle; and here the distinction that Homer draws between them in Book 3 has disappeared. The noise made by *both sides together* is compared to that made by three forces of nature.

Vergil's simile thus takes its place within his ongoing redaction of the *Iliad*. But the Iliadic context is relevant to the *Georgics* context as well. The sound made by the Trojan and Achaean warriors is the same as the

91. Servius *ad G.* 4.261 (Thilo 1887. 341) FRIGIDUS UT QUONDAM S. I. A. tres comparationes singulis impletae versiculis de Homero translatae sunt, quas ille binis versibus posuit (≈ Σ Bern. *ad loc.* [Hagen 1867.300]).

sound made by a hive full of unwell bees. Is Vergil suggesting an analogy between epidemic and warfare as comparable social disturbances? The thematic structure of the *Georgics* supports the view that he is. Books 1 and 3 end with images of civil war—which is characterized there as a cosmic disturbance—and of plague, respectively. The bee state is subject to both maladies; and this passage on the diseases of bees, like those on civil war and plague, gives instruction for diagnosing the illness by observing definite signs (253; cf. 1.351, 439, 461–471 and 3.440, 503). Thus here as elsewhere Vergil indicates the common ground between the heroic and georgic world.

From this discussion of the similes in Vergil's bee discourse several things have become obvious. First, we are clearly dealing with a series of similes that was designed with deliberate care. The first and the last are double similes: the swarming bees fall thicker than hail or acorns, and burst from the slaughtered bullock like rain or arrows. These frame a group of four similes in which brief comparisons of one or two lines (the dusty traveler and the sea-tossed boats) alternate with more developed ones (the Cyclopes and the triple simile of wind, sea and fire). The connections among the similes are telling. The fire of line 263, for instance, rages within a *fornax*, and Putnam has acutely observed that the word occurs only once elsewhere in the poem, among the portents attending the death of Julius Caesar:[92]

> quotiens Cyclopum effervere in agros
> vidimus undantem ruptis fornacibus Aetnam,
> flammarumque globos liquefactaque volvere saxa!
>
> *Georgics* 1.471–473

These furnaces prove to be those of the Cyclopes' volcanic forge, which reappears in another simile in Book 4. Thus a point of diction forms a pair of thematic links, on the one hand between the triple elemental simile of Book 4 and the gloomy finale of Book 1, and on the other between that simile and the Cyclops simile of Book 4. We may note a second, similar connection between the latter simile and another of the series. When the Cyclopes dip the glowing bronze in the basin, it sizzles upon contact with the cold water. When the bees begin to take form

92. Putnam 1979.269. A second connection with this passage is found in the description of the bees forming in the carcass of the bullock. At their first appearance, the nascent bees are *visenda modis animalia miris* (4.309). At Caesar's death were sighted *simulacra modis pallentia miris* (1.477).

within the bullock's carcass, they first appear without limbs, but begin to buzz as soon as they grow wings. Vergil evidently regarded these sounds as similar:

> alii taurinis follibus auras
> accipiunt redduntque, alii **stridentia** tingunt
> aera lacu
>
> *Georgics* 4.171–173

> trunca pedum primo, mox et **stridentia** pennis
>
> *Georgics* 4.310

Significantly, the sea also makes this sound in two passages. The first is the ominous description of the Lucrine lake in the *"Laudes Italiae"*:

> an memorem portus Lucrinoque addita claustra
> atque indignatum magnis **stridoribus** aequor,
> Iulia qua ponto longe sonat unda refuso
> Tyrrhenusque fretis immittitur aestus Avernis?
>
> *Georgics* 2.161–164

The second is in the triple simile of Book 4:

> ut mare sollicitum **stridit** refluentibus undis,
>
> *Georgics* 4.262

Here we may also note the connections between *indignatum...aequor* and *ponto...refuso* in Book 2 and *mare sollicitum* and *refluentibus undis* in Book 4. The connections thus forged between these similes and key passages from earlier book quickly become obvious. They do not stop here. The sound of the south wind in the triple simile is marked by a Vergilian *unicum*, *immurmurat* (4.261). Two other murmurs are heard in the poem, both in Book 1: the sound made by the runaway irrigation stream (109) and a sign of windy weather (359). We are reminded both of the way in which this ambitious transformation of Homer began and of the thematic associations of the sound with inclement weather in the Aratean "Weather Signs." We may compare the first simile of the bee discourse, where the bees are massed together (*glomerantur* 4.79) into a ball. The verb also denotes the gathering of a storm (1.323), the raging of fire in a treetop (2.311), and the proud gait of a war horse (3.117). Similarly the shaken oak in the second half of this simile recalls the conclusion of the "Aetiology of *Labor*":

heu! magnum alterius frustra spectabis acervum,
concussaque famem in silvis solabere quercu.

Georgics 1.158–159

It should thus be clear that the similes of Vergil's bee discourse are closely connected among themselves and that they serve to relate the context in which they occur to the chief thematic concerns of the poem as a whole.

A second point. These similes clearly embody a new and important step in Vergil's program of Homeric imitation. The epic simile, as Knauer rightly notes, plays but a small role in Books 1–3. Indeed, as our discussion shows, Vergil actually disguises Homeric similes as straightforward didaxis in those books. But here, in the first half of Book 4, we encounter this intricate series of similes, which flaunt their Homeric form while illustrating the variety of possible approaches to Homeric allusion. In concluding this discussion, we may remark upon one further aspect of Vergil's Homeric program in these lines. What has Homer to say about bees? There are only two bee similes in Homer. One seems to have no relevance to the *Georgics*.[93] The other, however, offers a fascinating commentary on Vergil's Homeric program. When the Achaeans muster for the assembly inspired by Agamemnon's dream, they too are compared to swarming bees:

ἠΰτε ἔθνεα εἶσι μελισσάων ἁδινάων,
πέτρης ἐκ γλαφυρῆς αἰεὶ νέον ἐρχομενάων·
βοτρυδὸν δὲ πέτονται ἐπ᾽ ἄνθεσιν εἰαρινοῖσιν·
αἱ μέν τ᾽ ἔνθα ἅλις πεποτήαται, αἱ δέ τε ἔνθα·
ὣς τῶν ἔθνεα πολλὰ νεῶν ἄπο καὶ κλισιάων
ἠϊόνος προπάροιθε βαθείης ἐστιχόωντο
ἰλαδὸν εἰς ἀγορήν.

Iliad 2.87–93

This happens to be the first extended narrative simile of the *Iliad*.[94] Although Vergil does not refer to it specifically, in a general way one

93. Two Lapiths, Polypoetes and Leonteus, defending the Achaean encampment, are compared by the Trojan Asius to wasps or bees fighting to protect their homes and children (*Iliad* 12.166–172). The simile is unusual, since insect similes usually illustrate number, as in the example discussed immediately below.

94. Three very brief similes can be found in Book 1. At line 47, Phoebus Apollo is said to move "like the night" (*νυκτὶ ἐοικώς*); at line 104, Achilles' anger causes "his eyes to resemble shining fire" (*ὄσσε δέ οἱ πυρὶ λαμπετόωντι ἐΐκτην*); and at line 359 Thetis leaves the grey sea "like a mist" (*ἠΰτ᾽ ὀμίχλη*).

might say that the entire series of similes in *Georgics* 4 simply reverses the terms of this Homeric comparison: Homer's warriors in a sense become Vergil's bees. Through this series of similes, and through the diction and imagery of Vergil's "technical" discourse, the bees are increasingly likened to Homeric warriors. With Vergil's imitation of Homer's triple simile, the process is completed. Again it is a matter of heroes gathering, now not for assembly, but for war. In *Iliad* 14, this simile describes the Trojan and Achaean forces as they rush together in battle; in *Georgics* 4, it is applied to the hive beset by disease. It is of course clear that in the thematic structure of the *Georgics* disease, unbridled lust, inclement weather, and war, especially civil war, are close equivalents. The sound of the bees, moreover, is mentioned only twice: when they swarm—i.e. when they go to war—and when they are sick. It is at this point, when their warlike nature succumbs to disease and death, that the bees are most frankly Homeric. It is at this point too, when the hive has been irrevocably lost, that we must learn to how to acquire a new one.

"Aristaeus" (Georgics *4.315–558)*

As I explained in Chapter 2, Vergil's method of recasting his source in the "Aristaeus" is, in formal terms, of a piece with the technique that we have observed throughout the *Georgics.* We have also seen that the poet makes an effort to integrate Homeric allusion with imitation of Hesiod, Aratus and Lucretius in earlier episodes. On the other hand, the fact that he uses Homeric raw material to create the poem's final episode remains, or should remain, surprising. Years ago, scholars were all too aware of the inconcinnity between this conclusion and the earlier books of the poem. Indeed, at one time the general tendency was to exaggerate this inconcinnity when passing qualitative judgment on the poem—just as we habitually ignore it today. In doing so, of course, scholars were merely—and quite understandably—following Servius' assertion that the original conclusion of *Georgics* 4, published in about 29 B.C., was different from the one that we have: that it consisted of or at least contained a passage in praise of Vergil's close friend, the poet and statesman Gaius Cornelius Gallus, and that the present ending was composed for a second edition of the poem at Augustus' request.[95] The request would have been made after Gallus' fall from favor and subsequent suicide in

95. Servius *ad G.* 4.1 (Thilo 1887.320); cf. *ad Ecl.* 10.1 (Thilo 1887.118).

26 B.C.[96] This information, scholars have felt, explains the difference between the conclusion and the rest of the *Georgics:* our ending was written several years after the original poem, at a time when Vergil had left didaxis far behind. A new and much greater project, the *Aeneid*, was already underway; the poet's mind was full of his new model, the poet *par excellence*, Homer; and, rather than return to his former preoccupations, he simply provided the earlier poem with a new ending consistent with his current interests.[97] For a time scholars even tried to prove that repetitions of lines occurring both in *Georgics* 4 and in the *Aeneid* indicated that the lines were original to the later poem, and provided material which Vergil was able to adapt quickly to a second edition of the *Georgics.*[98]

Critics who believed in this chapter of literary history might have responded more charitably to the solution that Vergil chose. The north tower of Chartres cathedral in all its flamboyance hardly ruins, and may even be felt to enhance, our experience of the austere original design. But it was once not impossible to deride the "Aristaeus" as if it were Marcel Duchamp's mustache on the Mona Lisa.[99] Of course it was inevitable that readers would begin to notice that the episode is not only beautiful poetry in itself, but in fact forms a highly suitable conclusion to the poem as a whole. This realization came about when critics started to appreciate what Seneca told Lucilius long ago, that Vergil wrote the *Georgics* not to instruct farmers, but to delight readers.[100] The poem, as all sensitive readers have always known and as scholars now acknowledge, did not serve as a how-to manual for Octavian's newly landed veterans, nor as a protreptic for jaded urbanites, nor as a patriotic reminder to the survivors of the civil wars that they were all children of the earth. Roman readers needed no such thing. They responded to the image of the farmer as deeply and instinctively as any American responds to images of the pioneer in fiction and in cinema. The Romans,

96. On Gallus' demise see Suetonius *Aug.* 66; Cassius Dio 51.9, 53.23.5–24.2; Jerome *Chron.*, *ad annum Abr.* 1984 (Helm 1913.162).

97. The scholarly controversy over the "*Laudes Galli*" is thoroughly and conveniently surveyed and discussed by Jacobson 1984.

98. See particularly Walter 1933, Mylius 1946, Büchner 1959.294–296 (= *RE* 8 A 1316–1318).

99. Sellar (1897.188–189) for example calls the episode "an undoubted blot on the artistic perfection of the work" imposed on Vergil by "the despotic will of the emperor."

100. The general principle remains valid despite the interesting and informative observations of Spurr 1986.

even in the most urbanized and sophisticated periods, could not think about farming merely as an occupation full of drudgery and devoid of edification. Even the age that produced a Catullus, the consummate poet of *urbanitas*, could find in the life of the farmer an ideal pattern of existence.[101] Vergil shows us both the harshest and the noblest aspects of the farmer's work: the backbreaking labor; the need to live in sympathy with the cosmos, both in its regular cycles and in its uniquely catastrophic events; pride in the mastery of nature, and humility in the face of its unbending laws; and most important, intimacy with the forces of birth and death. This is what the *Georgics* is about, and recent critics have done excellent work in showing why the "Aristaeus" is in fact a highly appropriate conclusion to such a poem. The corollary to their argument is that we have in the "Aristaeus" no hasty revision of Vergil's original conception, but an episode that is so fully integrated into the poem's thematic structure as to make Servius' statements about the "*Laudes Galli*" look unconvincing.[102] While I tend to agree with those who feel that Servius is basically mistaken about a second edition, I do not feel at present that the matter can or need be settled either way.[103] But in the light

101. Even in Catullus himself we find, along with derisive expressions like those applied to the poems of Volusius (*pleni ruris et infacetiarum* 36.19) and the poet Suffenus (*infaceto est infacetior rure* 22.14), e.g. the elaborate vindication of marriage via imagery drawn from viticulture (62.49–58).

102. Some scholars thus suggest that Servius' statement is confused and therefore simply meaningless, usually citing Anderson 1933 (although many of his arguments had been advanced by previous scholars, as Jacobson 1984.272 and 275 observes) and Norden 1934. Others try to explain Servius' confusion. According to one line of reasoning, the representation of Gallus in the concluding poem of the *Eclogue* book somehow gave rise to the idea that it was the *Georgics* that ended by honoring Gallus (Jacobson 1984.275–276). A more likely suggestion, which I myself favor for reasons that will become clear, is that Servius garbled a tradition that interpreted the "Aristaeus" episode as an allegorical tribute to Gallus (Nisbet 1987.189). Finally, some scholars have basically accepted Servius' testimony, arguing however that the "*Laudes Galli*" was a small tribute made in passing and easily excised or replaced with other material at Augustus' request; the latest representative of this view is Jacobson 1984.

103. If there was a first edition of the *Georgics*, it is now lost to us. If we had it, we might form an impression of it that differed a great deal, somewhat, or hardly at all from our impression of the version that we have. But even if we assume that the original *Georgics* was dramatically different from our version, there is no need to accept the notion that Vergil's revisions, even if he was moved to make them at the bidding of a tyrant, represent second best. Thomas makes a good point when he argues that "The *Georgics* was written at an average rate of about one line per day; but those who believe Servius' statements must assume that Virgil produced 250 of his most careful and allusive lines in a very short time" (1988.1.14 n. 35). He begs the question, however, of how long Vergil

of recent work, we can certainly proceed on the assumption that the "Aristaeus," which is unquestionably Vergil's own work, represents his final decision about how best to end the *Georgics;* that it is fully integrated into the thematic structure of the poem; and that any questions we may ask about it and any conclusions we may reach cannot be undermined by the specter of Gallus and Servius' testimony concerning a first edition.

With the thematic appropriateness of the "Aristaeus" no longer in doubt, we may turn to the real question. Why did Vergil take Homer's heroic epics as models for the conclusion of a poem based in large part on more obvious sources within the tradition of didactic epos? We can hardly be satisfied with the plausible but simplistic suggestion that he was already looking ahead to his great Homeric agon in the *Aeneid.* The teleological approach fails to recognize the artistic perfection reached in the *Georgics* itself. The question we must confront is, Why is the allusive structure of the entire *Georgics* arranged in such a way as to build so deliberately through the earlier books to this great Homeric conclusion? The pattern of allusion to Homer throughout the *Georgics* makes it clear that the "Aristaeus" represents more than a new direction in Vergil's thinking as he approached the end of his didactic essay. Rather, in thematic and allusive terms it is presented as the summation of that essay. Why did Vergil design the poem in this way?[104]

To answer this question, we must expand our approach. Up to this point, I have considered Vergilian allusion as a largely unmediated encounter between Vergil and his epic predecessors. I have appealed only occasionally to secondary sources, such as the scholia to Hesiod and Aratus, for clues as to why Vergil altered certain details in his sources, or how he conceived the relationship between different models. This critical strategy was dictated by necessity: the scholia to Hesiod and Aratus contain little information that is useful in a study of this kind. Opening the Homeric scholia, however, we enter another world. At

---

would have worked on his revisions, something about which we have no evidence. At any rate, Vergil's contemporaries evidently regarded the version we have as superior, to judge by how completely the hypothetical first edition, if it ever existed, has vanished.

104. Because the allusive structure of the *Georgics* is my object of inquiry, I cannot adduce my conclusions about it as evidence in subsidiary arguments. Nevertheless, I cannot refrain from adding *en passant* that the structure of Homeric allusion in the poem to my mind greatly corroborates the view that we have in the "Aristaeus" Vergil's original ending.

once more voluminous and more subtle in their treatment of the text, both the scholia and the collection of specialized treatises on various aspects of the Homeric poems constitute one of our best witnesses to the main stream of literary studies in antiquity. Properly interpreted, they are capable of giving us important and surprising insight into how ancient Greek and Roman readers understood their Homer. In our case, it is possible to use this scholarly tradition in conjunction with our previous analysis of Homeric allusion in the "Aristaeus" to shed light on Vergil's interpretation of the *Iliad* and the *Odyssey*. Specifically, by examining the remains of ancient scholarship on the Homeric episodes in question, I shall argue that in composing his homerizing "Aristaeus" Vergil was greatly influenced by the ancient tradition of allegorical exegesis that represented Homer not merely as a poet of heroic saga, but more importantly as a poet of natural philosophy.

The farfetched ideas of certain modern practitioners have given all allegorical exegesis a bad name. These need not detain us. By contrast, a few scholars—I will name only Armand Delatte, Félix Buffière and, most recently, Robert Lamberton—have done outstanding work to illuminate the important role of such criticism in antiquity. It cannot be doubted that many ancient readers, particularly of Homer, were in the habit of interpreting the adventures of Achilles and Odysseus as stories with a veiled historical, moral, philosophical, or mystical import. We see in Horace's second *Epistle* of Book 1, to Lollius, a clear example of this habit. Horace writes that he lately had the pleasure of rereading Homer on vacation at Praeneste (while his friend was toiling in the courts at Rome). He refers to Homer—*Troiani belli scriptorem* (1), he calls him—in such a way as to place him, conventionally, at the head of the grand tradition of heroic epos. The terms in which he proceeds to discuss Homer, however, are far different: *qui quid sit pulchrum, quid turpe, quid utile, quid non, | planius ac melius Chrysippo et Cantore dicit.* (3–4). The poet of the Trojan War is presented as a philosophical sage and a consummate ethical guide. Horace goes on to outline an allegorical reading of both Homeric epics as morality plays: Paris, Nestor, Achilles, and Agamemnon become emblems of perversity, reason, anger, and lust (9–12). By contrast (*rursus* 16) Ulysses is proposed as a type of the wise man and a model of right conduct (16–17). Horace closely paraphrases the opening lines of the *Odyssey* to suggest that the hero, as a student of human nature (*mores hominum inspexit* 19), was able to endure so many trials—he mentions prominently the threat of being turned into a dog or a swine by drinking Circe's potion (22–25), the act not of a *sapiens*, but of a *stultus atque cupidus* (23)—and to achieve his *nostos*. At length the lesson to be learned is made clear. We, the readers, are subject to the passions that Ulysses has mastered:

> nos numerus sumus et fruges consumere nati,
> sponsi Penelopae, nebulones, Alcinoique
> in cute curanda plus aequo operata iuventus,
> cui pulchrum fuit in medios dormire dies et
> ad strepitum citharae cessatum ducere curam.
>
> *Epistula* 1.2.27–31

Thus, in about thirty lines, Horace quickly digests the entire *Iliad* and *Odyssey* into a few moral concepts with an obviously didactic import.

Even if this poem is not wholly serious[105] it illustrates one way in which Vergil's contemporaries read and understood their Homer. If we consult either the scholia to the *Iliad* and *Odyssey* or a specialized treatise like Heraclitus' *Homeric Problems*, we discover that Horace's poem takes its place within a vast and diverse tradition of Homeric commentary that employs allegorical modes of interpretation. What is most impressive about this tradition is both how extensive it is, and the types of guidance that it may have provided to a *doctus poeta* engaged in an intertextual redaction of the Homeric poems.

Vergil certainly used several forms of this guidance in composing the *Aeneid*. Richard Heinze was the first to notice Vergil's use of Hellenistic scholarship in general[106] and with regard to allegorical exegesis in particular.[107] Years later, Knauer adduced a very specific example of the way in which the allegorical tradition informs Vergil's imitative technique.[108] The example concerns the famous "Song of Iopas" in *Aeneid* 1. The symbolic character of this song in relation to the story of Dido and Aeneas has been convincingly explained by Pöschl, who with other critics has seen in Iopas' subject matter evidence of Vergil's interest in

105. Cairns finds it marked by sophistication and wit, but notes that "it is based on a standard ancient way of interpreting Homer" (1989.86, with further references). For the view that the letter is a serious counterpoint to the generally humorous tone of the *Epistles*, see Fraenkel 1957.316.

106. Heinze 1915.496 s.v. "Homerscholien."

107. Heinze 1915.304–310, "Symbolische Götterszenen." Most important for our purposes is Heinze's argument that Mercury in *Aeneid* 4 is conceived against the background of Homer's Hermes interpreted as an allegorical representation of Λόγος (307–310).

108. Knauer 1964a.168.

Lucretius and in didactic poetry generally.[109] But we can be more specific than that. There is no doubt about the character of this song: its subject is natural philosophy. Lest we miss the point, it quotes directly from Vergil's statement about his own poetic ambitions at the end of *Georgics* 2:

> hic canit errantem *lunam solis***que labores,**
> **unde** hominum genus et pecudes, unde imber et ignes,
> Arcturum pluviasque Hyadas geminosque Triones,
> **quid tantum Oceano properent se tingere soles**
> **hiberni, vel quae tardis mora noctibus obstet.**
>
> *Aeneid* 1.742–746

> me vero primum dulces ante omnia Musae,
> quarum sacra fero ingenti percussus amore,
> accipiant caelique vias et sidera monstrent,
> defectus *solis* varios *lunae***que labores;**
> **unde** tremor terris, qua vi maria alta tumescant
> obicibus ruptis rursusque in se ipsa residant,
> **quid tantum Oceano properent se tingere soles**
> **hiberni, vel quae tardis mora noctibus obstet.**
>
> *Georgics* 2.475–482

Iopas in the *Aeneid* thus utters the ideal song, a didactic poem of natural philosophy, as Vergil describes it in the *Georgics*. This fact alone says a great deal about the character of both poems. But we must also consider, as always, the Homeric background of the *Aeneid* passage. Within the Homeric program of Book 1, Dido's banquet is modeled upon that of Alcinous in *Odyssey* 8; thus Iopas' performance before the Carthaginians does duty for the three songs that Demodocus performs for the Phaeacians.[110] The subjects of those songs—the quarrel of Odysseus and Achilles, the love of Ares and Aphrodite, and an episode of *Iliupersis* material—do not match that of Iopas' song, which concerns cosmology. This is not a problem. From Vergil we expect allusive variation. However: these variations are seldom unmotivated, and often, as we have seen, they contain the most interesting aspect of an imitation. With this principle in mind, we pose the question: what could have motivated Vergil to

109. Pöschl 1962.150–154; cf. Austin 1971.222–223 *ad Aen.* 1.742ff.

110. On this section of the *Aeneid* see Knauer 1964b.148–173. The fact that Vergil's banquet is modeled on Homer's was evidently recognized by ancient critics, as we may infer from Macrobius' reference at *Sat.* 7.1.14.

imitate the three "historical" and mythical songs of Demodocus as a poem of natural philosophy?

The answer lies in ancient Homeric scholarship. If we consult either the Homeric scholia *ad loc.* or Heraclitus' *Homeric Problems* 69, we discover that the second song of Demodocus, the "Love of Ares and Aphrodite," was commonly explained in antiquity as an allegorical poem of natural philosophy—that is, one in which Aphrodite and Ares represent Love and Strife as the principal creative forces in the physical universe:

> *Νομίζω δ' ἔγωγε καίπερ ἐν Φαίαξιν, ἀνθρώποις ἡδονῇ δεδουλωμένοις, ᾀδόμενα ταῦτα φιλοσόφου τινὸς ἐπιστήμης ἔχεσθαι· τὰ γὰρ Σικελικὰ δόγματα καὶ τὴν Ἐμπεδόκλειον γνώμην ἔοικεν ἀπὸ τούτων βεβαιοῦν, Ἄρην μὲν ὀνομάσας τὸ νεῖκος, τὴν δὲ Ἀφροδίτην φιλίαν.*
>
> Heraclitus *Quaestiones Homericae* 69.7–8

Of course, most of our testimonia to this exegetical tradition, Heraclitus included, are of late or uncertain date. In this particular case, however, Heraclitus himself makes a connection between his own exegesis and a famous passage of early Greek philosophy. On the basis of such connections, Buffière establishes a more general association of this and other physical allegories with pre-Socratic and especially early Pythagorean thought, and regards them as distinct from other types of allegorical exegesis.[111] Thus, while most of the surviving witnesses to this class of allegorical explanation cannot be dated, the interpretation of Aphrodite and Ares as a cosmic allegory can be traced back well before Vergil, even as far back as Empedocles' treatment of Love and Strife in his *Peri Physeos* We are therefore justified to infer that Vergil followed this tradition by substituting Iopas' overt cosmogony for Demodocus' "allegorical" one.[112]

111. Buffière 1956.168–172; cf. Delatte 1915.116.

112. Note also that Heraclitus' interpretation of the Phaeacian's character, a commonplace in ancient criticism (see the passages listed by Kaiser 1964.17–20 and, with regard to Phaeacian national character in general, Dickie 1983) is entirely congruent with that of Vergil's contemporary, Horace (see my remarks on pp. 257–258). Heraclitus' floruit is not definitely known, but is usually given as the reign of Tiberius, i.e. not much later than Vergil and Horace. The exegetical branch of the Homeric scholia are thought to have been compiled during the same general period (Lehnert 1896.69, Latte 1924–1925.171, Erbse 1960.171–173). In any case, the tradition that both Heraclitus and the scholia represent quite clearly derives in large part from earlier antiquity; and this is particularly true of allegories concerning the physical world, as we shall see. In what follows, I shall draw

The implications of such an allusion are enormous. We have grown accustomed to the idea that Roman poets imitate their predecessors in a way that strives for independence, and that this independence may involve more than a simple exercise in rhetorical *inventio*. The cult of *doctrina* that late Republican poets inherited from Hellenistic culture and which they passed on to the Augustans, challenged the imitator to interpret his model and to illustrate his interpretation through allusion. In the case of Iopas' song, Vergil's interpretation is far-reaching. Of Demodocus' three songs, he selects one; which is another way of saying, he rejects two. By selecting as his model a mythological love poem rather than the heroic and "historical" themes of the quarrel and the sack of Troy, Vergil outlines an important part of his epic program, a program inherited from the Hellenistic masters such as Apollonius of Rhodes. Vergilian epic will treat not only of "kings and battles," as a younger Vergil had himself had derisively put it in *Eclogue* 6, but will concern the life of love as well.[113] This theme of course has been quite prominent in recent criticism. But Demodocus' love poem appears in Vergil under a more serious guise, that of natural philosophy. This allusive variation not only reveals the poet's awareness that one particular Homeric episode had been interpreted allegorically; by acknowledging and accepting this interpretation, Vergil incorporates into his own work a tradition that regarded heroic epic not simply as a matter of, again, "kings and battles," but, when properly read, as a profound meditation on the very nature of the universe. In this respect as well, Iopas' song speaks subtly but eloquently to a crucial element in Vergil's own poetic program.

The fact that Iopas' song embodies the poetic ideal to which Vergil himself claims to aspire in *Georgics* 2, as I mentioned above, can tell us a lot about both the *Georgics* and the *Aeneid;* the fact that it is the product of an allegorical interpretation of Homer gives us a crucial insight into

---

upon these sources as witnesses to a critical tradition already available to Vergil in essentially the form that we have it.

It is intriguing to find that Servius *ad Aen.* 1.742 (Rand 1946.304) contrasts Iopas' recitation of a cosmology with Clymene's song about Venus and Mars in *Georgics* 4 (345–346)—i.e. the subject of Demodocus' second song in *Odyssey* 8—and that he does so as if he were answering an objection to Vergil's choice of topics (as noted by Austin 1971.222 *ad Aen.* 1.742 ff.). Could it be that Vergil had been criticized on this point as an instance of κακοζηλία with respect to his Homeric model?

113. Even this development may be reflected in Homeric criticism: note the emphasis on love and desire in Horace's moral allegory of the *Iliad* (*Epist.* 1.2.6–16). See pp. 257–259, and cf. Lausberg 1983.234–235.

Vergil's poetic technique and literary ideals. Oddly, allegorical exegesis has been all but ignored in the only book-length study of Vergil and Homeric scholarship.[114] But the importance of physical allegory in the *Aeneid* has received increasing scholarly attention in recent years.[115] No one, however, has yet investigated the possibility that an interest in physical allegory informed Vergil's imitation of Homer in the *Georgics* as well.[116] The evidence presented so far suggests that allegorical readings of the Homeric poems were available to Vergil, that he used them in composing the *Aeneid*, and that they may also have influenced his imitation of specific passages of the *Georgics*. When one turns to the conclusion of the poem, the "Aristaeus," this possibility becomes very strong indeed.

The central motif of the "Aristaeus," the bugonia *aetion*, which connects this climactic episode with the "technical" material that precedes it, readily lends itself to allegorical exegesis. We can see how ancient readers might have reacted to this story from the way in which it enters ancient Homeric criticism. At *Odyssey* 13.96–112 Homer describes the harbor in Ithaca inhabited by the sea god Phorcys. It is framed by twin cliffs, and at the back of it stands an olive tree beside a cave sacred to the Naiads. In the cave are stone mixing bowls and amphoras in which bees deposit honey. There is also a stone loom on which the nymphs weave purple cloth, and entrances on the north and south sides, the first for mortals and the second for the gods. This cave of the nymphs became the subject of a famous allegorical monograph by Porphyry of Tyre, the third century A.D. Neoplatonist philosopher. In this treatise, amidst a jumble of other mystical lore, Porphyry indicates that the bees symbolize the souls of the dead, and specifically those that were to be reborn and to live just lives.[117] He further connects this notion with the bugonia motif.[118] It is therefore clear that for Porphyry, the bugonia is an allegory of metensomatosis, a religious idea central not only to his own Neoplatonic thought, but to many other eschatological systems of antiquity.

114. Schlunk 1974.

115. See Murrin 1980, esp. ch. 1, "The Goddess of Air," pp. 3–25, which illustrates Vergil's application of the common allegorical etymology of Hera as ἀήρ; Innes 1979; Wlosok 1985; Hardie 1985, esp. pp. 90–95; Hardie 1986.

116. The possibility is raised in passing by Knauer (1964a.168 n. 2) with regard to Clymene's song (*G.* 4.345–347), but not pursued in his subsequent study of the *Georgics* (1981.890–918).

117. Porphyry *Antr.* 18–19.

118. Ibid.

Whether any such explanation of the Homeric passage was available to Vergil is not definitely known; but it seems likely that it was. In the first place, Porphyry cites in this same passage a fragment of Sophocles that presupposes a similar connection between the souls of the dead and bees (*βομβεῖ δὲ νεκρῶν σμῆνος* fr. 879 [Radt 1977.568]). While this passage does not guarantee that the exegetical tradition represented by Porphyry extended back to Sophocles' or even to Vergil's time, it does establish the antiquity of a connection between bees and *ψυχαί* among the poets themselves. Second, the details of Porphyry's exegesis fit the thematic structure of the "Aristaeus." The death of the bees and their recovery, the image of new life bursting forth from the carcass of a young bull, is clearly a symbol of the rebirth of a society that has, literally or figuratively, died. Third, Vergil was to apply bee imagery to the just souls awaiting rebirth in the underworld of *Aeneid* 6:

> interea videt Aeneas in valle reducta
> seclusum nemus et virgulta sonantia silvae,
> Lethaeumque domos placidas qui praenatat amnem.
> hunc circum innumerae gentes populique volabant:
> *ac veluti in pratis ubi apes aestate serena*
> *floribus insidunt variis et candida circum*
> *lilia funduntur, strepit omnis murmure campus.*
> horrescit viso subitu causasque requirit
> inscius Aeneas, quae sint ea flumina porro,
> quive viri tanto complerint agmine ripas.
> tum pater Anchises: "animae, quibus altera fato
> corpora debentur, Lethaei ad fluminis undam
> securos latices et longa oblivia potant.
> has equidem memorare tibi atque ostendere coram
> iampridem, hanc prolem cupio enumerare meorum,
> quo magis Italia mecum laetere reperta."
>
> *Aeneid* 6.703–718

Most commentators have felt that the simile was inspired by a tradition of representing souls as bees,[119] but here too the circumstances are close to Porphyry's allegorization of Homer: Vergil's souls, waiting to be born again as the greatest heroes of Roman history, are like bees as they buzz around the plains of Elysium on a summer day. Can anyone read this passage without thinking of the bees, the *Quirites* of *Georgics* 4, whose

119. Norden 1957.305–307 *ad Aen.* 6.706 ff.; Austin 1977.217 *ad Aen.* 6.707 ff.; Cook 1895.19, 1914.469 n. 7; Briggs 1980.75–77.

society is wiped out by a disease thematically identical to civil war and then reborn through the miracle of the bugonia?

In his decision to make apiculture the last "technical" subject of his poem, Vergil was surely guided by the tradition represented by Sophocles and Porphyry. After the total, hopeless destruction of the georgic world in the cattle plague of Book 3, the bees clearly raise the issue of whether rebirth is possible. This issue is the central concern of the "Aristaeus"; and when one examines the ancient scholarship on the Homeric episodes that Vergil imitates in the "Aristaeus," one discovers that all the episodes in question were connected with the allegorical tradition. In fact Vergil's interest appears to be in a special class of allegorical stories concerning natural philosophy; for of those episodes that are so explained by ancient critics, only a few deal specifically with the theme of *cosmogony*, the birth of the universe. They are:

1. the insurrection of Hera, Athena, and Poseidon against Zeus, who was aided by none of the gods except Thetis and Briareus (*Iliad* 1.396–406)
2. the shield of Achilles (*Iliad* 18.478–613)
3. the story of Proteus and his daughter Eidothea (*Odyssey* 4.351–572);
4. the love of Aphrodite and Ares (*Odyssey* 8.266–366).

Remarkably, Vergil's "Aristaeus" alludes to all of these episodes, either by structural imitation or by direct quotation. The relationship between the Vergilian theme of the rebirth of a society and the Homeric theme of the birth of the natural order has been established in Hardie's recent work on the *Aeneid*.[120] When we now find that the concluding episode of the *Georgics* is full of allusions to Homeric episodes that were explained in antiquity as allegorical cosmogonies, the natural inference, I think, is that Vergil was guided in his selection of these episodes by the tradition of allegorical commentary on the *Iliad* and the *Odyssey*.

Let us examine the details. In Chapter 3 I discussed the Aristaeus as a conflation or combination of various Homeric scenes and characters, strictly from the standpoint of imitative technique. I further suggested that this combination represents a reworking of a Homeric theme that I defined as "life, death, and rebirth." Such an interpretation may be seen as parallel to Pöschl's analysis of the symbolism contained in "The Song

120. Hardie 1986.

of Iopas." But, as does Iopas' song, the "Aristaeus" embodies the fruits of an exegetical tradition that regarded Homer as a poet of natural philosophy. We may begin to understand how this is so by considering a rather unusual and striking feature of Vergil's imitation. The "Aristaeus" is particularly notable for the scope of its allusive reference: on my interpretation, it is meant to represent an important theme that pervades both Homeric poems. Obviously such a short narrative can achieve so much only by employing what I have called allusive synecdoche and, especially, by combining or conflating action, characters, and motifs found in both Homeric poems. Within this ambitious conflation of Homeric elements, however, one thing stands out. Besides the Homeric episodes in question, two pairs of Homeric characters are conflated into two Vergilian ones: Achilles from the *Iliad* and Menelaus from the *Odyssey* are conflated into Aristaeus; while Thetis from the *Iliad* and Eidothea from the *Odyssey* are conflated into Cyrene. One character only—and somewhat ironically, perhaps, in view of his ever-changing nature—maintains his individual identity in both poems. I refer to Proteus, who explains in the *Odyssey* what Menelaus must do to achieve his *nostos*, and in the *Georgics* why Aristaeus' bees have perished and how he may get new ones.[121] We can explain Vergil's appropriation of this character entire in direct thematic terms. Proteus represents Menelaus' hope for a *nostos*, i.e. in Odyssean terms, for a return to life; for in the *Odyssey*, loss of homecoming is consistently equated with death and annihilation, its achievement with life and rebirth. In the thematic structure of *Georgics* 4, the bugonia holds out the possibility of rebirth, just as the story of Orpheus and Eurydice acknowledges the forces of oblivion.

This interpretation, the product of an imaginative comparison of Vergil and his source, may well have been inspired by Homeric exegesis. Homer's Proteus makes possible Menelaus' *nostos*, his "rebirth." Vergil, following the tradition known to Homer's allegorical commentators that bees represent souls awaiting rebirth through bugonia, connects his Proteus with the revelation of this technique to the beleaguered Aristaeus. But the specific importance of Proteus, the factor that leads Vergil to maintain his identity alone among the Homeric characters that served as models for the players in this episode, is another

121. In fact, the explanation of how Aristaeus may replenish his hive through the bugonia falls ultimately to his mother Cyrene (*G.* 4.530–547), although she had told him earlier (396–397) that Proteus would both explain the *causa* of her son's situation and direct him towards a good *eventus*.

question. I suggest that Vergil has maintained this character's Homeric identity along with his narrative function as a kind of clue to the reader. Proteus, more perhaps than any other Homeric character, lends himself to allegorical interpretation. As Buffière comments, his very name suggests the primordial, particularly the primordial matter out of which the organized universe was ultimately composed.[122] When Homer calls him *γέρων ἅλιος νημερτής* (4.384), Heraclitus explains the choice of epithets allegorically:

> Τὸ *μὲν γὰρ* ⟨**γέρων**⟩ *οἶμαι τῆς ἀρχεγόνου καὶ πρώτης οὐσίας σημαίνει γεραίτερον, ὥστε ἀποσεμνῦναι τῇ πολιᾷ τοῦ χρόνου τὴν ἄμορφον ὕλην.* **Ἅλιον** *δ' ὠνόμασεν οὐ μὰ Δί' οὐ θαλάσσιόν τινα δαίμονα καὶ κατὰ κυμάτων ζῶντα, τὸ δ' ἐκ πολλῶν καὶ παντοδαπῶν συνηλισμένον, ὅπερ ἐστι συνηθροισμένον.* **Νημερτὴς** *δ' εὐλόγως εἴρηται· τί γὰρ ταύτης τῆς οὐσίας ἀληθουργέστερον, ἐξ ἧς ἅπαντα γεγενῆσθαι νομιστέον;*
>
> *Quaestiones Homericae* 67.2–4

"As to his being a *γέρων*, I believe that this means older than original, primordial substance, so that the poet dignifies unformed matter with the grayness of age. And Homer has called Proteus *ἅλιος* not at all because he is a kind of *marine* divinity who lives under the waves, but because he is *composed* of or *gathered* together out of many and diverse elements"—a willful derivation of the adjective not from ἅλς, "sea," but from ἁλής, "crowded, massed"—"and *νημερτής* is well said; for what could be more productive of truth than the substance out of which we must suppose that everything came into being?" To summarize the rest of the argument, in the various forms that Proteus assumes, Heraclitus finds symbols of the four Empedoclean elements into which the primordial matter disposed itself; in the vatic song that the sea god sings after these transformations, the principle of mind appears. Finally, Proteus' daughter, who introduces Menelaus into these mysteries and thus in a sense presides over them, is none other than Eidothea; and while it is clear that her name in fact refers to her divine beauty, Heraclitus, in another act of willful etymology, argues that it reveals her as the "goddess of forms."

The fairly obvious case of Proteus should alert the reader to the possibility that allegorical exegesis motivated Vergil's choice of other Homeric episodes. Upon inquiry we find that the episode with which

122. Buffière 1956.181.

he introduces the Proteus imitation also depends on Homeric episodes that were explained as physical allegories. Aristaeus' encounter with Cyrene, we should remember, is modeled on Achilles' two meetings with Thetis. In the first of these meetings, the hero asks his mother to intervene with Zeus to restore his lost τιμή. When Thetis complies with this request, the scene shifts to Olympus for the remainder of *Iliad* 1. Here I must mention two interesting myths that are related at significant points in this narrative. First, when Achilles asks Thetis for help, he cites as the reason for her influence with Zeus the assistance that she and Briareus rendered him some time in the past in a battle against three rebellious divinities, namely Hera, Athena, and Poseidon. Second, the book concludes with the gods' amusement at Hephaestus' own account of his fall from Olympus. These two myths bracket the Olympian episode introduced by Achilles' interview with Thetis; and both were explained in antiquity as elemental allegories.

In the insurrection myth Heraclitus sees an allegory based on pre-Socratic physics; he cites Thales, Anaximander, Heraclitus *philosophus*, and Empedocles in particular. The divinities involved in this myth stand for the elements of the physical world. Zeus, who represents the dominant substance, or fire, is conspired against by the other elements: Hera represents air (ἀήρ; and here again creative etymology comes into play), Poseidon the "moist substance," or water, for the obvious reason, and Athena earth, since she is the maker of all things and the divinity of craft.[123] In order to achieve some balance among the elements, there was a need for a rational principle to impose itself upon them; and this was the role of Thetis, who, Heraclitus observes, is aptly named: for she established a harmonious *arrangement* (the Greek word is ἀπόθεσις) among the four elements. Finally, the monstrous Briareus represents the brute force that reason frequently requires to actualize its schemes. Thus the story of the rebellion is revealed as an allegorical cosmogony.

The story in which Zeus hurls Hephaestus from Olympus, according to Heraclitus,[124] signifies the acquisition of fire by humankind. Human beings once lived without fire, he says, but at some point learned to capture sparks from the sky by setting bronze devices in the beams of the midday sun. He then refers to the myth of Prometheus' theft of fire,

123. *QH* 25. The logic of Athena as earth is not compelling, nor is the identification found in other myths concerning her; but in this passage ancient critics were determined to see an allegory of the four elements, as the three varying explanations of Eustathius (van der Valk 1971–1987.1.188–191 *ad Il.* 1.396–406) attest.

124. *QH* 26; cf. Cornutus *DND* 19, Σ *ad Odysseam* 8.284, 300 (Dindorf 1855.381, 383).

which he explains as another allegory of the same event. I may also refer to a farther-reaching explanation of the Homeric myth unknown to Heraclitus, but preserved in the *Iliad* scholia, which sees Hephaestus' fall into the ocean as a transformation of elements into one another, fire into air into water.[125] The scholiast misses a chance to squeeze in land by invoking the island of Lemnos, where the oafish god actually landed, but no matter. What is clear is the general way in which ancient scholars attempted to explain these myths, which taken at face value they found either opaque or embarrassing.[126]

What have these stories to do with Vergil's poem? The *Georgics* passage does not allude to the myth of Thetis and Briareus, nor to that of Hephaestus' fall, but only to the Iliadic episode that introduces these myths, Achilles' first interview with his mother in Book 1. At first sight, it is difficult to see how allegory might come into play. But we must remember that Vergilian allusion takes many forms. It is, for instance, characteristic of this poet to imitate episodes from earlier narratives not just for the events that they themselves contain, but also, as we have seen, with reference to some other narrative function.[127] This, I believe, is the key to the passage under discussion here. What the reader must realize is that Aristaeus' interview with Cyrene is similar to Achilles' first interview with Thetis not only dramatically but functionally. Both episodes, though not allegorized themselves, *introduce* passages that ancient critics regarded as allegorical cosmogonies. The Homeric episode introduces a change of scene to Olympus and a pair of physical allegories, the myth of Briareus and Thetis and that of Hephaestus. Vergil's imitation introduces only one such episode, that of Aristaeus' contest with Proteus.

125. Σ A *ad Il.* 15.189 = Σ B *ad Il.* 15.21 (Dindorf 1875–1877.2.69–70, 4.73–77).

126. On the role of Homer's critics in provoking these interpretations from his defenders see Buffière 1956.9–78; Lamberton 1986.15–22.

127. An obvious example: in *Aeneid* 1, Vergil models Aeneas' arrival in Africa on three distinct experiences of Odysseus. At lines 94–101, Aeneas "quotes" Odysseus' words as Poseidon tries to overwhelm him en route to Scheria (*Od.* 5.299–312). At lines 198–207, he makes a speech modeled on the words that Odysseus addresses to his men after arriving on Aeaea (*Od.* 10.189–197). Finally, at lines 314–409, the hero encounters his mother and divine patron, Venus, in a scene fraught with significant thematic reversals vis-à-vis its model, Odysseus' meeting with Athena upon reaching Ithaca (*Od.* 13.221–351). The three allusions have a greater than local significance; they are, to use Knauer's term, *Leitzitate* (1964a.145, etc.), indications of a more general correspondence between the actions introduced by these episodes in both the *Aeneid* and the *Odyssey*. As such, they invite the reader to question whether Aeneas will find in Dido an Arete, a Circe, or a Penelope.

What I believe confirms this interpretation is that the pattern is repeated. When Achilles for the second time complains to Thetis of his fate, the scene again shifts to Olympus, where Thetis takes it upon herself to intercede on her son's behalf, this time with Hephaestus; and the book concludes with a description of Achilles' divine shield. The shield, of course, was famously interpreted in antiquity as an image of the universe, and its construction, therefore, as a cosmogony.[128] It is clear from Heraclitus' introduction just how obvious he felt this particular allegory to be; and a long tradition of learned imitation and scholarly commentary has borne out his interpretation. Of course, we needn't equate Hephaestus with the Platonic *demiurge*, nor agree that Homer's real purpose here was to describe the creation of the universe. On the other hand, we can hardly avoid interpreting the shield in some sense as an image of the Homeric world, just as the description of the goatherd's cup in Theocritus' first *Idyll* is an image of the bucolic world; and, again, just as the shield described in *Aeneid* 8 assimilates the world of Achilles, the consummate Achaean warrior, to that of Aeneas, the prototypical Roman hero. Hardie sees in this last passage an impressive example of how allegorical exegesis of Homer influenced Vergilian imitation.[129] The historical scenes depicted on Aeneas' shield—a cycle of events culminating in the Battle of Actium—outline the establishment of a stable political order based on the supremacy of Augustus. By Hardie's argument, Vergil's decision to substitute these scenes for those found on Achilles' shield, in view of the overwhelming tendency of ancient critics to read the Homeric passage as an allegorical cosmogony, equates the establishment of Augustus' regime—the rebirth of the Roman state that had perished in the plague of civil war—with the creation of a stable and orderly universe under the rule of Zeus/Jupiter.

The point of this argument is that, in the *Aeneid*, Vergil's imitation of the Homeric shield suggests that he was familiar with the allegorical exegesis of that passage as a cosmogony. In the *Georgics*, he does not imitate the shield passage itself. He does, however, imitate the episode that introduces the shield passage, Achilles' *second* interview with Thetis, in combination with a similar episode, the *first* interview between these characters. We have already seen that the first of these interviews in the *Iliad* introduces both a change of scene to Olympus and a pair of stories

128. Heraclitus *QH* 43; Eustathius *ad Il.* 18.482 (van der Valk 1971–1987.4.219–221); cf. Reinhardt 1910.61; Mette 1936.36; Hardie 1986.340–346.

129. Hardie 1986.336–376.

that had been the subject of allegorical interpretation. Now we find the pattern repeated in the second. Thus in combining these two interviews into the single encounter between Aristaeus and Cyrene, Vergil is imitating not only the dramatic content of his Homeric models, but also their narrative function of introducing episodes of allegorical cosmogony. Instead of using the same stories that Achilles' interviews with Thetis introduce in the *Iliad*, however, Vergil, by a typical procedure of allusive variation, "replaces" them with a single episode, itself an allegorical cosmogony, from the *Odyssey*. Thus the chief components of Vergil's Homeric imitation play the same roles as their Iliadic and Odyssean models: Aristaeus' meeting with Cyrene, just like Achilles' two meetings with Thetis, introduces an episode of allegorical cosmogony; and the episode that is introduced centers, just like its Odyssean counterpart, on the vatic and ever-changing figure of Proteus. These are the central passages of the Homeric poems from the point of view of natural philosophy, because they reveal Homer as a poet concerned not just with the legends of heroic saga, but with the very grandest of philosophical themes, the creation of the universe. In selecting these episodes as the basis of his synecdochic imitation of Homer's epics, Vergil clearly aligns himself with the tradition that regarded Homer in this light.

So much for the main components of the "Aristaeus." What is equally interesting is the way in which Vergil alludes in this same episode, apparently quite casually, to other Homeric passages as well, because here too his choice of models is influenced by allegorical criticism. When Cyrene first hears Aristaeus' complaint, she has been listening to the song of her sister Clymene, whose theme is of great interest:

> curam Clymene narrabat inanem
> Volcani, Martisque dolos et dulcia furta,
> aque Chao densos divum numerabat amores.
>
> *Georgics* 4.345–347

Clymene, that is, has been telling the story of Hephaestus, Ares, and Aphrodite—the same story that Demodocus tells in *Odyssey* 8—but has included it in a theogony instead of in a heroic song. This detail seems clearly to indicate Vergil's awareness of the allegorical tradition surrounding the Homeric episode. It also anticipates the allusive technique that we find in the "Song of Iopas" in *Aeneid* 1, which I have already discussed.[130] Vergil's treatment of the reference is consonant with what we have observed previously: the Homeric myth is made part of a Hesiodic theogony. Further, both of these august archaic themes are cast in a form that renders them suitable to the interests of first-century poets and readers. Vulcan's *curam inanem*, Mars's *dulcia furta*, Chaos, the source of countless *divum amores:* here Vergil wittily treats of cosmogony and theogony as mere branches of love poetry. Thus, as in Book 1 but with a different emphasis, he joins Hesiod and Homer under the umbrella of a sensibility formed by the poetry of Alexandria and of the Neoterics.[131]

Again, later in the "Aristaeus," Cyrene prays on her son's behalf to "Father Ocean and my sister Nymphs":

> Oceanumque patrem rerum Nymphasque sorores
>
> *Georgics* 4.382

This line is modeled on three of Homer's:

> Ὠκεανόν τε, θεῶν γένεσιν, καὶ μητέρα Τηθύν
>
> *Iliad* 14.201=14.302
>
> Ὠκεανοῦ, ὅς περ γένεσις πάντεσσι τέτυκται
>
> *Iliad* 14.246

The last cited of these lines is specifically adduced by Heraclitus (*QH* 22) as evidence that Homer was already aware of the ideas usually associated with pre-Socratic philosophy. Moreover, the context in which the lines are found is extremely relevant: they occur in the "Deception of Zeus" (*Iliad* 14.153–351), a passage that ancient scholars explained as an allegory of spring. The union of Hera, who appears in her familiar elemental capacity as ἀήρ, and of Zeus, who is identified with the fiery αἰθήρ, represents allegorically the union of those elements, according to one interpretation, is what produces the season of spring.[132] Another view represents the union of air and ether as a more fundamental event, even as the source of everything that exists.[133] Both of these interpretations suit Vergil's purposes admirably. In Book 2, he introduces in the

130. See pp. 258–261.
131. Cf. p. 216.
132. Heraclitus *QH* 38–39.
133. [Plutarch] *De vita et poesi Homeri* 96 (Bernardakis 1896.380–381)

hyperbolic "Praise of Spring" (315–345) the motif of the *hieros gamos*, to which the ancient Homeric commentators also refer in discussing "The Deception of Zeus."[134] Moreover, Vergil himself, like Lucretius before him, explicitly draws the connection between spring weather and the conditions that must have prevailed at the world's beginning (*G.* 2 .336).[135] Thus his brief reference to the "Deception of Zeus" in Book 4 looks is if it too were motivated by an effort to relate the main themes of the *Georgics* to the tradition of interpreting Homer as an allegorical poet of natural philosophy.

This view of Vergil's "Aristaeus" does not require us to look back over his entire program of Homeric imitation in an effort to relate all of it to the allegorical tradition. It does, however, put that program in an entirely new light. What are we to conclude about this remarkable design? Three things, I think. First, that the Homeric allusions of Books 1–3 and those of the bee discourse in Book 4 build toward the "Aristaeus," partially adumbrating the themes that are central to that climactic episode. Second, that Vergil united in the "Aristaeus" allusions to the central themes of the Homeric poems as determined by the allegorical tradition that explained Homer as a poet of natural philosophy. Indeed, as an interpretive effort in its own right, the "Aristaeus" and the *Georgics* as a whole must be placed squarely within this tradition. Third, by imitating Homer in this way, Vergil shows that he has really not departed from the literary tradition defined by his sources in the earlier books of the poem. Instead, he indicates the common ground between Homer and other poets of nature, and thus identifies Homer's heroic epos with the didactic epos of Hesiod, Aratus, and Lucretius.

134. Σ D *ad Il.* 15.18 (Dindorf 1875–1877.4.73–74), with further reference to Hesiod's *Theogony*.

135. See p. 103.

PART III

The Georgics *in Literary History*

CHAPTER 7

Vergil's Early Career

Literary History in the *Georgics*

The argument of the preceding chapters presents the *Georgics* as an unusually elaborate allusive design. It illustrates, I hope, the inadequacy of reading the poem as an imitation of any single model. Rather, the poet makes a more ambitious claim about his poem's relationship to the literary past. More than a Roman version of *Works and Days* or an updated *De Rerum Natura*, the *Georgics* embodies a coherent literary-historical statement about the entire tradition that produced it. Moreover, by its tendentious reading of its poetic forebears, both individually and in their relations one to another, it manages to define itself as the consummate representative of that tradition. So pronounced is this feature of Vergil's allusive artistry that it seems appropriate to speak not of the poetic tradition that created the *Georgics*, but rather of the tradition that the poem itself, as an encrypted literary-historical essay, calls into being.

The form in which Vergil chose to present his vision of the literary past raises further questions. I mentioned earlier that we appeared to have in the *Georgics* a deliberate design, in which one model succeeds another.[1] I would emphasize the word "deliberate." Even in recent times good scholars have found it possible to interpret the most sophisticated allusions as evidence of how the force of tradition works upon the passive imagination of a poet. It is commonly held, for instance, that those *Eclogues* bearing the strongest relation to specific passages of Theocritus must necessarily be earliest, and that allusions to other authors betoken

1. See pp. 131–134.

some "new" interest on the part of the poet.[2] To return to the *Georgics*, the old view that the homerizing Book 4 was composed after the *Aeneid* was under way has returned in a new guise, namely, that the "new direction" represented by the "Aristaeus" and continued in the *Aeneid* is the result of Vergil's coming to grips with the essence of Homer for the first time only after completing *Georgics* 1–3.[3] Such notions beg important questions of artistic unity, and fly in the face of what we know about Vergil's deliberate working methods. It makes much more sense, particularly in light of the evidence that I have presented so far, to see that arrangement of models in the *Georgics* as a matter of choice reflecting the poet's conscious design—the imposition of his will on the literary past, and not the reverse.

The question is, What does this design mean? Beyond our primary conclusion, that Vergil is presenting Homer as an important part, perhaps even the source, of the didactic tradition more frankly represented by the poem's three other major epic models, what, if anything, can we infer from the sequence that the poem describes, from Hesiod and Aratus through Lucretius to Homer? This question may not finally be answerable; but it is worth making the attempt, if only to point out conclusions that we can not legitimately draw. It will be possible, in any case, to show what significance Vergil's arrangement of models has for modern literary historians. In these final two chapters I will try to use what we have learned so far to situate the *Georgics* in the literary history of first-century Rome.

The Neoteric Background

The most influential critics, particularly in the English-speaking world, have for some time been accustomed to a literary-historical model that sees the mid-first century B.C. as a watershed.[4] It was then, they argue,

2. Representative of this view is Coleman 1977.16–17 and e.g. 205 (on *Eclogue* 6: "The Callimachean form of Vergil's opening *recusatio* to Varus (see 3–4n.) now assumes its full importance: Callimachus, not Theocritus, is now the dominant influence in his poetic thinking. He is nearing the end of his career in pastoral."

3. Knauer 1981.900, for instance, traces the differences that he perceives between the earlier books and Book 4 to Vergil's "growing acquaintance with Homer." Cf. his remarks on pp. 912–914.

4. The views that I summarize here are essentially those of Clausen and his "school" (see pp. 14–17). These scholars take a rather extreme position, particularly with regard to Callimacheanism, and my own ideas are closer to those who emphasize other influences together with that of the Alexandrians, as will become clear. I have framed this discussion

that Latin poetry matured, became independent of inhibiting social structures like the *patrocinium*, assimilated fully the lessons of the Hellenistic scholar-poets, and found a voice with which to express even the most exquisitely subtle concerns of a sophisticated culture more completely and authentically than it had done before. This literary coming of age is seen as the contribution of a small group of poets known as the Neoterics, a group defined differently by different scholars but, for most practical purposes, represented for us chiefly by the one poet whose work survives in bulk, Catullus. Careful and intensive research into the nature of this poetry has produced in many the conviction that, despite certain superficial dissimilarities, the poetry of the next generation—the so-called Augustan poets—remained Neoteric in its essential character; that it was, in the words of one scholar, "a natural growth in the soil prepared by Catullus."[5]

This conception of how Latin poetry developed has been built up gradually and with care by many scholars over many years. It is, in my view, largely correct, certainly an essential piece of the puzzle. And yet, given the prevalent understanding of the Neoteric program, it is difficult to explain the characteristic forms and themes of Augustan poetry in exclusively Neoteric terms. The Neoteric revolution, the argument goes, was in favor of Callimachus, but, just as important, was against a native tradition represented by Ennius and his followers—chief among them, to speak only of surviving poets, being Catullus' near-contemporary, Lucretius—a tradition that based itself, ultimately, on frank imitation of Homer. The Neoterics, it is argued, followed Callimachus in rejecting this sort of poetry; and, in order to prove that it was the Neoteric tradition that exerted the essential influence on the Augustans, some scholars have found it necessary to argue that their supposed rejection of Homer, Ennius, and Lucretius was maintained by the Augustans as well.[6]

---

as, to some extent, a reply to the more extreme position for two reasons. First, it is by far the clearest and most articulate analysis of the literary-historical forces at work in Vergil's world, even if some of its clarity is owed to a certain reductivism. Second, though I am generally in sympathy with scholars who take a broader view of those forces, no one, in my opinion, has successfully challenged the radical Alexandrian position, or made it more compatible with other views. (Clausen 1987 and Thomas 1985 themselves both recognize the need to do so; but in the end, they continue to emphasize Alexandria.) If we really want to understand Vergil and his contemporaries, this seems to me the necessary next step.

5. Ross 1975.2, 163.

6. Ross, for example, has consistently refused to see substantive Lucretian influence on Vergil (1975.25–26, 29–30; Feeney 1988.476). Likewise Thomas (1988.1.4) minimizes Lucretius' importance as a model for the *Georgics*.

Thus many have remarked on the *a priori* unlikelihood that the poet of the *Eclogues* would someday become the poet of the *Aeneid.*[7] If one accepts the current view, Vergil's masterpiece—a long, heroic epic written in frank imitation not only of Homer, but of Ennius and even Naevius as well—presents a serious problem. At first glance, Vergil's very decision to attempt such a poem looks like a repudiation of his earlier Neoteric principles. But on the other hand, the degree of Hellenistic influence on the poem is well known. No serious scholar will deny the influence of Apollonius and, in a more general way, of Callimachus on the *Aeneid.* "Repudiation" therefore seems to strong a word. And yet a sense of uneasiness remains. On the one hand, the *Aeneid* unarguably does participate in both a Homeric/Ennian and a Callimachean/Apollonian tradition.[8] On the other hand, this fact is difficult to understand fully in light of current scholarship.

To the extent that a poem of overwhelmingly Homeric form and content violates Callimachean and Neoteric principles, the *Aeneid* is an un-Callimachean, un-Neoteric poem. But it may be that Vergil in writing did not so much reconcile opposing forces as take advantage of similarities not readily apparent to us. I suggest that one can read in the allusive program of the *Georgics* the clear indications of these similarities. By studying this program we rediscover, given Vergil's place both in the contemporary poetic scene and in literary history, what use he might have had for Homer. But before we complete our investigation of the *Georgics*, let us turn to examine the *status quo ante* as it was perceived by Vergil himself.

The *Eclogues*

Vergil's career, which progressed steadily from the lower to the higher epic genres and embraced the *tria genera dicendi* of rhetorical theory,

7. See especially the incisive remarks of Zetzel 1983b.97 on the magnitude of the step even from the *Georgics* to the *Aeneid.*

8. Some scholars deliberately blur the distinction between these two traditions. Newman 1967.61–98, for instance, argues correctly that Ennius and Lucretius must themselves be understood in terms of Hellenistic literary ideas rather than as representatives of a hypothetical "native Roman tradition." Instead he distinguishes Ennius' Hellenistic tendencies from the Alexandrianism of the Neoterics, although he does not always do so with sufficient clarity (see especially pp. 95–96); and his concept of "Alexandrianism" itself (pp. 437–454), particularly in his later work (Newman 1986), is poorly defined.

became a paradigm of the ideal poetic career.[9] One must therefore be careful of succumbing to the idea that Vergil's development was preordained.[10] It is very unlikely that Vergil was so prescient as to have accurately mapped out his career in advance. Nevertheless, there is a remarkable developmental unity in his literary-historical observations throughout his works. What follows here is a selective survey of those observations, which I hope will illuminate his relationship to the epic genre and to Homer in particular.

The Pollio Eclogues[11]

In the *Eclogues* Vergil follows an elaborate program of allusion to Theocritus. The choice of model indicates the poet's Alexandrianism, while the style of the collection has many affinities to Neoteric poetry. It is of course common for Latin poets to define the essential qualities of their work both by celebrating specific examples of their literary ideals and by contrasting their work with poetry belonging to other traditions. Such definitions might be made in open declaration, or by using a more veiled approach. The *Eclogue* poet, too, makes clear his orientation by contrasting his work with examples from the higher genres, including tragedy and heroic epos. But it is not always clear in such instances that the generic distinction is congruent with a stylistic one. Indeed, there are cases in which the poet contrasts his own work with that of another in terms of their differing generic backgrounds, while nonetheless celebrating their mutual adherence to Alexandrian and Neoteric ideals. I can illustrate the procedure with reference to the ways in which Vergil honors one of his patrons [12]

9. Curtius 1973.232; cf. 201 n. 35

10. Cf. Zetzel 1983b.93, who discusses Vergil"s development in the context of a more general literary-historical revisionism on the part of the Augustan poets.

11. On the connections between these *Eclogues* in general see Desport 1952.229–242; Berg 1974.158–162. On *Eclogues* 3 and 4, see Segal 1977 = 1981.265–270.

12. I use the traditional word without meaning to imply that the poem was offered by Vergil in return for services rendered, that he was a paid servant, or anything of the kind. From my remarks it will become clear that I agree with those who see the choice of addressee as governed by poetic considerations as much as or, in some cases, more than by social ones. I have benefitted from discussion of this point with my colleague, Matthew Santirocco, who is engaged in a study of literary patronage in the Republic and early Empire; until this project is completed, we shall have to content ourselves with his papers on Horace, Tibullus, and Propertius (1982 and 1984). See also Zetzel (1982).

Asinius Pollio was an influential literary figure in the late Republic. Catullus in *Carmen* 12 makes a playful indictment of Pollio's brother,[13] Asinius Marrucinus, for (supposedly) pilfering a napkin. The addressee is sharply contrasted with his kinsman. Marrucinus does not use his left hand *belle* (2), stealing napkins is not *salsum* (4), he is addressed as *inepte* (4), and the whole matter denounced as *sordida...et invenusta* (5). If Marrucinus refuses to believe Catullus, then he can ask Pollio, who would give anything to change this behavior (6–8): *est enim leporum* | *differtus puer ac facetiarum* (8–9). Thus the two kinsmen are opposites, Marrucinus lacking those qualities of *urbanitas* that Catullus so admires, Pollio imbued with them. Now, the terms in which Catullus praises Pollio and blames Marrucinus are the same as those of his literary criticism. Moreover, not only things but people in Catullus are frequently endowed with the qualities of a good or a bad poem; and, since we know that Pollio was a poet,[14] it makes sense to assume that this witty poem of trifling invective was written, at least in part, to express Catullus' approval of Pollio the poet.

Catullus was not alone in his admiration of Pollio. Cinna wrote a propempticon to Pollio, a poem which in late antiquity was arguably his best known work.[15] The young man in turn became influential in fostering poetic careers: he mentions Cornelius Gallus in a letter to Cicero touching on a literary curiosity (*Fam.* 10.32.5), and is honored as a patron by Vergil in the *Eclogues*. It is here that we can glimpse something of the character of his own poetry.

Eclogue 3 contains a rustic singing contest between the shepherds Damoetas and Menalcas. Exactly halfway through the match the name of Pollio appears (86). Its central position lends emphasis to what is anyway a startling occurrence, the intrusion of a living person into the imaginary rustic discourse of the shepherd poets.[16] More than startling,

13. Or cousin (i.e. *frater patruelis*)? We know very little about this man, and the possibility that he was a cousin rather than a brother of Pollio seems to me at least as likely as the assumption that their father assigned them different *cognomina* (Fordyce 1961.129).

14. Schanz–Hosius 1927–1935.2.26–27.

15. We have five fragments of this poem, as compared with three of *Zmyrna* and six from various other poems.

16. Contemporary figures are named again in *Eclogues* 4, 6, 9, and 10, but the mention of Pollio at 3.86 is the first such reference in the *Eclogue* book; the sculptor Alcimedon (3.37 and 44, not otherwise known) and the third-century astronomer Conon (3.40) are in a different category.

Pollio's appearance is in fact openly disruptive: it changes the course of the singing match and the tone of the poem. Before he is mentioned, the contestants exchange six pairs of couplets; beginning with his name, they exchange another six pairs.[17] The first half of the contest is full of pleasant motifs of pastoral love: lusty nymphs, love offerings, assignations, and so forth. The second half is full of foreboding: a warning to those picking flowers and berries to beware the snake that lurks in the grass; a dangerous river bank; excessive heat that may dry up the sheep's milk.[18]

What causes this change? Or, rather, why does Vergil associate it with the sudden appearance of Pollio's name? If we consider again the structure of the singing match, we notice that each half begins with an honorific preamble:[19]

> D. Ab Iove principium, Musae: Iovis omnia plena;
> ille colit terras, illi mea carmina curae.
> M. Et me Phoebus amat; Phoebo sua semper apud me
> munera sunt, lauri et suave rubens hyacinthus.
>
> *Eclogue* 3. 60–63
>
> D. Pollio amat nostram, quamvis est rustica, Musam:
> Pierides, vitulam lectori pascite vestro.
> M. Pollio et ipse facit nova carmina: pascite taurum,
> iam cornu petat et pedibus qui spargat harenam.
> D. Qui te, Pollio, amat, veniat quo te quoque gaudet;
> mella fluant illi, ferat et rubus asper amomum.
> M. Qui Bavium non odit, amet tua carmina, Maevi,
> atque idem iungat vulpes et mulgeat hircos.
>
> *Eclogue* 3.84–91

As the contest begins, Damoetas follows Aratus in beginning with Zeus/Jupiter. In fact, he translates Aratus quite closely before introducing an element all his own:

17. Cf. Van Sickle 1978.129 n. 65. Others divide the singing match into three parts: see Büchner 1959.172 (= *RE* 8A 1192); Leach 1974.176. For a more complex scheme see Segal 1967.296 = 1981.252.

18. Cf. Leach 1974.177: "The search for subjects culminates in references to the patronage and favor of Pollio, and here the singers' rivalry takes an explicitly unpleasant turn."

19. Cf. Klingner 1967.56; Putnam 1970.129–130.

ἐκ Διὸς ἀρχώμεσθα. τὸν οὐδέποτ᾽, ἄνδρες, ἐῶμεν
ἄρρητον· μεσταὶ δὲ Διὸς πᾶσαι μὲν ἀγυιαί,
πᾶσαι δ᾽ ἀνθρώπων ἀγοραί, μεστὴ δὲ θάλασσα
καὶ λιμένες· πάντη δὲ Διὸς κεχρήμεθα πάντες.

Phaenomena 1–4

Aratus does not make the claim that Zeus cares for his song. Damoetas' assertion would be arresting in any case, but the departure from his model makes it more so. A second point is the fact that Damoetas addresses the Muses. This is not startling in itself, but it is another departure from Aratus.[20] When we look ahead to the second invocation, the reason behind these allusive variations becomes clear. Pollio also loves Damoetas' poetry; for him too the Muses (here Pierides) are invoked. Thus the structure of the poem exploits the commonplace of equating one's patron with Jupiter/Zeus. By developing these verbal and thematic echoes, Vergil makes it clear that Pollio in this poem plays the role of a patron.

But what sort of patron? First, the patron of the bucolic poet: the Pierides are to raise a heifer for their reader (85). The image is Callimachean[21] as its subsequent appearance in *Eclogue* 6 will make clear. But Pollio is not just a patron: he is also a poet who makes *nova carmina* (86). Nothing here seems inconsistent with the Neoteric message of the previous couplet; if anything it is confirmed. But after observing that Pollio is himself a "new poet," Menalcas bids someone—presumably again the Muses—raise a bull, one that is ready for the ring. Now the conventions of the singing contest require that the second singer try to "cap" the first, and this explains the progression from Damoetas' *vitulam* (85) to Menalcas' *taurum* (86)—but only in part. The heifer is a perfect symbol for bucolic poetry. By bidding the Pierides raise a heifer for Pollio in his capacity as a reader, Damoetas is in effect dedicating a poem to him. Menalcas' response is obviously in much the same

20. *Musae* (60) is often taken as genitive, e.g. by Coleman 1977.117 *ad loc.*, who cites Cicero's translation of Aratus (*A Iove Musarum primordia*, *Aratea* fr. 1 [Soubiran 1972.158]) as a parallel. But when Vergil borrows from a Greek source via an earlier Latin version, he habitually alters the Latin in some small way. This is especially true when the alteration demonstrates Vergil's knowledge of what lay behind his source—in this case, the proemium of *Works and Days*, Aratus' own model (see p. 163). Thus Cicero's translation actually suggests that *Musae* in Vergil is vocative (cf. my remarks on *aërius*, pp. 222–223).

21. *Aetia* 1, fr. 1.23–24 (Pfeiffer 1949.5).

vein; but when he mentions Pollio as a poet in his own right and then bids the Muses raise for him the more aggressive bull, is he not hinting at a difference between Pollio's interests and his own bucolic discourse?[22]

If this interpretation is correct, some rereading is required. We have been told that Pollio loves Damoetas' Muse, rustic though she be (84). At first we note nothing more than the conventional deprecation of his lowly genre and uninflated style by the pastoral and Neoteric poet. But, after finding that Pollio himself writes poetry of a loftier strain, we may read more into the conventional disclaimer: he loves Damoetas' rustic Muse, even though she differs in this respect from his own.[23] Second, when we hear that Pollio writes *nova carmina* (86), our natural inclination is to think that he is being praised as a Neoteric poet; but this "new poetry" is evidently something very different from Vergilian pastoral, and perhaps something new to Pollio as well.

This emphasis is instructive for what follows. Pollio, we next learn, is going somewhere, and all his friends should follow (88–89). There ensues the famous reference to Bavius and Maevius (90–91), and then a series of cautions (92–103) before the contest ends (in a draw, 108–111) with a pair of riddles (104–107). Not all the details disclose their meaning readily or clearly. In general, however, they all seem, at least on one level, to be concerned with the writing of poetry.

It seems necessary, given the lines that precede it, to interpret Pollio's journey as literary: he is striking out into the higher genres.[24] Such a reading is perfectly consonant with the earlier reference to Pollio's *nova carmina* (86) and, in fact, enhances it. The Neoteric stress on novelty is familiar from other contexts.[25] It may even by this time have come to sound rather jejune. Here though there may be a more than just a hint of the unexpected about Pollio's poems. It is difficult to be very specific since we actually know very little about them; what we do know is that Pollio was hailed by his contemporaries and remembered in aftertimes as a tragedian.[26] This in itself may be enough. It was

22. Compare Leach 1974.178, who, in keeping with her general emphasis on the antagonism between the two shepherds, sees Menalcas' couplet as a riposte exposing the overweening ambitions of Damoetas.

23. Cf. Segal 1967.295–296 = 1981.251–252.

24. For poetic composition as a journey see Becker 1937.68–85; Wimmel 1960.103–111.

25. Most notably Catullus *Carmen* 1.1 *lepidum* ***novum*** *libellum*.

26. Pollio is named as the tragic poet of his day by Horace (*Sermo* 1.10.42–43), and is later mentioned (though with no great approval) by Tacitus' Aper (*Dialogus* 21.7), who

certainly an important factor for Vergil. In the proemium to *Eclogue* 8, another singing match, Vergil does not name Pollio, but a reference to his Sophoclean tragedies marks him as the most likely addressee.[27] The context alludes both to Pollio's *facta* (8) and to his *carmina* (10), the one military and the other tragic. Then an important clue: *a te principium, tibi desinam* (11). The Aratean allusion reminds us of the similar passage in *Eclogue* 3, with its parallelism between Pollio and Jupiter. Finally, an invitation: the poet bids his patron accept the poem that he requested, and let his temples be bound by pastoral ivy and heroic laurel—the latter a second reference to both Pollio's *facta* and his *carmina*—together (11–13). Here a deliberate contrast is drawn between Pollio's poetry and Vergil's: the entire passage is a type of *recusatio* in which the *Eclogue* poet excuses himself from celebrating Pollio's military feats, and contrasts his own poetic endeavor with Pollio's high tragic style. We may even observe that Pollio's tragedies are made the equivalent of his military deeds in a surprising way: they are cited as a suitable subject for panegyric (cf. 7–8 with 9–10).

This is the man of whom the shepherds speak in *Eclogue* 3. Here, of course, they ignore Pollio's military endeavors, referring only to his poems; but nothing they say should cause us to question the congruency between these two spheres that Vergil exploits in *Eclogue* 8. Nothing, that is, except the phrase *nova carmina* (3.86). If we take this phrase in a very narrow sense, then we shall expect it to mean that Pollio writes Neoteric *nugae*.[28] But of course, the "Neoteric program," if we can use such a term, was never confined just to this. Catullus' range is sufficient indication of its scope in the generation before Vergil. The loss of Gallus and Pollio, poets who loom so large in the *Eclogues*, is an immense frustration; but everything we know of their work attests to the fact that they were central figures in the Neoteric movement, and at the same time deeply committed to the suitable expression of serious themes.[29] In Pollio's case, this commitment took the form of composing Sophoclean tragedies, an endeavor that seems on its face completely at variance with the Callimachean attitudes of first-century Rome. Callimachus famously

---

finds him reminiscent of the pre-Neoteric tragedians Accius and Pacuvius.

27. See pp. 65–67.

28. The view held by, e.g., Coleman 1977.86 *ad loc.*

29. On Gallus see especially Ross 1975.18–38 on the Gallan background of *Eclogue* 6 as a poem of "science."

ridiculed the bombast of tragedy,[30] and Propertius (2.34.41–42) and Ovid (*Amores* 3.1) were to make tragedy a specific generic and stylistic antitype of their elegies. And yet Callimachus is not above echoing Euripides in his most important apologia,[31] nor was the tragic form unknown to the Alexandrian Lycophron of Chalcis; while Ovid's poem must be understood as an act of self-definition performed by a poet who had written tragedy with notable success.[32] What we must remember, but too often forget, is that Callimachean polemics cannot be reduced to simple categorical injunctions against writing long books, epic verse, or imitations of Homer; Callimachus himself did all three.[33] What counts, always, is the manner of the composition. The same holds true for the Neoterics, whose program, if indeed it did begin with narrow emphasis on small, non-epic, unheroic forms, nevertheless continually expanded. It is in this context that Vergil mentions Pollio: as a poet engaged in expanding the horizons of the new poetry.

There can be no doubt that the reference to Pollio, here as in *Eclogue* 8, is laudatory. When Damoetas claims him as a patron (3.84–85), Menalcas responds by hailing him as a New Poet, a member of that charmed inner circle of Roman letters to which Catullus once and now Vergil himself belonged (86–87). Damoetas then praises the direction that Pollio's poetry has taken by recommending that others follow his lead and wishing them fabulous success (88–89). That this succession of beatitudes is proffered sincerely is guaranteed by the next couplet. It matters little whether Servius is correct in seeing historical personages behind the masks of Bavius and Maevius.[34] Menalcas heaps on them his unrestrained contempt, wishing them ill in terms that exactly invert the success that Damoetas desires for Pollio's followers. The point of mentioning this ill-omened pair is precisely the contrast that they provide with Pollio and his friends, including Vergil. Damoetas and Menalcas are

30. In his *Iambi* (fr. 215 Pfeiffer 1949.211, with testimonia).

31. Pfeiffer 1949.8 *ad Aetia* 1, fr. 1.36. Of the tragedies he is supposed to have written (*Suda* 10.227 s.v. Καλλίμαχος [Adler 1928–1935.3.19–20]; cf. *Iambus 13* [fr. 203 Pfeiffer 1949.205–209]), nothing actually survives.

32. Quintilian *IO* 10.1.98.

33. On the *Aetia* as a *magnum opus* see Knox 1986.23 n. 7. In addition to Callimachus' best-known epic, the *Hecale*, we have the remnants of a *Galatea* (frr. 378–379 Pfeiffer 1949.304–306). On his imitation of Homer see von Jan 1893.

34. Servius *ad Ecl.* 3.90 (Thilo 1887.40–41); cf. Horace *Epod.* 10.1 and Porphyrio *ad loc.* (Holder 1894.204).

simple shepherds. Their praise and blame is explicit and open. Their exclusion of Bavius and Maevius from the society of poets whom they admire underlines the evident approval accorded to Pollio.

At the same time, Pollio's tragedies serve as a foil for Vergil's own bucolic poetry. The imagery of the distichs that follow the appearance of Bavius and Maevius is at once forboding and literary in reference.[35] Gathering flowers is of course a common image for poetic composition; here, however, there is said to be danger from snakebite (92–93). Writing poetry is not, apparently, the carefree pursuit it is elsewhere (in Theocritus, for instance) presumed to be. The specific intent of this cautionary remark is not immediately clear; but a similar note is sounded in the next couplet. The sheep, Menalcas says, must not go too near the treacherous bank; even the more powerful ram has fallen in (94–95). This contrast between ewe and ram recalls the earlier one between heifer and bull (84–87), a context in which we saw a distinction between Vergil's pastoral songs and Pollio's more "ambitious" endeavors. Here the dichotomy is reasserted in conjunction with the potent image of the river, which runs on into the next couplet. There Damoetas enjoins Tityrus to drive the grazing nanny goats away from the river; the singer himself (*ipse* 97) will wash them in the spring at the proper time (96–97). The contrast between river and spring is familiar to readers of Callimachus' second *Hymn.* "I admire not the poet whose song is not as great as Ocean," says Envy to Apollo (106). Apollo replies by kicking Envy, and by contrasting the muddy stream of the great Euphrates unfavorably with the pure trickle from a sacred spring (107–112). Tityrus, of course, besides being a kind of emblem of the *Eclogues*, is in several passages a stand-in for Vergil himself in his capacity as the *Eclogue* poet. By ordering Tityrus away from the river and offering to take the goats to the spring himself, is Damoetas not giving pastoral form to this Callimachean idea? His purpose in doing so, I believe, is to justify pastoral poetry—the *Eclogues*—in Callimachean terms, partly, perhaps, with reference to carping *Invidia* in the persons of Bavius and Maevius, but more importantly as regards the "higher" poetry of Pollio. We need not assume that Vergil is doing anything so petty as accusing his friend of being un-Callimachean; Pollio's Neoteric credentials have already been

35. Cf. Klingner 1967.57, however, who relates the unpleasant imagery to Bavius and Maevius; Putnam 1970.130, with reference to the potential for self-destruction inherent in the bucolic world; Leach 1974.179–180, who relates it to Damoetas and Menalcas' quarrel.

submitted (86). Rather, he is exploiting literally the imagery of the hymn in support of his own slighter genre. By doing so he anticipates the motif that provides closure to this, the longest of the *Eclogues.* When Palaemon finally calls a halt to the contest (and the to poem itself), he says *claudite iam rivos, pueri; sat prata biberunt* (111). Overabundant irrigation is incompatible with the bucolic epos.

Menalcas answers Damoetas' warning to Tityrus with more foreboding, expressing fear that excessive heat will cause his sheep's milk to dry up (98–99)—possibly an image of inspiration run out. He orders the workers to collect the sheep and, presumably, drive them to shade. We think of the shade that everywhere is the distinguishing emblem of the pastoral world, of pastoral poetry. Damoetas then returns to the image of the bull, not, this time, tossing his head and pawing the ground, like Pollio's aggressive bull (86–87), but standing thin and lovesick in a verdant pasture (100–101). One is reminded of the commonplace that asserts the power of love even over the strongest heroes, a figure for the superiority of the humbler genres to the more exalted, serious ones. Menalcas responds with an image of lambs emaciated not from love, but from a magic spell (102–103)—again, perhaps, an oblique reference to poetic *Invidia.* Then, finally, the riddles. Damoetas first poses a cosmic riddle:[36] Where in the world does the expanse of heaven extend no more than three cubits? The solution, for our purposes, is immaterial. What is important is that the terms of the riddle recall astronomical motifs elsewhere in the poem.[37] Most notably, at the center of each of the chased cups that Menalcas puts up as his stake in the contest stands an astronomer, Conon, and another figure whose name the shepherd fails to remember (40–42). The allusion to Aratus with which Damoetas begins the contest (60–61) is a second example. In light of what we have observed so far, these repeated references to astronomy take on a poetic relevance. Menalcas responds to Damoetas' wager by noting that he already has a pair of cups by the same artisan depicting, not Conon or any other astronomer, but Orpheus (44–46). Thus astronomy and poetry are in a sense equated, or, better perhaps, united in the serious poetry of nature represented by Orpheus.[38] Others have seen Damoetas' allusion to Aratus, himself a poet of nature and astronomy, as "grandiose"

36. For the term see Ohlert 1912.83–105.

37. Cf. Segal 1967.297 = 1981.253.

38. For Orpheus as a Vergilian emblem for scientific poetry see Ross 1975.23–27, 93–96, and *passim;* 1987.227–229.

or "pretentious."[39] Segal in particular, who sees in Damoetas' invocation "a distinct departure from [Vergil's] immediate source" (Theocritus *Idyll* 5.80–83), notes that "this divergence is only part of a larger attempt at a redefinition of pastoral in loftier terms."[40] While I agree that a large part of Vergil's pastoral program is a generic transformation, there are two sides to this coin. Just as Vergil exalts the genre vis-à-vis Theocritus, he is also at pains to distinguish it from other, "inherently" more exalted genres.

Menalcas' riddle uses different terms to express a similar idea. Where in the world, he asks, do flowers grow inscribed with the names of kings (106–107)? Here there is more consensus: the flower that Menalcas has in mind is the hyacinth, the petals of which were marked, according to the ancients, with the Greek letters "AI."[41] Variant *aetia* explained the mark as a reference to the name of Ajax or to Apollo's cry of grief for Hyacinthus.[42] As with the first riddle, however, it is not the solution that interests us, but rather the terms in which Menalcas poses it. The paradoxical image of a flower inscribed with the name of a king perfectly embodies the contrast that I have been tracing throughout this contest poem between the bucolic and heroic modes—the contrast between Vergilian pastoral and Pollio's tragedy—which is subsumed within the larger unity of the aesthetic ideals that both poets share. As such, the image of the hyacinth resolves the dichotomy that informs the entire poem while establishing a welcome and necessary individuality for both the *Eclogue* poet and his literary friend.

Though some details remain obscure, it seems to me that one point is clear. Vergil is playing the familiar game of distinguishing himself from his patron, but this time in purely literary terms. Pollio, himself a Neoteric poet, has turned to the higher genres. Vergil, Pollio's admirer, has devoted himself to the quintessentially humble bucolic mode. By the terms in which he characterizes Pollio's shadowy endeavor, he expresses in part the superiority of his own choice; but Vergil is clearly not adducing Pollio as a reactionary or as an anti-Callimachean: *Pollio et ipse facit*

39. Segal 1967.295 = 1981.251; Leach 1974.177.

40. Segal 1967.295 = 1981.251.

41. Euphorion fr. 40 (Powell 1925.38), Ovid *Met.* 10.215.

42. The flower first grew either from the blood that Ajax spilled in suicide or from that of Hyacinthus when Apollo accidentally killed the boy with a discus. Both versions were current, and Ovid, typically, relates each of them in the *Metamorphoses* (10.162–219, 13.382–398). The minor variants preserved by Junius Philargyrius *ad Ecl.* 3.106 (Hagen 1902.70–71) are not to the point.

nova carmina. Rather, he is a poet whose instincts and abilities Vergil thoroughly admires, but one who writes within a different generic tradition. What we are observing in both *Eclogues* 3 and 8 is no more than the expression of generic "rules," a game whereby Vergil defines his bucolic endeavors as against other, more exalted poetic types. Disapproval of Pollio is out of the question, just as it is in Horace's careful definition of his lyric program in opposition to the epics of Varius and Vergil in *Odes* 1.3 and 1.6.[43]

One other poem, *Eclogue* 4, is also addressed to Pollio. Its position in the *Eclogue* book relative to poems 3 and 8 is in itself significant. In *Eclogue* 3, Pollio appears primarily as a poet, as we have seen. In the following poem, he is emphatically a consul (*Ecl.* 4.3, 11). Finally, *Eclogue* 8 unites the two spheres of literature and action, praising Pollio for both his tragedies and his Illyrican conquests. Thus the juxtaposition of poems 3 and 4 confronts the reader with Pollio's contrasting pursuits, literature and warfare, which are united only in *Eclogue* 8. But while they present a contrast on one level, the two interests are parallel: heroic poetry and heroic deeds.

We may therefore expect to find significant contrasts between *Eclogues* 3 and 4, and this is in fact the case. *Eclogue* 3, one of the most conventionally "Theocritean" poems in the collection, is self-consciously "humble" in its style. The colloquial *cuium pecus* of line 1, quoted by Menalcas at *Eclogue* 5.87 and famously derided by Numitorius,[44] immediately sets the tone that Vergil wants to assume and establishes a distance between the pastoral mode and Pollio's more exalted tragedies. *Eclogue* 4, although it begins by emphasizing the collection's Theocritean program, has very little to do with Theocritus or with the humble pastoral style in general.[45] Instead, the poet calls for a grander style (*paulo maiora canamus* 1). If the song is to be pastoral, it will at least be pastoral worthy of a consul (3)—worthy, in fact, of Pollio (12), the same consul who, in his capacity as tragic poet, had formed the opposite pole to Vergilian pastoral in the preceding poem.

The basic contrast between the two poems is thus pronounced. The same is true in matters of detail. Charles Segal in particular has shown that the imagery of *Eclogue* 4 consistently reverses that of *Eclogue* 3.[46]

43. See pp. 332–334.
44. *Antibucolica* fr. 2 (Morel–Büchner 1982.135).
45. See pp. 58–59.
46. Segal 1977 = 1981.265–270; cf. Desport 1952.237, Schmidt 1972.161, Berg 1974. 158–162.

What is significant about these reversals for our purposes is how they reformulate the relationship that they had expressed in *Eclogue* 3 between Vergilian pastoral and Pollio's higher style. In all cases, images that had been used to express the extreme differences between bucolic and heroic poetry are resolved into the harmonious imagery of the Golden Age. As the singers approach the midpoint of their contest, Damoetas mentions the danger with which wolves threaten sheep (*Ecl.* 3.80), anticipating the notion to be developed in the second half of the contest that excessive literary ambition is inimical to pastoral poetry. But in *Eclogue* 4 this motif is reversed: in the new age ushered in by Pollio's consulate, the herds will not need to fear such predators (*nec magnos metuent armenta leones* 4.22). The fact that these predators are called "great" echoes the contrast between great and slight poetry in *Eclogue* 3. Next, as we have seen, at the midpoint of the contest Pollio appears. Damoetas offers him a bucolic heifer, Menalcas a heroic bull (*Ecl.* 3.84–87). Again in *Eclogue* 4 the bull's aggressive qualities (*iam cornu petat et pedibus qui spargat harenam* 3.87), like the threat to livestock from predators, are, almost literally, softened in the new age as the farmer loosens the yoke from the bulls' necks (*robustus quoque iam tauris iuga* ***solvet*** *arator* 4.41). Then, in the lines that hail Pollio's literary program as a bold new endeavor, Damoetas wishes luck to whoever follows Pollio by using an actual Golden Age motif—may honey flow for him and even the bramble bear exotic spices: *mella fluant illi, ferat et rubus asper amomum* (*Ecl.* 3.88–89). In *Eclogue* 4, these wishes will become actual features of the world ushered in by Pollio's consulate (*Assyrium vulgo nascetur amomum* 4.25; *durae quercus sudabunt roscida mella* 4.30). Finally, *Eclogue* 3 resumes its catalogue of dangers that threaten the pastoral poet: the snake in the flower bed (92–93), the treacherous riverbank (94–95), and the heat that can dry up the sheep's milk (98–99). All of these, too, are recalled and transformed in *Eclogue* 4. Whereas poetic flower gatherers formerly had to beware of snakes, now, in a telling sequence of motifs surely designed to recall the preceding poem, the cradle of the *puer mirabilis* automatically pours forth blossoms; and the serpent dies (*ipsa tibi blandos fundent cunabula flores. | occidit et serpens*... 4.23–24). Nor must the livestock be rounded up to avoid oppressive heat; they come home unbidden with their udders full (*ipsae lacte domum referent distenta capellae | ubera* 4.21). And what of the ram who has slipped on the treacherous bank and fallen into the river? He reappears in another guise, as verbal repetition makes clear:

> parcite, oves, nimium procedere; non bene ripae
> creditur. **ipse aries** etiam nunc **vellera** siccat.
>
> *Eclogue* 3.94–95
>
> nec varios discet mentiri lana colores,
> **ipse** sed in pratis **aries** iam suave rubenti[47]
> murice, iam croceo mutabit **vellera** luto.
>
> *Eclogue* 4.42–44

Segal sees in these imagistic games "a self-conscious progression from small to great themes, from the rustic Muse loved by Pollio as a private citizen to the *Sicelides Musae* who will help sing the *grandia* which will come to pass under Pollio as an important public figure."[48] Although I would qualify slightly what he says about the rustic and Sicilian Muses, this interpretation in general is surely right. The themes of *Eclogue* 4 are the grandest of the *Eclogue* book. The stylistic progression that he sees between *Eclogues* 3 and 4 is an important aspect of Vergil's effort to re-create the pastoral tradition that he inherited from Theocritus. We can view this process as the crossing of genres, to use Kroll's familiar term,[49] with this proviso: we are dealing not with a mechanical process that can be understood in reductivist terms—bucolic or heroic, pastoral or tragedy, Callimachean or not. Rather, Vergil uses Pollio for the multivalent significance of his poetic and military careers. When projected against this multifaceted figure, our straightforward image of pastoral is refracted in many directions, and becomes a complex meditation on generic and stylistic self-definition.

Eclogue *6*

A similar function has been generally imputed to the poem addressed to Varus, *Eclogue* 6, as well. While I agree with this assessment, I believe that we will find on closer inspection that the poem forms an interesting complement to the Pollio *Eclogues*, serving as an essay on the future of poetry in first-century Rome.

The fact that the poem in question is part of a cycle of bucolic poems has caused readers some difficulty. Like its companion piece, *Eclogue* 4, it is in many ways not a bucolic poem at all, not least in that it has very lit-

47. Cf. *Ecl.* 3.63 *suave rubens hyacinthus.*
48. Segal 1977.163 = 1981.270.
49. Kroll 1924.202–224.

tle to do with Theocritus. Some have ignored or attempted to obscure this fact. It is, however, inescapable, and Vergil—in a characteristically understated way, to be sure—insists on it quite definitely. In the first two lines of *Eclogue* 6, he proclaims himself the singer of the *Eclogues:*

> prima Syracosio dignata est ludere versu
> nostra neque erubuit silvas habitare Thalea.

This opening explicitly recalls that of *Eclogue* 4:

> Sicelides Musae, paulo maiora canamus:
> non omnes arbusta iuvant humilesque myricae;
> si canimus silvas, silvae sint consule dignae!

The passages are clearly linked by the singer's claim to divine inspiration (*Thalea* : *Musae*), by similar diction (*dignata* : *dignae; silvas* : *silvas*, *silvae*), and by acknowledgement of Theocritus of Syracuse as the "model" of the *Eclogues.* With regard to the last point, however, we have had occasion to comment on the playful duplicity that these passages share with Vergil's allusion to his Hesiodic program in the phrase *Ascraeum carmen* (*G.* 2.176). The reader will recall our discovery that the phrase signals Vergil's belated farewell to the Hesiodic imitation of *Georgics* 1, just as the phrases *Sicelides Musae* and *Syracosio versu* mark *Eclogues* 4 and 6 as temporary departures from an overall plan of Theocritean imitation in the *Eclogues.*[50]

The parallelism between the two poems continues. Despite Vergil's invocation of Sicilian muses in *Eclogue* 4.1, his purpose in invoking them is, as we have seen, explicitly to sing a different kind of song, *paulo maiora* (1). Similarly, the tone of *Eclogue* 6.1–2 suggests the perspective of a poet who has finished with pastoral poetry (note the perfects *dignata est* and *erubuit*),[51] even one with a new project in mind.

What sort of project? An answer is suggested by the placement of this poem in the collection. The structural principle, which was so important to all writers of this period, has been applied to the *Eclogues* in a variety of ways. For my purposes, one need only allow that this collection, like the *Georgics* and the *Aeneid*, may reasonably be divided into halves. I attach to such a division only one point of importance, namely, that Vergil seems in all three of his major works to have regarded the

50. See pp. 58–60.
51. Hence the supposition that these are among the latest of the *Eclogues.*

start of the second half as the appropriate place to meditate upon the direction that his work is taking.[52] In the *Georgics* and the *Aeneid*, this meditation points in the direction of grander themes than had yet been essayed. At the beginning of *Georgics* 3 we find the impressive "Marble Temple" passage (1–48), which has been plausibly interpreted as a prefiguration, however approximate, of the heroic poem that Vergil would eventually write.[53] Even in the *Aeneid* itself we find the second, "Iliadic" half of the poem distinguished from the earlier, "Odyssean" half by the famous utterance, *maius opus moveo* (7.45). In the *Eclogues*, of course, *Eclogue* 6, as the sixth poem out of ten, corresponds to the opening passages of *Georgics* 3 and *Aeneid* 7 and is therefore the most "appropriate" place in the collection for the poet to disclose, in however veiled and tactful a fashion, his ambition to try his hand at the higher genres.

When we read the beginning of the poem, the opposite message seems to come through. On the one hand, Vergil alludes to his bucolic program by way of momentarily dismissing it; but this act is followed by his famous Callimachean *recusatio* (3–12). Tityrus, the speaker of these lines, once tried to sing *reges et proelia* (3), heroic poetry; but Apollo Cynthius advised him *deductum dicere carmen* (3–5). He therefore excuses himself from composing the *laudes* and *tristia bella* of his addressee, Varus (6–7),[54] concluding his graceful exordium with the modest hope that this *Eclogue*, proffered in place of the grander poem that Varus had putatively requested, will yet bring honor to its dedicatee. By representing itself as a more humble sort of poetry than heroic epos, this poem seems to resemble *Eclogue* 3 and to move in the opposite direction from both *Georgics* 3 and *Aeneid* 7 on the one hand, and from *Eclogue* 4, where the poet announces his specific intention to sing *paulo maiora*, on the other. In *Eclogue* 6 he speaks of such an endeavor as an earlier, unsuccessful experiment. On the other hand, by raising the issue of heroic epos at the beginning of *Eclogue* 6, the poet seems to invite his reader to consider the poem in that light, as an alternative to heroic epos in particular.

A comparison of Vergil's *recusatio* with its model will bear this interpretation out. The similarity between this passage and the prologue of Callimachus' *Aetia* is strong, and since the discovery of the Callimachus fragment, scholars have tended first to emphasize, and then to take for

52. Cf. Thomas 1985.62–63.
53. Thomas 1988.2.36–37 *ad G.* 3.1–48.
54. His exact identity is in question: see Coleman 1977.177 *ad Ecl.* 6.7.

granted the notion that the opening of *Eclogue* 6 makes exactly the same point as the beginning of Callimachus' poem. This is to give Vergil less than due credit for the care that he lavished on all his imitations. It is of course clear that the *Eclogues* as a whole are the product of a poet who was very much at home with Callimachean poetics, which had been popular at Rome among poets of the previous generation and was not unknown before then. In other words, there is no reason to think that Callimacheanism in Latin poetry was a Vergilian innovation. The fact that this *Eclogue* begins with a ringing allusion to one of Callimachus' most important programmatic passages does not imply that Vergil is announcing something new and unfamiliar. Catullus made much of Callimachus, and it is a reasonable inference that his friends Calvus and Cinna did so as well. The same goes for Cornelius Gallus, who plays a leading role in this *Eclogue*. If Callimachus had been so important in Latin poetry during the two decades that preceded the *Eclogues*, what would be the point of Vergil's merely parroting the *Aetia* prologue at the center of the collection? The issue as I see it is not so much that Vergil announces in this passage a Callimachean poetics as that he chooses this particular method and occasion to do so. Both the method and the occasion suggest another interpretation of the passage, one that parallels our interpretation of the Pollio *Eclogues:* namely, that the poem is concerned not only to distinguish pastoral poetry from the higher genres, but also—and this is just as important—to speculate on the place of both in contemporary literature.

Let us turn to the passages themselves and observe the manner in which Vergil reinterprets his source.

> *καὶ γὰρ ὅτε πρώτιστον ἐμοῖς ἐπὶ δέλτον ἔθηκα*
> *γούνασιν,* 'Α[*πό*]*λλων εἶπεν ὅ μοι Λύκιος·*
> "*.......*]...*ἀοιδέ, τὸ μὲν θύος ὅττι πάχιστον*
> *θρέψαι, τὴ*]*ν* Μ*οῦσαν δ' ὠγαθὲ λεπταλέην.*"
>
> Callimachus *Aetia* 1, fr. 1.21–24
> (Pfeiffer 1949.5)

> cum canerem reges et proelia, Cynthius aurem
> vellit ad admonuit: "pastorem, Tityre, pinguis
> pascere oportet ovis, deductum dicere carmen."
>
> *Eclogue* 6.3–5

The basic similarity should not blind us to Vergil's important departures from his model. Some of these may amount to little more than *oppositio in imitando*, a form of *aemulatio* intended to display the imitative poet's independence and learning. Thus for Apollo's Callimachean cult title, *Lycius*, Vergil substitutes another, *Cynthius*, while suppressing the god's common name altogether.[55] Again, where the Callimachean Apollo uses a very general form of address (ὠγαθέ 24), Vergil's Tityrus is pointedly hailed as a shepherd (*pastorem* 4). Thus the motif of the *Dichterweihe* is anchored to Vergil's pastoral endeavors, while the contrast between Callimachus' "victim" and "Muse" gains point. A third variation may have a more specific purpose. Both poets are visited by Apollo as they begin to compose. Callimachus emphasizes that poetic composition involves writing: the experience took place, he says, "when first I placed my tablet on my knees" (21–22).[56] For Vergil's Tityrus, composition is an oral art: he does not write, but sings (*canerem* 3). This detail too looks to the conventions of Theocritean pastoral, but derives ultimately from reflections of the poet's craft in archaic poetry, a conception adopted by poets of all sorts in Vergil's Rome.[57] Finally, Callimachus does not tell us what he wanted to write on the occasion of first taking up his tablet; he merely states that Apollo appeared with timely advice that the poet made the basis of his art. The point of Apollo's contrast between a fat victim and a slender Muse is clear enough: the poet has just been speaking of those who hold it against him that he writes only short poems and avoids heroic themes:

> ...οὐχ ἓν ἄεισμα διηνεκὲς ἢ βασιλ[ήων
> πρήξι]ας ἐν πολλαῖς ἤνυσα χιλιάσιν
> ἢ προτέρ]ους ἥρωας, ἔπος δ᾽ ἐπὶ τυτθὸν ἑλ[ίσσω
>
> Callimachus *Aetia* 1, fr. 1.3–5
> (Pfeiffer 1949.1–2, with supplements)

This is the contrast that in Vergil becomes that between *reges et proelia* (3) and *agrestem...Musam* (8). Vergil, however, alters the Callimachean situation, slightly but significantly. Where Callimachus does not specify what he would have written without Apollo's warning, Vergil's Tityrus emphasizes this point: he would have, was in fact already trying to sing

55. Cf. Clausen 1976.

56. The detail is typical of the emphasis on poetry as a written medium shared by the Hellenistic poets: see Bing 1988.10–48.

57. Newman 1967.45 relates the emphasis on singing in Augustan poetry both to the developing *vates* concept and to Callimachus, citing, in fact, several occurrences of ἀειδεῖν in the *Aetia* prologue.

(N.B. the conative force of the imperfect *canerem*) about "kings and battles." Now, Callimachus gives us no hint that he ever aspired to such themes. Vergil's Tityrus need not have done so. We should not ignore so pointed an allusive variation, nor should we assume that Vergil's purpose in this reference is to make exactly the same point about such poetry as Callimachus and others had already made.[58]

It is often assumed that Tityrus' refusal here to compose a panegyric account of Varus' military exploits in itself constitutes a Callimachean stance. Nothing could be further from the truth. This application of the *Aetia* prologue is either Vergil's invention or that of an unknown predecessor. The poet's effort to establish distance between himself and his addressee, particularly when mentioning an addressee might be taken to imply the poet's obligation to him in some form, such as that of *cliens* to *patronus*, is a characteristically Roman motif.[59] The Alexandrians, by way of contrast, made the traditional claims about the benefits that poets conferred on their powerful patrons, but were lavish and abject in their general attitude. Callimachus is no exception to this rule. Apart from the panegyric of Berenice that framed Books 3 and 4 of the *Aetia*, we find effusively hyperbolic praise of Ptolemy in the *Hymns*. Vergil's more tactful use of the equivalence between patron and Jupiter in *Eclogue* 3 stands in sharp contrast to, for instance, Callimachus *Hymn* 1.79–90 or Theocritus *Idyll* 16 or 17.[60] There is even a piece of ancient evidence suggesting that the *Aetia* prologue contained the motif of Arsinoe as a tenth Muse.[61] If this is true, then Vergil's contrasting treatment of Varus at the beginning of *Eclogue* 6 can be seen as another variation on its model.

It certainly should have been possible in Vergil's day to take for granted among certain readers—those for whom the *Eclogues* was composed, at any rate—a general familiarity with Callimachus' literary theories and even a close acquaintance with some of his poetry. Indeed, were this not

58. Cf. Wimmel 1960.132–141; Newman 1967.121.

59. It appears prominently in Catullus' dedicatory poem to Cornelius Nepos: see especially Cairns 1969. The strategy later becomes an important element in all the personal genres: for examples see Bright 1978.38–65; Zetzel 1982; Santirocco 1984.

60. This restraint was, of course, destined to be forsaken with Augustus' gradual ascendancy. He appears as *τρισκαιδεκατὸς θεός* in place of Jupiter at *G.* 1.18–42 (Wissowa 1917.100–104), and poetic references to his divinity are common thereafter. Along these lines, Clausen 1972 even argues that the *iuvenis deus* motif of *Eclogue* 1 is evidence for a later date than the scholiastic tradition would allow. This point loses some force, however, in the light of Mayer 1983.20.

61. Pfeiffer 1953.102 *ad Aetia* 1 fr. 2a.7.

so, passages such as the one under discussion would fail to make half their effect. Probably the Neoterics had done a lot to popularize Callimacheanism among the litterati. But even they were hardly preaching an unknown gospel. Callimachus appears as a model for one of the early Roman erotic epigrammatists, and the obviously Hellenistic and even Alexandrian tone of much early Latin poetry ensures that Callimachus was not unknown in Rome long before Catullus. It has been argued that what was known of him were his more conventional works, and that his most distinctive contribution to literature—his programmatic poetry—was largely unknown or ignored.[62] And yet it is generally acknowledged that Callimachus stands behind perhaps the most important single passage of early Latin epic. I refer, of course, to the prologue of Ennius' *Annales*, the famous "*Somnium*," to which Lucretius alludes and in which Ennius declares himself Homer reborn. This passage, as everyone agrees, was modeled on Callimachus' "*Somnium*,"[63] in which the poet draws a comparison between Hesiod's meeting with the Muses on Mt. Helicon, when they taught him the *Theogony* and *Works and Days*, and a dream which Callimachus himself claims to have had in which the same Muses, accompanied by Queen Arsinoe, answered his questions about the causes of things; and it is this question-and-answer format that was the structure of the first two books of *Aetia*. Ennius clearly imitated this passage in his own prologue to the *Annales* while inverting Callimachus' conceptual scheme. Where Callimachus had rejected long epics on heroic themes and held up Hesiod as a conceptual model for his own poetry, Ennius presented himself as actually possessing the soul of Homer, taught by the Muses an epic of Roman history.[64]

We have been asked to believe that this ambitious transformation made no lasting impression on Roman poetry, and that Callimachus had no influence until he was rediscovered in the mid-first century B.C. by Catullus and his friends. Thus Vergil's reference to Callimachus in *Eclogue* 6 is to be read with practically no Roman context whatsoever. This account strikes me as unlikely in the extreme. Little can be made of the silence with regard to Callimachus that stretches between Ennius and Catullus, since we have virtually no poetry from that time. Indeed, it is all the more impressive that among our scanty remains there exists

62. Clausen 1964.187–188.

63. *Aetia* 1 fr. 2 (Pfeiffer 1949.9–10).

64. On Ennius' dream see Skutsch 1985.148–150, with further references; Clausen 1964.185–187.

evidence that Roman poets did know and use Callimachus. As for the programmatic poetry, we cannot know how many other poets alluded in their own verse to the *Aetia* prologue; but there is some indication that Callimachus' *Dichterweihe* was a passage well known even to poets and audiences far removed from the rarefied aesthetics of Catullus and his contemporaries. In the *Pseudolus*, Plautus puts the following soliloquy into the mouth of the character for whom the play is named:

> neque exordiri primum unde occipias habes
> neque ad détexundam telam certos terminos.
> sed quasi poeta, tabulas quom cepit sibi,
> quaerit quod nusquam gentiumst, reperit tamen,
> facit illud veri simile quod mendacium est,
> nunc ego poeta fiam: viginti minas,
> quae nusquam sunt gentium, inveniam tamen.
>
> *Pseudolus* 399–405

The comparison is facetious in the extreme, the language unencumbered with the jargon of a first-century *doctus poeta;* and yet Plautus seems clearly to be parodying Callimachus, and doing so in a context that suggests its familiarity to a Roman audience.[65] The parody, despite its seemingly general import, and notwithstanding the raucous conditions under which Roman comedies were performed, is quite detailed. Plautus' poet, like Callimachus, is signified as such by the act of taking up his *tabulas* (≈ δέλτον). The context, again like that of the *Aetia* prologue, is concerned with the process of *becoming* a poet (*poeta fiam* 404). Even the Hesiodic background of the Callimachean *Somnium* is cleverly parodied in the irreverent business about truth and falsehood (402; cf. *Theogony* 27). This is of course not a context in which one expects to find sophisticated allusion to the most important programmatic passage of Callimachus; but there it is, presented as something altogether familiar to a general Roman audience, well over a century before it was supposedly made a permanent part of Latin poetry by Vergil and Tityrus in *Eclogue* 6.

On this evidence, we must see in Vergil's *recusatio* a reference to Callimachus, but one that is qualified by other references in Latin poetry

65. What is even more surprising is that this allusion probably antedates that of Ennius. The *Pseudolus* was produced by M. Iunius Brutus, *pr. urb.* in 191 B.C., at the Megalesia (Livy 36.36.4; *didascalia in Pseud.*), whereas Ennius, on Skutsch's reckoning (1985.6), did not begin the *Annales* until about seven years later.

stretching back to Ennius and Plautus. Time and again we have seen in Vergil's allusions a tendency to create dialogue, not merely between his own work and that of a single predecessor, but often one that involves several predecessors. In this context, in which the poet meditates on the nature of epic poetry in particular, it is Ennius as well as Callimachus that commands our attention.[66] The fact that Ennius had established his authority as the poet of the *Annales* by means of Callimachus' similar claim in the *Aetia* was surely not lost on Vergil. In fact, it makes sense to read Vergil's imitation of Callimachus as a pointed response to Ennius' earlier imitation as well. The *Aetia* prologue consists of two main parts, two different scenes of initiation. The first is Callimachus' encounter with Apollo Lycius.[67] The second is his dream encounter with the Muses.[68] Ennius, of course, imitated this double structure, first telling of a meeting with the shade of Homer, and then of a second with the Muses.[69] Thus his Callimachean allusion is varied by the substitution of Homer both for Hesiod (as his epic model) and for Apollo (as the main figure of his first *Dichterweihe*). But the double structure and the imitation of the second *Dichterweihe* involving the Muses are both fairly close to the original. Vergil instead dispenses with the double structure, "transfers" the motif of the Muses to a much later point in the poem, and casts Gallus, like Callimachus, in the role of the neo-Hesiodic poet. Tityrus himself, in contrast to Ennius, follows closely only the first part of Callimachus' initiation, the encounter with Apollo. Thus, we may say, Vergil has revised Ennius' imitation of Callimachus both by focusing on Apollo instead of the Muses and by tracing Gallus' poetic genealogy back to Hesiod instead of Homer.

This is one aspect of the poem's allusive dialogue, the creation of its Callimachean, Neoteric, and Hesiodic background. The other aspect,

66. Although it is intriguing to find that in line 1 Tityrus claims a special relationship with Thalea, the comic Muse. Is this a sly acknowledgement of Pseudolus' burlesque *Dichterweihe?*

67. Fr. 1.21-28 (Pfeiffer 1949.5–6).

68. Fr. 2 (Pfeiffer 1949.9–10).

69. Skutsch 1985.147–148 argues against the actual presence of the Muses in Ennius' dream, but admits (p. 148 n. 7) the difficulty of excluding them altogether. The fantastic nature of the scene in my view tells against Skutsch's contention that "the wraith of Homer, just risen from the *Acherusia templa*, can hardly have kept company with the Muses," nor is there any reason why Gallus, if he is the immediate model of *Eclogue* 6.64–73, or Vergil would have refrained from Alexandrianizing such an Ennian scene—particularly since they would have been conscious of Ennius' own use of Callimachus.

its Ennian and Homeric background, is outlined by Vergil's pointed departures from Callimachus. In abandoning the double structure of Callimachus' initiation, Vergil also distances the Callimachean experience from himself. The Apollo encounter is related by a poet named Tityrus. Though we may see this Tityrus as a mask for the poet Vergil, the credulous errors of the ancient biographical tradition warn us to respect the distance that this mask establishes between Vergil and the positions taken by the poem's narrator. The *Musenweihe* motif is placed at an even greater remove from Vergil. It is presented to us not as his experience, or even as Tityrus', but rather as that of Gallus. Critics are therefore obviously correct to infer a strong connection in Vergil's mind between Gallus and Callimachus—particularly since Gallus is commissioned to tell the *origo* (≈ *αἴτιον*) of the Grynean grove; but we may not conclude that Tityrus, or still less that Vergil, sees himself as occupied with an identical mission.

I therefore suggest that Tityrus' remarks on his bucolic achievements and his subsequent *recusatio* are in fact not offered simply as the manifesto of a rising neo-Callimachean poet, but that they express a deep concern with the richness and complexity of the entire epic tradition as it viewed by a Roman poet who had trained in the Callimachean poetics of the previous generation and had distinguished himself in this tradition. To put it another way: Tityrus presents himself as a bucolic poet with a serious interest in heroic epos. The question that he poses concerns the relationship between what he has and what he might have accomplished in these diverging branches of the epic tradition.

This question is clearly articulated in the main part of the poem, the "Song of Silenus" (27–86), an astounding composition that has challenged critics for generations with its eclectic mix of subjects and styles; but, while we may not understand it fully, a few important points now appear certain.[70] First let us outline the structure of the piece. As Silenus begins to sing (26), Tityrus, before detailing the substance of the song, sets the mood by describing its effect: the Fauns and beasts frolicked in time with the music, and the stiff oaks became pliant and swayed their tops; neither Parnassus prides itself so much on Apollo, nor Thrace on Orpheus (27–30). The tone here contrasts sharply with that of the previous scene. There Silenus, nursing a hangover, is bound with his own reveler's garlands by two playful fauns and a nymph, and concedes to the

70. The most illuminating discussion remains that of Ross 1975.18–38, although I disagree with him on a number of important points, as will become clear.

former the often promised song that follows as a reward for "capturing him," while pledging to the latter, true to his obscene nature, a suggestive "something else" (13–26). The august, Apollonian/Orphic character of the song itself is therefore wholly unexpected. Both the proem with its *recusatio* and this introductory vignette lead us to expect a pastoral song; but what we get proves to be anything but.

The song proper (31–81) can reasonably be divided into four distinct sections, to which is added a coda (82–86). In each of these sections Tityrus reports on the shifting subjects of Silenus' song; and in the first three of these sections, the subjects are reported with increasing vividness and immediacy, so that the persona of Tityrus, the primary narrator, appears to dissolve.

The first section (31–40), which continues to distance the tone of the song from that of the introductory mime, deals with a highly impersonal but supremely august topic, natural philosophy—or, more specifically, cosmogony. There is no consistent adherence to a specific philosophical school, any more than in Anchises' eschatological discourse in *Aeneid* 6. In both passages, what matters is literary reference. The language of the *Eclogue* passage is unmistakably Lucretian,[71] and the science, at first, appears to be as well: *namque canebat uti magnum per inane coacta | semina...fuissent* (31–32). The appearance of atoms and void indicate an Epicurean universe, while the very rhythms of the verse honor the system's greatest expositor, Lucretius. But Lucretius in such passages does not specify what particular atoms he envisions "ruining along the illimitable inane"; atoms, simply atoms, flying through the void is his basic image for the structure of the universe. Here, however, Vergil departs from his model both imagistically and philosophically: *semina terrarumque animaeque marisque... | et liquidi simul ignis* (32–33). Here Epicurean atoms are turned into particles of the four Empedoclean elements, earth, air, water, and fire; and Vergil's detailed treatment of these elements is marvelously allusive. By naming only the first three elements in line 32, he seems to adopt a typically Roman conception of the universe as consisting of three components, earth, sea, and sky, a conception which Lucretius himself exploits.[72] And yet, his way of mentioning these three parts of the world contains something distinctly un-Roman. The triple repetition of *-que* in this line has been taken as a deliberate Grecism, an imitation of the similar, but more common treatment of τε

71. Ross 1975.25 attempts—wrongly, I think—to minimize this fact.
72. Clay 1977.27; cf. *DRN* 1.278, 340, etc.

in Greek epic.[73] This mannerism, then, points us towards a Greek source. The delayed inclusion of *ignis* in line 33 next forces us to reread the previous line as referring not to land, air, and sea, but to the elemental earth, air, and water: the Empedoclean elements, in fact. Even the delay in mentioning fire corresponds to an emphasis on the role of this element in Empedocles' system.[74]

This mixed Epicurean/Empedoclean universe presents a philosophical problem; but it makes good sense allusively.[75] In Book 1 of *De Rerum Natura*, Lucretius discusses at some length problems with three "rival" scientific theories about the fundamental structure of the universe. First Heraclitus is criticized for asserting that everything is made of fire (635–704); next is Empedocles and his quadripartite theory (705–829); and last is Anaxagoras, with his concept of *homoeomeria* (830–920). In dealing with Heraclitus and Anaxagoras, Lucretius is openly scornful, and not just of their scientific positions.[76] The essential aspects of each philosopher's thought and manner of expression is made a foil not only for Epicurus' atomistic system, but also for Lucretius' own poetic program. Thus Lucretius adopts a markedly polemical stance against Heraclitus and Anaxagoras, deriding both of them for stylistic infelicity as much as for their scientific views. His treatment of Empedocles, however, is entirely different.

Although Lucretius registers his disagreement with Empedocles' scientific theories, towards the man himself he shows a reverence that he normally reserves for Epicurus alone. The four-element theory, moreover, is linked imagistically with the topography and *thaumata* of Empedocles' homeland, Sicily (716–725). The island contains many natural and man-made wonders, Lucretius continues, but nothing more wonderful than Empedocles himself (726–730). These lines are worth quoting:

> nil tamen hoc habuisse viro praeclarius in se
> nec sanctum magis et mirum carumve videtur.
> carmina quin etiam divini pectoris eius
> vociferantur et exponunt praeclara reperta,
> ut vix humana videatur stirpe creatus.
>
> *De Rerum Natura* 1.729–733

73. Ross 1975.26 n.1.
74. Stewart 1959.184 and 201 n. 26, with further references.
75. For what follows see especially Tatum 1984, with further references.
76. Cf. Lenaghan 1967; Schrijvers 1970.84–85; Kollmann 1971.

The passage is remarkable in several ways. Not only is the theory that Empedocles begat similar to the island that bore him, but his discoveries are like the man himself in that both are *praeclarus* (729, 732). This is high praise, coming as it does after Heraclitus' indictment on a charge of obscurity, and only shortly before Lucretius asserts the clarity of his own verse.[77] Divinity, too, is a quality to which only the Epicurean sage approximates and only Epicurus himself is said actually to have achieved.[78] The climactic element, however, is Empedocles' poetry; and this, I think, explains his importance to Lucretius. We have already seen him similarly invoke Homer and Ennius for literary purposes while repudiating their eschatology. Here too Empedocles' philosophy is rejected, while his importance as a poetic predecessor is tacitly acknowledged. It has even been suggested that the lines suggest a closer connection between Lucretius and Empedocles than the fact that both are philosophical poets. The island of Sicily, Lucretius states, contained nothing more brilliant, more holy, more wonderful, *or more dear* than this man (729–730). The final term is surprising, since it points to a much more intimate feeling for Empedocles than seems strictly called for. But in Latin it makes sense: nothing more *carus* than Empedocles. When we remember that Carus is Lucretius' cognomen, it seems quite possible that the poet wishes to make a punning connection between his great literary model and himself.[79]

Thus in *De Rerum Natura* there is an important parallelism between Lucretius and Empedocles qua poets; and it is as poets in the same tradition, not as rival philosophers, that Vergil alludes to both of them in *Eclogue* 6.

There is yet a further element in these lines that requires notice. The opening and the close of this first part of Silenus' song quote yet another epic cosmogony, the one that Orpheus sings in Apollonius' *Argonautica:*[80]

> namque **canebat uti** magnum per inane coacta
> semina **terrarumque animaeque marisque** fuissent
>
> *Eclogue* 6.31–32
>
> ἤειδεν δ' ὡς γαῖα καὶ οὐρανὸς ἠδὲ θάλασσα
>
> *Argonautica* 1.496

77. *DRN* 1.921–922, 933–934.
78. *DRN* 3.322, 5.8, 6.68–78.
79. Kollmann 1971.89 n. 46
80. Stewart 1959.186 traces the discovery of this parallelism back to Ursinus 1568.

incipiant silvae cum primum surgere, cumque
rara per ignaros errent animalia montis

Eclogue 6.39–40

οὔρεά θ᾽ ὡς ἀνέτειλε, καὶ ὡς ποταμοὶ κελάδοντες
αὐτῇσιν νύμφῃσι καὶ ἑρπετὰ πάντ᾽ ἐγένοντο

Argonautica 1.501–502

Realizing that Vergil is alluding to a song put into the mouth of Orpheus, we recall the Orphic qualities that Tityrus attributes to Silenus' song (30). The recurrence of the Orpheus theme guarantees the force of the allusion. But this allusive mixture of an Empedoclean Lucretius with an Orphic Apollonius makes little sense in terms of prevalent literary history. Apparently the preconceptions with which we customarily view the passage are somehow wrong.

At this point some background is in order. Lucretius' manifest importance in these lines has embarrassed some critics, who have attempted minimize his importance here. Their reasoning is that the poem as a whole is consistently Callimachean and Neoteric, and that this character leaves no room for Lucretius. Accordingly, their interpretations stress the un-Epicurean character of Silenus' physics and the precedent of the Orphic cosmogony in Apollonius of Rhodes. But, as I have shown, the Empedoclean element is itself an important aspect of Lucretius' poem, and I have also suggested that there is common ground between Lucretius and the Apollonian Orpheus as well. This suggestion will seem strange to those who view Apollonius as an orthodox Callimachean and Lucretius as a follower of a debased Homeric and Ennian tradition rejected by the Neoterics. But, in the first place, even if modern scholars are correct in viewing Apollonius as an orthodox Callimachean, it is by no means clear that Vergil did so. Indeed, the tradition available to him probably characterized the two poets as enemies who differed sharply on literary principles.[81] In the second place, even if the Neoterics did indeed set themselves against Ennius' annalistic epigoni—poets like Hostius, Volusius, and others—they show a much more respectful attitude towards Ennius himself;[82] and there is absolutely no need to suppose that their hostility towards Volusius was also meant for Lucretius, who represented a different tradition altogether.

81. Pfeiffer 1949.307 *ad* fr. 382 expresses his opinion that the supposed enmity between the two was current in ancient times.

82. See especially Zetzel 1983a.252.

It is commonly supposed that Catullus' contempt for Volusius, which he expresses so forcefully in *Carmina* 36 and 95, applied to all others like him, meaning all followers of Ennius; and Lucretius is undoubtedly the most important surviving follower of Ennius. But Lucretius followed Ennius in a very different way than Volusius and the other annalists did. Although we lack a well-developed sense of the context in which these poems were produced, we can distinguish strong alternative traditions deriving from Ennius, mythohistorical on the one hand and philosophical on the other. The former has been more prominent in modern literary histories; but Lucretius' poem, in terms of its most basic characteristics, was hardly an unprecedented undertaking. The tradition of didactic epos inaugurated by Ennius is, indeed, merely another aspect of the inveterately Hellenistic character of Latin literature; and it continued to flourish beyond the time of Lucretius. To focus on that period, we know of several philosophical poems more or less contemporary with *De Rerum Natura*. Some of these probably took the form of a translation or paraphrase, like Sallustius' *Empedoclea*;[83] we may compare the tradition of *Aratea*, to which Cicero and, later, Germanicus belong. Others presented themselves as independent compositions, even if based explicitly on one of the Greek philosophical systems: Egnatius' *De Rerum Natura* is a case in point.[84] It is thus an accident of fate that has deprived us of the proper context of Lucretius' poem, and we are mistaken to lump it together uncritically with poems like Volusius' *Annales*.

Lucretius gives us an important clue concerning his relationship to his great poetic models, Ennius and Homer, but it seems to have been ignored. The clue is found in the famous programmatic passage in which Lucretius invokes his two most important predecessors. This well-known passage makes a number of points, many of them familiar to all students of Latin literature: Lucretius implicitly aligns himself formally and stylistically with Homer and Ennius while rejecting on philosophical grounds Ennius' claim to be *Homerus redivivus*. But at least one remarkable feature in Lucretius' description of Ennius' encounter with Homer's shade seems to have gone completely unnoticed. Lucretius argues that there is great ignorance about the nature of the soul, whether it is born

83. Cicero contrasts this poem unfavorably with that of Lucretius in a well-known letter to Quintus (2.10.3) Lucretius.

84. Mentioned by Macrobius (*Sat.* 6.5.2, 12; see Schanz–Hosius 1927–1935.1.314). Varro Reatinus wrote a poem of the same title (Schanz–Hosius 1927–1935.1.560).

with us or rather insinuates itself into our bodies when we are born, and similarly whether it perishes along with us, goes to dwell in the underworld, or insinuates itself into other creatures when we die (*DRN* 1. 112–116). The third possibility, Lucretius notes, is the one favored by Ennius, who was the first to bring down from lovely Helicon that evergreen crown famous among the peoples of Italy, although he also asserts in his eternal verses that there is an underworld where neither our bodies nor our souls dwell, but rather shimmering images of a mysterious kind (117–123). It was from this infernal realm of ghostly images, Ennius claimed, that the shade of ever-vigorous Homer appeared to him and, weeping, began to discourse *on the nature of the universe* (*rerum naturam expandere dictis* 126)!

What a strange thing for Homer to do! Or, what a strange view Lucretius has of what we know was a tradition of heroic, martial epos. But of course Lucretius is reading his sources and representing them to us in a way that suits his own poetic and literary-historical agenda, which could hardly be more different from those of a Volusius. An important theme of *De Rerum Natura* is the vanity of political and military power, the very themes glorified by both Ennius and Homer. Such themes pale in comparison to Lucretius' material: the nature of the universe and of the soul, the origin and destruction of the cosmos, the rise and fall of civilization. And yet Homer is the poet who has shown all poets for all time how to write on a theme of great importance, and Ennius has shown Latin poets how to write like Homer. Lucretius therefore appropriates the style of these two lofty models while reinterpreting their subject matter to suit his own purposes.

On what does Lucretius base his interpretation? In the case of Ennius, little more is involved than selective reading. The *Annales* evidently began with a passage of what one might call natural philosophy: the surviving fragments and testimonia deal mainly with the Pythagorean doctrine of metensomatosis, which Ennius made the basis of Homer's authority to sing the events of Troy and of his own authority to compose in the meter and manner of Homer. Although the exact relationship between this prologue and the main narrative of the poem remains conjectural, the philosophical discourse must have been in some sense logically prior to the mythohistorical discourse that followed. In the case of Homer, of course, more imagination was required; but we have already seen a possible source of inspiration for Lucretius' view of Homer. When he characterizes Homer as a poet of natural philosophy,

Lucretius must be following an exegetical tradition that is basically similar to the one that Vergil was to use in composing the *Georgics.* What is more, Lucretius' very brevity and almost offhand manner in so referring to Homer can only suggest that he expected readers to be familiar with a Homer who could hold forth about the nature of the universe—i.e. readers fully accustomed to the idea that the *Iliad* and the *Odyssey* were, or at least could be interpreted as, poems of natural philosophy.[85]

If we understand these facts about *De Rerum Natura*, then the allusive structure of *Eclogue* 6 does not seem so confused. In fact, it makes excellent sense if we understand that Vergil is alluding here through Lucretius and Apollonius to an epic tradition not of "kings and battles," but one of natural philosophy, a tradition found writ large in *De Rerum Natura* and writ small in the Orphic cosmogony of the *Argonautica*. It is with this exalted tradition that Silenus' song begins.

Recognition of these elements in section 1 is important for everything that follows. Section 2 (41–63) effects what has seemed to many an abrupt transition: *relictis prudentibus rebus de mundi origine, subito ad fabulas transitum fecit*, as Servius puts it.[86] The transition is marked by the introduction of mythical personages, such as Pyrrha, Saturnus,

85. We may wonder about Lucretius' summary of Ennius' dream, and particularly about his testimony that Homer discoursed to Ennius on natural philosophy. Does Lucretius impose on the original passage his own interpretive view of Homer as a philosophical poet, or does he faithfully report on Ennius' own characterization of Homer? To put it another way, was allegorical understanding of the *Iliad* and *Odyssey* a recent arrival in first-century Rome, or was it merely another aspect of the Homeric tradition brought to Rome by Ennius, and thus as familiar to Latin poets as Ennius' other great Homeric borrowing, the epic meter? The question cannot be settled. Ennius' Homer probably delivered a Pythagorean discourse designed merely to establish the authority of both Ennius and Homer over the events of which they sang, without making Homer into a philosopher in general; the second step would then have been Lucretius' contribution. On the other hand, the early allegorical tradition is associated with the Pythagoreans (Lamberton 1986.31–43); and its later popularity among the Stoics (DeLacy 1948; Buffière 1956.137–54) could have some relevance to Ennius as well. Such views could easily have reached Rome as early as the visit to Rome by Crates of Mallos in 168 B.C., by which time the poet was probably dead; but Ennius was very conversant with Hellenistic culture, as his adaptation of Callimachus' "*Somnium*" and poems like the *Euhemerus* show, and with Stoicism in particular. For a different opinion concerning Stoic allegoresis, see Long (forthcoming).

86. Servius *ad Ecl.* 6.41 (Thilo 1887.71).

and Prometheus (41–42).[87] It is eased somewhat by the nature of the first three myths, which concern the earliest period of the world: the creation of humankind after the flood, the Golden Age, and the punishment of Prometheus for stealing fire from the gods (41–42). The myths then take on an erotic color: Hylas (43–44), Pasiphaë (45–47 and 52–60), the daughters of Proetus (48–51), Atalanta (61), and the sisters of Phaethon (61–62). Most of the myths are mentioned so briefly that little can be said for sure about their exact purpose here; that they receive a highly Neoteric treatment is obvious, particularly in the case of Hylas and the allusion to Calvus' *Io* in the Pasiphaë/Proetides sequence.[88]

The progression both of themes and of poetic models found in sections 1 and 2 of Silenus' song is provocative, but not immediately intelligible. Like Servius, we notice mainly the contrast between natural philosophy and mythology of a tragic and erotic cast. There is also a contrast between two different traditions of Latin epic, the archaizing and the Neoteric schools as represented by Lucretius and Calvus respectively. But both contrasts may be only apparent.

If we think of the underlying Greek models of the two sections, several unifying elements begin to appear. Thematically, the obvious importance of Empedocles in section 1 calls to mind the opposition of the basic forces, love and strife, that animate his four-element system. Could Vergil have seen a connection between this theme at the cosmic level and the theme of *erotika pathemata* in myth, following Empedocles' own allegorizing use of the Homeric "Ares and Aphrodite?"[89] He was certainly to exploit the philosophical songs of Clymene in *Georgics* 4 and (in a different way) of Iopas in *Aeneid* 1.[90] It may well be that Silenus' song also relies on a literary and exegetical tradition that regarded myth as a cryptic form of philosophical discourse.

The importance of Apollonius in these lines corroborates this view. In section 1, we noted how Vergil began by combining Lucretian diction with allusion to the Orphic cosmogony in Apollonius' *Argonautica*. Others have commented that, beyond the lines that are actually quoted, Orpheus' song shows certain formal similarities to that of Silenus, notably in the progression from inanimate things to sentient creatures.[91]

87. The metonymy Nereus = *mare* appears in line 35, but no specific reference is made either to this character or to any particular myth.

88. Skutsch 1901.32–33; Stewart 1959.188–190.

89. Empedocles fr. 17 (Diels–Kranz 1951.315–318).

90. See pp. 258–262.

Thus the influence of Apollonius may in a general way be felt throughout the passage. In section 2, we again note that Apollonius' influence is signaled early. After the introductory myths of Pyrrha and Deucalion, the Golden Age, and Prometheus, the first of the erotic myths to which Vergil turns is that of Hylas. Commentators tend to seize on this fact as at least one element of Theocritean pastoral in this otherwise very unpastoral poem. It is no doubt right to see here some relevance to Theocritus, even if his "Hylas" (*Idyll* 12) is hardly representative of the truly pastoral idylls.[92] Indeed, if Theocritus is relevant here, it is most likely because his "Hylas" gave Vergil an opportunity to point out what his great model for bucolic epos could accomplish outside of pastoral. But even more important, in my opinion, is the Hylas episode of Apollonius. In the first place, Theocritus is generally quite unimportant in this poem, while the *Argonautica* is an essential ingredient in its allusive scheme. Even more telling is the fact that Vergil's treatment of the myth agrees more closely with that of Apollonius. Where Theocritus focuses on the pathos of the tale from both Hylas' and Heracles' points of view, Apollonius instead looks to the role of the other Argonauts (cf. Vergil's *nautae* 43) and to the *aetion* concerning rites observed in later times at the fountain where the boy was lost (cf. *quo fonte relictum* 43). Thus we are dealing not with an inert allusion to a famous pastoral tale, but with a a very active double reference that challenges the conventional boundaries of bucolic epos.

The presence of Apollonius here serves two purposes. First, it represents an element of unity between the first two sections of Silenus' song; second, it helps to establish that the poem is not merely unpastoral, but that it is actually, in some sense, about the higher modes of epic. What seems to be on Vergil's mind are those figures in literary history that define for him the epic tradition that he has inherited, or that he wishes to create. Lucretius and Calvus, it seems clear, represent different strands of the tradition in Rome; while Empedocles and Apollonius stand as the important Greek predecessors. What seems to emerge from all this is an interest in philosophical epic, both in its straightforward form, as in Empedocles and Lucretius, and in mythological garb, as in Apollonius and, apparently, Calvus.

91. Stewart 1959.186.

92. For a different view of the relationship between epyllion and idyll, see Stewart 1959.190.

Silenus' allusion to Calvus' mythological epyllion of bizarre eroticism in what began as a philosophical poem strikes the modern reader as odd. We must remember, though, that we are dealing here with Vergil's decidedly willful interpretation of the literary past. In this interpretation, Apollonius, too, on the strength of a single episode, joins the ranks of Empedocles and Lucretius. Our reaction may also have something to do with our almost total ignorance about Calvus' poem. It is extremely interesting that the scholia to another passage cite Calvus in particular as Vergil's source for certain details concerning the myth of the Golden Age:

> ANTE IOVEM et reliqua: Dicunt Iovem commutasse omnia, cum bonus a malo non discerneretur, terra omnia liberius ferente, quod Calvos canit.
>
> Σ Bern. *ad G.* 1.125 (Hagen 1867.186)

The scholiast is certainly not referring to one of the poet's *nugae* on the order of Catullus' polymetrics, still less to an epigram. It is therefore surprising that we hear nothing else of the poem in which Calvus handled this theme, unless we are overlooking the obvious. It is not impossible that Calvus brought this theme into his *Io*, perhaps in something like Catullus' manner of deploring the degeneracy of modern times in his own epyllion. Whether it was in the *Io* or elsewhere, however, the appearance of such a theme in Calvus, and the fact that we have evidence of Vergil's interest in his treatment, helps to explain his presence and importance in *Eclogue* 6. It also helps to prepare for the remarkable lines that follow.

In the third section of the poem (64–73), as also in *Eclogue* 3, the reader is startled to find an actual, living Roman poet made the subject of song by one of Vergil's characters. The *Dichterweihe* of C. Cornelius Gallus picks up the Callimachean allusion of section 1 and traces it to its Hesiodic roots, while the description of the pipes with which Gallus is bidden to sing the origin of the Grynean grove recall again the earlier prominence of Orpheus. The treatment of these two figures as the ultimate ancestors of Gallus' poetry expresses a very specific view of the epic tradition. It matters little in this context whether Gallus wrote only elegies or also composed hexameter poetry.[93] Here Vergil's friend is

93. Although poets often observed this formal distinction for specific purposes, it nevertheless seems clear that in the broadest sense they considered elegy a species of epic: see Wilamowitz 1924.1.231, 2.96; Knox 1986.10. With respect to Gallus see Ross 1975.39–46.

clearly presented as heir to the epic tradition that it is the purpose of Silenus' song to delineate. As in the case of Calvus, one might be surprised to find Gallus the love elegist identified with this tradition, and scholars once found it necessary to invent a whole range of unattested non-erotic poetry by him in order to account for his prominence here. Ross, however, has carefully traced the logic of the "poetic genealogy" outlined here to show how we must understand this passage: Gallus the elegist is presented as the heir to an Orphic/Hesiodic tradition of "scientific" poetry.[94] By referring so prominently to Hesiod in his guise as *Ascraeus senex* (70), Vergil avails himself of the Hellenistic conceit that casts Hesiod in the role of an alternative to Homer, i.e. as the founder of an ancient and august tradition of epic poetry that concerns not kings and battles, but *res prudentes* (including cosmogony), the loves of gods and mortals, and, generally, the origins of things. In this connection, the continuation of the Orpheus theme here is also crucial. The Orphic qualities attributed to Silenus' song in lines 27–30 color our perception of everything in it; and the references in sections 2 and 3, both verbal and structural, to a cosmogony sung by the character Orpheus in Apollonius' *Argonautica*, further the process.[95] This thematic development suggests that even Lucretius—on this point I must differ with Ross—and Apollonius, along with every other author whom we can identify through allusion, is being presented as an Orphic poet. Read in this way, *Eclogue* 6, I suggest, defines those elements within the epic tradition that fit in with Vergil's sense of modernist propriety and seriousness. Lucretius, Empedocles, Apollonius, Calvus, Gallus, Hesiod, and Callimachus all appear under the aegis of Orpheus, a mysterious but potent symbol for the power of poetry to disclose the secrets of the universe, and thus to control nature.

In this interpretation Empedocles, Lucretius and Apollonius are willfully excluded from the Homeric tradition. Thus *Eclogue* 6 embodies a quite tendentious view of literary history. This is also true of the way in which it virtually ignores Homer's place in the epic tradition. While accepting Empedocles and Apollonius, Lucretius and Calvus, Gallus and Hesiod as Orphic poets, Vergil explicitly rejects those other followers of Homer, the epigoni of Ennius who gratify the powerful by writing annalistic poetry—Varus will find others to sing his praises (6–7). When

94. Ross 1975.31–38.

95. Note the ambiguous phrasing of *Eclogue* 6.69–71, and cf. the tempting interpretation of it by Ross 1975.23.

Homer at last is acknowledged, he comes almost as an afterthought. After Gallus' Hesiodic *Dichterweihe*, the increasing vividness with which Tityrus had presented Silenus' song abruptly reverses itself. The narrative voice retreats into rhetorical questions that amount to a simple *praeteritio* of the singer's final themes: Why should I mention Scylla, or the story of Philomela (74–81)? This reversal proves an effective means of closure; but the way in which Vergil has fashioned this passage produces an additional effect. On the surface, the question "Why should I mention Scylla and Philomela?" does little more than enter another pair of items into the catalogue of Silenus' topics.[96] Previously, however, Silenus himself was more emphatically the singer: *namque canebat* (31), *hinc...refert* (41–42), *his adiungit* (43), *tum* (*canit*) (61, 62, 64). As the poem progresses, the reader comes to feel more and more that Silenus himself is the primary narrator. Here, however, Tityrus calls attention to himself as the singer of the song we hear. *Quid loquar* (74), "Why should *I*, Tityrus, sing of Scylla and Philomela?" This emphasis on Tityrus problematizes the *Eclogue* poet's attitude towards these myths. For this reason, they bear further scrutiny.

We will proceed ὕστερον πρότερον, Ὁμηρικῶς. The disastrous love stories of Silenus' song focus on the plights of various heroines. This feature is in keeping with Neoteric practice, and derives ultimately from Hesiod's *Catalogue of Women*.[97] In fact, all but one of the heroines mentioned in *Eclogue* 6 make their first literary appearance in Hesiod rather than in Homer.[98] The exception is the last mentioned, Philomela. Her story appears first in the *Odyssey*.[99] She is not mentioned by Hesiod. This may be a coincidence, but it certainly reverses a consistent pattern. Scylla, the heroine with whom she is paired, also receives unusual treatment, and the lines in which Tityrus characterizes her are well worth quoting:

96. Coleman 1977.198 *ad Ecl.* 6.74 sees a resumption of earlier themes ("Silenus now reverts from allegory to mythology").

97. This poem was regarded in antiquity as a continuation of the *Theogony* (West 1966.48–51). Does the structure of Silenus' song allude to this fact?

98. All but one of them in the *Catalogue of Women:* Pyrrha frr. 2–4 (Merkelbach–West 1967.4–5); Pasiphaë fr. 145.2 (?) and 12 (? p. 71); the Proetides fr. 131 (p. 64); Atalanta frr. 72–76 (pp. 46–49); the sisters of Phaethon fr. 311 (*incertae sedis* pp. 161–162). On Scylla see n. 100.

99. *Odyssey* 19.518–523, a simile.

quid loquar aut Scyllam Nisi, quam fama secuta est
candida succinctam latrantibus inguina monstris
Dulichias vexasse rates et gurgite in alto
a! timidos nautas canibus lacerasse marinis...?

Eclogue 6.74–77

Scylla of Megara, daughter of Nisus, is an eminently Neoteric heroine, one who falls in love with the man besieging her father's kingdom and so betrays it to him, but fails to win his love. Ultimately, she and Nisus undergo metamorphosis, becoming the ciris and the sea-eagle that forever pursues it. This is not, however, the story that Vergil tells. Instead, he outlines the tale of Scylla of Zancle, who, though she rejected the advances of the sea god Glaucus, nevertheless incurred the wrath of Circe, who was in love with Glaucus; and, through Circe's drugs and spells, she became the monster that infested the Straits of Messina in the Homeric *Odyssey*. The conceit of conflating alternate myths told of a single character, or completely different myths involving homonymous characters, is common.[100] Though the poet of the *Ciris* chastises others for making the very mistake that Vergil commits here, the conceit appears to be a somewhat paradoxical way of demonstrating one's *doctrina*. Other learned and Neoteric elements appear in these lines: the appeal to tradition (*fama* 74), antonomasia (*Dulichias...rates* 76), a modish exclamation (*a!* 77), a tactful gloss (*canibus* 77). And yet Tityrus' telling of the story—particularly his mention of Odysseus' ships—points us, for the first time in this poem, to the *Odyssey*. "Why should I tell this story from Homer?" he seems to ask, or almost "What need do I have of Homer?" The very idea seems beyond him, the material incompatible with his talents and training. And yet, the stories of Scylla and Philomela are parts of Silenus' song.

What these allusions mean beyond this is unclear, and may never be clear: this is a poem that cannot be pressed too hard. What we can say by way of summary is this. Vergil as a bucolic poet and as a Neoteric shows the concern that we would expect to define his work within those traditions. He also shows, in this poem as in *Eclogues* 3 and 4, a pronounced interest in expanding those traditions, or in reinterpreting them so as to open new possibilities. Neither Silenus' song nor the poem as a whole can easily be labeled as "pastoral" or "Neoteric" as

100. The phenomenon is discussed by Lyne 1978a.125 *ad Cirim* 54–91.

those terms are normally conceived. *Eclogue* 6 embodies a tradition of "Orphic" poetry—a poetry at once of science and of myth—that ranges far beyond the boundaries of Theocritean pastoral, drawing on poets as diverse as Ennius and Callimachus, Lucretius and Apollonius, Calvus and Empedocles, Gallus, "Hesiod," and even Homer. In a sense, the new possibilities glimpsed by Tityrus are realized in this self-enacting tour de force. But it is difficult to resist Cartault's suggestion that in this shifting, tantalizing sequence of allusive artistry, Vergil is "dreaming of his poetic future."[101]

The *Georgics*

If *Eclogue* 6 really was conceived as a kind of prospectus, it turned out to be a very misleading one. This comes as no surprise. The *Georgics* too contains at its midpoint a meditation on Vergil's future career as an epic poet, the "Marble Temple" (*G.* 3.1–48), itself a very inaccurate adumbration of the *Aeneid.* As I have stated before, we must beware of regarding Vergil's remarkable career as preordained, or as something that unfolded according to a master plan. If these programmatic centerpieces have some dim relevance to the future, their main purpose remains that of commenting on the project that is already under way.

It is true that many of the same elements that made up the epic tradition in "The Song of Silenus" are clearly present in the *Georgics.* This observation applies both to major motifs and to details. The poem labels itself an *Ascraeum carmen* (*G.* 2.176; cf. *Ecl.* 6.70 *Ascraeo seni*), reuses the motif of ritual binding in order to gain mastery over a prophet (Proteus), and alludes to the stories of both Scylla and Philomela, which close Silenus' song, in a symmetrical way, towards the ends of Books 1 (404–410) and 4 (511–515) respectively. Perhaps most significantly of all, the enigmatic figure of Orpheus reappears at the heart of the poem's final episode. In the *Georgics* as a whole, however, the relationship among the literary sources mentioned in *Eclogue* 6 has changed, some receding in importance, others assuming a greater prominence than before.

What first draws our attention is the role of Hesiod in *Georgics* 1. We cannot be sure of Vergil's utter originality in the way that he incorporates *Works and Days* into his own poem. Calvus, as I have noted above, had already made use of the Golden Age motif in a way that ancient scholars

101. Cartault 1897.286, cited with approval by Berg 1974.183.

felt anticipated Vergil's "Aetiology of *Labor.*"[102] Vergil himself, moreover, cites Gallus as Hesiod's rightful heir among the Romans. And yet, in two respects, Vergil's Hesiodic program probably was unprecedented. First, it seems unlikely that Vergil would undertake so extensive an imitation of *Works and Days*, and then call attention to his achievement at *Georgics* 2.176, if he were not the first to attempt an encounter on this scale with the authentic Hesiod. Certainly it is impossible to suggest who might have produced a comparably Hesiodic poem before the *Georgics*, since there is no indication that Roman poets before Vergil were at all interested in Hesiod per se.

This brings me to my second point. In so far as earlier Roman poets, such as Calvus and Gallus, did show interest in Hesiod, their interest was probably confined to the symbolic Hesiod of the Alexandrians. The reference to Gallus' "Orphic" Hesiod in *Eclogue* 6 certainly makes the most sense in these terms. Even among the Hellenistic poets a sincere interest in Hesiod is unusual; their fascination with Homer left little room for another. Thus the Hesiodic injunctions of Milon's song in Theocritus' *Idyll* 10 are in sharp contrast with the sophisticated imitation of Homer that pervades the *Idylls*. Hesiod was useful to these poets, of course, as an archaic spokesman for the vatic nature of poetic utterance, and as an alternative model to the Homeric ἄεισμα διηνεκὲς ἐν πολλαῖς χιλιάσιν; but when it comes to detailed imitation of specific actions, episodes, and even words, it is to Homer much more often than Hesiod that even Callimachus turns. Aratus stands out as the most Hesiodic of the Hellenistic poets,[103] as Callimachus saw, though he too devoted his scholarly energies primarily to Homer,[104] and it was even debated which of the two great archaic masters he followed in his poetic *oeuvre*.[105] It was Aratus, too, whose poetry exerted such a fascination on first-century Rome; and when we stumble upon a hint that Calvus imitated a scene in *Works and Days* that had also, and famously, been imitated by Aratus in *Phaenomena*, we should probably expect to see the latter's influence. Similarly, what room is there in Gallus' work for a Hesiodic imitation on the scale of *Georgics* 1? When Vergil places his friend in the tradition of the *Ascraeus senex*, he is certainly thinking of the

102. See p. 310.

103. His teacher Menecrates of Ephesus wrote poems on agriculture and beekeeping (*Suda* 1.3745 s.v. Ἄρατος [Adler 1928–1935.1.337–338]), but these are unlikely to have been more Hesiodic than the *Phaenomena* or, say, the poetry of Nicander.

104. See the discussion of Pfeiffer 1968.121–122.

105. See p. 217.

Alexandrian Hesiod, the reverentially cited but seldom imitated alternative to Homer who served as an emblem for epic on a small scale, carefully crafted, and dealing at least in part with wisdom, science, and the origins of things rather than with kings and battles.

Some scholars have argued that this is the Hesiod of the *Georgics* as well, and certainly the literary principles of which this Hesiod is a symbol inform the poem from beginning to end. But we must not allow this fact to obscure Vergil's remarkable originality in tapping the true source, the ancient poet of *Works and Days*, Hesiod himself. Latin poetry, we are taught, begins with a translation of Homer, is born again in Ennius' Homeric agon, and in the first century B.C. is dominated by second-rate homerizing poetasters. One naturally infers from this that imitation of archaic models had always been the norm. But this is far from the truth. In most genres from the very beginnings Latin literature shows a much greater affinity with the more nearly contemporary Hellenistic world. Tragedy, it is true, follows the masters of the high Classical period; but Roman comedy derives primarily from Attic New Comedy, and in the late thirties Horace can claim that no one reads the masters of the fifth century.[106] His predecessor in satire, Lucilius, draws heavily on Hellenistic literature.[107] The surviving fragments of Laevius, Catulus, and the rest follow Hellenistic models. Even Cicero and Hortensius, the foremost orators of their generations and, like Catulus, men of affairs, were "new poets" in their youth. In Cicero's case, this fact is often regarded as an aberration. But it is at least as likely that his early poetic experiments, the *Glaucus*, the *Alcyones*, the *Limon*, and the *Aratea*, were the sort of thing that came naturally to any young Roman educated in Hellenistic literary culture. Quite possibly Cicero's later admiration of Ennius was a bit unusual, something normally confined to deliberate archaizers like Lucretius.

It was therefore only too natural for Vergil to think of Hesiod as a kind of emblem for a certain approach to poetry, as he does in *Eclogue* 6. It was quite another thing to come to terms with the authentic Hesiod, as he does in the *Georgics*. This is one of the revolutionary achievements of the later poem, this rediscovery of the archaic figure concealed behind the modernist rhetoric. Indeed, one of the main purposes of *Georgics* 1 is to produce a confrontation between, on the one hand, the genuine,

106. *Sermo* 1.10.16–19.
107. Puelma Piwonka 1949.

archaic Hesiod and, on the other, the symbolic, Alexandrian Hesiod in the person of Aratus. The double imitation of both *Works and Days* and *Phaenomena* expresses Vergil's debt to both Hesiods while throwing into high relief what he owes to each.

The structure of Silenus' song makes Hesiod into a kind of conceptual symbol that embraces and sums up everything of value in the epic tradition. After Gallus' reception by Apollo and the Muses, little room is left for Homer in the world of a Hesiodic poet. In the *Georgics*, by contrast, Hesiod is only a starting point; and he is not the same Hesiod whom we remember from *Eclogue* 6. Instead, he is the actual poet of *Works and Days*. This change is highlighted throughout Book 1 as this Hesiod is brought into repeated contact with his great follower, Aratus. The process both illustrates the congruencies between the two authors and brings their basic differences into sharp contrast. Vergil puts this contrast to effective poetic use in modulating the tone of the book, which alternates between sunny optimism and grim pessimism, as he explores the dialectic that exists within the tradition of Hesiodic poetry. On the other hand, he finds common ground between these different poems not only in the fact that the *Phaenomena* is an imitation of *Works and Days*, but with reference to a third element: both Hesiod and Aratus are presented through allusion as imitators of Homer. This means that the *Georgics* makes an important addition to the literary-historical scheme outlined in *Eclogue* 6. There the tradition of scientific epos is traced back to Hesiod and, through him, to Orpheus. In the *Georgics*, Hesiod is only the starting point, while the climactic position formerly reserved for the *Ascraeus senex* is now granted to the *fons et origo* of all poetry—including that of the Hesiodic tradition—Homer.

Hesiod is not the only poet whom the *Georgics* casts in a new role. In the "Song of Silenus" Lucretius is the allusive starting point, just as Hesiod is in the *Georgics*. But if in the *Georgics* Lucretius gives place to Hesiod in this respect, his importance in the poem's allusive structure is if anything greater than it had been in *Eclogue* 6. In "The Song of Silenus" Lucretian language and imagery signals one aspect of the opening cosmogony's literary orientation, while in *Georgics* 2 and 3 repeated and detailed allusion points to *De Rerum Natura* 5 and 6 as Vergil's main source on the creation and eventual dissolution of the world. No longer the starting point, Lucretian material has become in a way the core of the poem, outlining a thematic opposition between life and death within a structure that emphasizes the centrality of that theme. It is in his ambitious imitation of Lucretius' poem of science that Vergil makes explicit the tension that is his main concern throughout the poem. Further, Lucretius is freed in the *Georgics* from the procrustean framework that made him in *Eclogue* 6 part of a tradition stretching back through Hesiod to Orpheus. Instead he is acknowledged as a follower of Homer. But this change in Vergil's outlook towards Lucretius affects his conception of Homer as well. *Eclogue* 6 seems to ignore or to suppress Lucretius' explicit invocation of Homer as a poet of natural philosophy. In the *Georgics* this interpretation of Homer is given full play, not only in the concluding "Aristaeus" episode, but in the common ground that the poet continuously labors to establish between Homer and the poets of the didactic tradition. To put it simply, he no longer views Homeric epos as merely the recitation of kings and battles, nor as a potential, if unlikely, source of stories dealing with disastrous love affairs; rather, he adopts Homer as his inspiration for poetry on the most sublime of all possible themes, the poetry of nature.

Thus Lucretius' characterization of Homer is an essential gloss on the allusive structure of the *Georgics*. If imitation of Hesiod and Aratus can be seen as a route to Homer, the same is evidently true of Lucretius, but in a different sense. After rediscovering Hesiod through the medium of his Hellenistic imitators, Vergil turns to the Roman branch of a different Hellenistic tradition, one that looked more frankly to Homer than to Hesiod as its conceptual model, one that viewed Homer not as a poet of kings and battles per se, but of myth as the reflection of a higher truth.

This revolutionary combination, one of the great achievements of the *Georgics*, may have been the single most important factor in the genesis of the *Aeneid*, which could certainly not have been written in its existing form without some reconciliation between these traditions. In the view of most current literary history, these traditions are basically antithetical, their combination almost a contradiction in terms. This, I think, is the reason why Lucretius in particular tends nowadays to get short shrift from those who see Vergil throughout his career as a "second-generation Neoteric" and who regard Callimacheanism as practically the sole motivating force behind the best poetry produced in Rome during the period that begins with Catullus and ends with Ovid. But this approach makes facile assumptions about the nature of Vergil's attitude towards Homer, Ennius, and Lucretius and leaves unexplained the philosophical element his poetry. We have seen that Vergil owes his mature conception of Homer to the Hellenistic scholars who followed the lead of philosophers such as Empedocles and Pythagoras. Poetically, however, his debt is to that branch of the epic tradition most splendidly embodied in Lucretius' *De Rerum Natura.* The effect of this poem on Vergil has been variously assessed; but in my estimation, its impact on Vergil's interpretation of Homer—an impact clearly measurable in the different outlooks that Vergil maintains in *Eclogue* 6 and the *Georgics*, and one that made possible the imitative program not only of the *Georgics*, but (in a different way) that of the *Aeneid* as well—was that of making Homer important to poets of Neoteric sensibility in a way that he had not been before. On this view, Lucretius' influence on Vergil has to be counted among the truly decisive forces that have determined the character of European literature.

In his Homeric program, Vergil does not content himself with the games of his Alexandrian predecessors; though he plays them, and plays them well, in his design they add up to something much more than the sum of their parts. In "The Song of Silenus" only the smallest place can be found for Homeric epos, which is mentioned only briefly and as an afterthought, warped almost beyond recognition into a sophisticated, "Neoteric" song of erotic disaster and metamorphosis. The assumption that underlies this reference seems to be that if Vergil is to imitate Homer, it must be in the Alexandrian fashion, which one scholar has aptly described as an attempt, "in den Bahnen Homers so unhomerisch zu sein wie möglich."[108] The possibility of Homeric imitation is limited to those passages that have something in common with the normal themes of Alexandrian and Neoteric verse. Moreover, it is probably significant that only the *Odyssey* is suggested as a likely source of suitable material, instead of the more exclusively martial *Iliad.* In the *Georgics*, however, a new attitude is signaled by the first Homeric allusion in the poem, which, as I have argued, involves one of the most sublimely heroic passages of, precisely, the *Iliad;* and, thereafter, allusion to both Homeric poems, but particularly to the *Iliad*, runs like a red thread through the *Georgics.* To be sure, the treatment that the Homeric material receives from Vergil the craftsman is not unlike what we see in *Eclogue* 6. The passages to which Vergil alludes are carefully chosen, with frequent representation of non-heroic material from the similes, and are reshaped to meet the requirements of a very different poetic style. We see this with

108. Herter 1929.50.

great clarity in the "Aristaeus," where Homeric material is given the form of a Neoteric epyllion. And yet, the point is no longer how to fit Homer into an alien aesthetic scheme, nor even to be as un-Homeric as possible in handling Homeric material. Instead, the earlier allusions to Hesiod, Aratus, and Lucretius look ahead to the Homeric "Aristaeus," which is based on a conception of Homer as a poet not of kings and battles, but of nature and philosophy. Viewed in these terms, Homer supplants even Lucretius as Vergil's guide to the secrets of the universe. Thus Vergil's predecessors no longer, as in *Eclogue* 6, establish a norm into which Homeric epos, if it is to be usefully available to Vergil, must by whatever means possible be made to fit; rather, they serve as a series of more familiar steps by means of which the ultimate poet can finally be approached.

A final token of Vergil's development may be seen in the figure of Orpheus. Unlike the symbol that he is in *Eclogue* 6, Orpheus appears in the *Georgics* as an actual character, the hero of the tale told by Proteus to Aristaeus. Still possessed of his ability to control nature through song, this Orpheus nevertheless cannot control himself, and so fails in his attempt to restore his beloved Eurydice to life. This feature of the story is probably Vergil's innovation; in any case, it represents an important qualification of the extraordinary power that Orpheus usually possesses. At the climax of *Eclogue* 6, Gallus receives a set of pipes used formerly by Hesiod and Orpheus, the very ones with which he used to bring the rigid ash trees down from the hills. Gallus is initiated into poetry of the most potent kind. At the climax of the *Georgics*, this marvelously powerful singer fails to save his wife and, dragging out a sterile existence in lamentation, eventually brings about his own destruction. The tale is affecting, and poignantly told; but the figure of Orpheus seems somehow diminished. It is not he after all, but Aristaeus, the georgic hero in cast in a Homeric mold, who succeeds in reversing the natural sequence of life and death.

In fashioning Orpheus' failed attempt, Vergil draws again on the poet who has by now become his chief model; and this too constitutes a reversal. In *Eclogue* 6, Orpheus represents a kind of absolute: he is the ultimate ancestor of all the poetry to which Vergil alludes there, even of Hesiod and Callimachus. But as this triumphant Orpheus gives way to the tragic figure of *Georgics* 4, even his experiences become derivative. As he descends to the depths of the world below, relying on the magic of his song, Orpheus disturbs the infinite multitude of shades gathered there:

at cantu commotae Erebi de sedibus imis
umbrae ibant tenues simulacraque luce carentum,
quam multa in foliis avium se milia condunt,
Vesper ubi aut hibernus agit de montibus imber,
matres atque viri defunctaque corpora vita
magnanimum heroum, pueri innuptaeque puellae,
impositique rogis iuvenes ante ora parentum,
quos circum limus niger et deformis harundo
Cocyti tardaque palus inamabilis unda
alligat et novies Styx interfusa coercet.

Georgics 4.471–480

His experience here is modeled on that of Odysseus in his "*Nekyia:*"

*τοὺς δ᾽ ἐπεὶ εὐχωλῇσι λιτῇσί τε, ἔθνεα νεκρῶν,
ἐλλισάμην, τὰ δὲ μῆλα λαβὼν ἀπεδειροτόμησας
ἐς βόθρον, ῥέε δ᾽αἷμα κελαινεφές· αἱ δ᾽ ἀγέροντο
ψυχαὶ ὑπὲξ Ἐρέβευς νεκύων κατατεθνηώτων.
νύμφαι τ᾽ ἠΐθεοί τε πολύτλητοί τε γέροντες
παρθενικαί τ᾽ ἀταλαὶ νεοπενθέα θυμὸν ἔχουσαι·
πολλοὶ δ᾽ οὐτάμενοι χαλκήρεσιν ἐγχείῃσιν,
ἄνδρες ἀρηΐφατοι βεβροτωμένα τεύχε᾽ ἔχοντες·
οἳ πολλοὶ περὶ βόθρον ἐφοίτων ἄλλοθεν ἄλλος
θεσπεσίῃ ἰαχῇ· ἐμὲ δὲ χλωρὸν δέος ᾕρει.*

Odyssey 11.34–43

If any guarantee were needed that Vergil is thinking of this Homeric passage, one could note that he borrows from the *Georgics* in establishing the Homeric background of Aeneas' catabasis (*Aeneid* 6.305–312). By modeling Orpheus' experience on that of Odysseus, Vergil effects a poignant contrast: the Homeric hero, like Vergil's Aristaeus and, later, his Aeneas, was successful in his symbolic confrontation with death. Indeed, reading the *Georgics* for the first time, and noticing the reference to the *Odyssey*, a reader might well take it as an allusive anticipation of the successful outcome that he probably expects to Orpheus' story. Upon learning the story's tragic end, however, we recognize that where Odysseus, Aristaeus, and eventually Aeneas all triumph, Orpheus must pathetically fail.

Later, at the very moment of his failure, Orpheus' experience again reminds us of Homer. Just as Eurydice begins to disappear again into the shades of death, Orpheus tries in vain to embrace her:

dixit [sc. Eurydice] et ex oculis subito, **ceu fumus in auras**
commixtus tenuis, fugit diversa, neque illum
prensantem nequiquam umbras et multa volentem
dicere praeterea vidit; nec portitor Orci
amplius obiectam passus transire paludem.

Georgics 4.499–503

Odysseus has the same experience when he attempts to embrace the shade of his mother:

ὣς ἔφατ᾽, αὐτὰρ ἐγώ γ᾽ ἔθελον φρεσὶ μερμηρίξας
μητρὸς ἐμῆς ψυχὴν ἑλέειν κατατεθνηυίης.
τρὶς μὲν ἐφορμήθην, ἑλέειν τέ με θυμὸς ἀνώγει,
τρὶς δέ μοι ἐκ χειρῶν σκιῇ εἴκελον ἢ καὶ ὀνείρῳ
ἔπτατ᾽.

Odyssey 11.204–208

Here Orpheus and Odysseus are united in their grief: Eurydice and Anticleia are lost forever, and the heroes must return without them. Again, however, we remember that Odysseus rises above his grief, returns to Ithaca and life, and, unlike Orpheus, does not succumb to self-pity or to any other deadly indulgence. If we listen closely to Vergil's language, however, we will catch another reference. When the shade of Patroclus appears to Achilles, he too tries in vain to embrace it:

ὣς ἄρα φωνήσας ὠρέξατο χερσὶ φίλῃσιν,
οὐδ᾽ ἔλαβε· ψυχὴ δὲ κατὰ χθονὸς **ἠΰτε καπνὸς**
ᾤχετο τετριγυῖα· ταφὼν δ᾽ ἀνόρουσεν Ἀχιλλεὺς....

Iliad 23.99–101

The visit of Patroclus' spirit sets in motion the concluding action of the *Iliad*, the exquisitely symbolic sequence of events through which Achilles grows, not as a warrior, but as a man until finally he can put aside his anger and receive with courtesy and respect the father of his most bitter enemy, the man who took Patroclus from him, Hector. Even Achilles, the relentless, is to this extent able to come to terms with death. But for Orpheus there is no such possibility. When he fails to conquer death, death conquers him.

Finally, after his return to earth, alone in his grief, Orpheus again follows where Homer has led. For seven months he dragged out his lament, like the nightingale bewailing the loss of her brood:

> qualis populea maerens philomela sub umbra
> amissos queritur fetus, quos durus arator
> observans nido implumis detraxit; at illa
> flet noctem, ramoque sedens miserabile carmen
> integrat, et maestis late loca questibus implet.
>
> *Georgics* 4.511–515

Luscinia, not *philomela*, is the usual Latin for "nightingale." The Greek word is never used elsewhere except as the name of the heroine. This fact, together with the simile form, refers the reader to the Homeric simile that makes use of her myth. Just after Odysseus' identity is revealed to Eurykleia, his aged nurse, Penelope returns to speak to her husband, thinking that he is just an ill-used traveler. She speaks of her grief, comparing herself to the ever-lamenting nightingale, reciting the *aetion* of its behavior:

> ὡς δ' ὅτε Πανδαρέου κούρη, χλωρηῒς ἀηδών,
> καλὸν ἀείδῃσιν ἔαρος νέον ἱσταμένοιο,
> δενδρέων ἐν πετάλοισι καθεζομένη πυκινοῖσιν,
> ἥ τε θαμὰ τρωπῶσα χέει πολυηχέα φωνήν,
> παῖδ' ὀλοφυρομένη Ἴτυλον φίλον, ὅν ποτε χαλκῷ
> κτεῖνε δι' ἀφραδίας, κοῦρον Ζήθοιο ἄνακτος....
>
> *Odyssey* 19.518–523

Thinking back to this passage, the reader of the *Georgics* realizes that Penelope's remarks describe Orpheus' situation even better than her own. The myth refers directly to Philomela's own role in bringing about her unhappiness by murdering her son, just as Orpheus has brought about Eurydice's second death by failing to keep his gaze away from her. But Vergil in his simile shields Orpheus from blame; his mistake was venial, if only the shades below knew forgiveness (*G.* 4.489). Vergil's Philomela bewails the brood that the *durus arator* has deliberately removed from their nest. This alteration of a model is, as usual, far from arbitrary. The motif of a plowman destroying or despoiling a bird's nest recalls an earlier passage of the *Georgics* (2.207–211), and the desire to produce a thematic echo might in itself have been enough to inspire this change.[109] But earlier in the *Odyssey* the reunion of the hero and his son produces in them a grief like that of birds whose children were stolen by farmers:

> *κλαῖον δὲ λιγέως, ἀδινώτερον ἤ τ' οἰωνοί,*
> *φῆναι ἢ αἰγυπιοὶ γαμψώνυχες. οἷσί τε τέκνα*
> *ἀγρόται ἐξείλοντο πάρος πετεηνὰ γενέσθαι.*
>
> *Odyssey* 16.216–218

The conflation of these two Homeric similes creates an effect of typically Vergilian ambivalence. Just as the allusion to Penelope's simile charges Orpheus with responsibility for his own actions, the detail imported from the other makes him a mere victim of arbitrary forces beyond his ken. In both Homeric similes, however, we notice an ironic inconcinnity. The grief of Odysseus and Telemachus is like that of birds that have lost their young; but here father and son, who thought they had lost each other, are reunited. Similarly Penelope grieves continually for the loss of her husband; but she is about to be reunited with him, is, in fact, speaking with him, though she does not know it, at this very moment. When applied to Orpheus, the similes lose this inconcinnity: Orpheus has lost Eurydice, for the second time, and irrevocably.

We may interpret Vergil's more "appropriate" use of these similes as an aspect of his Homeric *aemulatio*. More important, however, is the simple fact that so much of Orpheus' experience in the *Georgics* derives from Homer. In speaking of the differences in Vergil's attitude towards Hesiod in *Eclogue* 6 and in the *Georgics*, I spoke of the reversal whereby the *Ascraeus senex*, the symbol of all the epic poetry that Vergil admired, is subsumed into a scheme in which all derives from Homer. By modeling the central experiences of his Orpheus on those of characters from the *Iliad* and *Odyssey*, Vergil seems to be making a similar statement about this even more potent poetic symbol. In *Eclogue* 6, as David Ross has put it, Orpheus stands for Vergil's conception of "a poetry of 'science,' rather than of a particular school or sect."[110] In *Georgics* 4, even Orpheus is subsumed under the penumbra of the quintessential poet of science, as of all else, Homer.

109. Thomas 1989.2.233 *ad G.* 4.511–15 regards the internal reference as the main point here.

110. Ross 1975.26.

CHAPTER 8

Conclusion: After the *Georgics*

The *Aeneid*

IN MANIBUS TERRÆ. The main business of this essay is now complete. The allusive scheme of the *Georgics* and the significance of such a design in the light of Vergil's earlier career have been examined. It remains only to take a very brief look at what in later Augustan literature we can ascribe to the new direction in his own work signaled by the *Georgics*.

As I have already noted, just as Vergil's didactic essay in the *Georgics* differs from the one that he had first sketched in *Eclogue* 6, so his achievement in heroic epos is not what his brief for such a project in the *Georgics* leads one to expect. And yet here, too, elements of continuity can be observed. This is not the place to embark upon a detailed discussion of the *Aeneid* from this or any other point of view. Rather than indulge this urge, I will content myself with a few observations that address one or two problems, both traditional ones and those raised by the outcome of my inquiry into the *Georgics*.

It remains surprising that the poet of the *Eclogues* should ever have come to write the *Aeneid*. But what we have learned about both the *Eclogues* and the *Georgics* sheds some light on this development. An important part of Vergil's allusive program in the *Eclogues* is to elevate the tone and thematic purview of Theocritean pastoral, while at the same time distinguishing his *molle atque facetum epos* from the *forte epos acer* of poets like Pollio. These aims are of course basically incompatible, and it is partly the tension between them that gives the *Eclogues* their unique flavor. The central poems of the collection are perhaps the most interesting in this regard; and the fifth, which I have not discussed, strikes a particularly fine balance between the opposing tendencies of pastoral form and elevated theme. *Eclogues* 4 and 6, in their concentration on themes of the grandest possible sort—heroism, Sibylline prophecy, the *magnus annus*, orphic shamanism, cosmogony, and so forth—threaten to leave pastoral far behind. These sophisticated generic and stylistic games are based on a variety of other models, but have little to do with Homer. There are only the faintest indications that Homer was a part of Vergil's thinking at this time, although some of Homer's imitators surely were. Our main impression, however, is of the contrast between Theocritus' reliance on Homer throughout the *Idylls* and Vergil's virtually complete independence of Theocritus' master.[1]

In the *Georgics* we see that Vergil's plan was not to write heroic epos in a Homeric vein, but rather, through allusion, to interpret Homer as a poet of natural philosophy, or better perhaps, to exploit this well-established scholarly interpretation of the *Iliad* and *Odyssey* in his own allusive program. This is an obvious *volte-face* from the position of the *Eclogues*, and it is difficult to explain; I have suggested that it is related to the growing influence of Lucretius on the aspiring didactic poet. Here I would stress again a factor that was so important in the *Eclogues*. The Homeric program of the *Georgics* serves the dual purposes of elevating Vergil's notionally "humbler" didactic discourse to a plane nearer that of heroic epos, and of defining more clearly the tradition in which the *Georgics* itself participates. On the latter count Vergil follows a somewhat different procedure from what we have seen in the *Eclogues*. Instead of drawing open distinctions between the *Georgics* and examples of the more elevated heroic epos, Vergil actually "shows" through allusion that the ultimate heroic poet, Homer, was in fact a poet of nature.

The question that faces us now is, How much of this method that had served Vergil so well in the *Georgics* carries over into the much more considerable Homeric imitation of the *Aeneid?* The answer to this question has been anticipated in part. In Chapter 6 I introduced my discussion of Homeric allegory in the "Aristaeus" by noting that allegorical exegesis had already been found to underlie certain Homeric allusions in the *Aeneid*. I also referred to a number of studies that testify to a growing awareness that this tradition is an essential ingredient of *Aeneid* criticism. The most ambitious critic of this type, Philip Hardie, argues that the *Aeneid* "draws extensively on techniques and themes

1. Vergil's transformation of Theocritus' para-Homeric Polyphemus (*Idyll* 11) into the thoroughly un-Homeric Corydon (*Eclogue* 2) illustrates this difference very well.

which can broadly be classed as 'cosmological,'" i.e. techniques and themes that "derive either from philosophical discussions concerning the nature of the physical universe, or else from mythical cosmology of a pre-philosophical age."[2] Elsewhere, Hardie states that "Vergil's conception of the epic as an imitation of the cosmos can be derived directly from the ancient idolatry of Homer, which made of him a universal genius."[3] Another scholar, Antonie Wlosok, in discussing Macrobius' Neoplatonist reading of the *Aeneid* as an allegory possessing a surface layer of mythological narrative that conceals an inner philosophical core, points to the tradition of allegorical interpretation of Homer and to passages of both the *Aeneid* and other Augustan poetry to show that such interpretive procedures were familiar to Vergil and his contemporaries. She concludes that "Vergil does not aim at making two levels of meaning constantly present and constantly in harmony, but rather he creates a congruence of sense which is only partial. In this way he is actually calling the readers' attention at important points to the possibility of a cosmic interpretation in addition to the surface level. This amounts to telling the reader from time to time that there is a double meaning.... Macrobius, in my view, understood Vergil correctly."[4]

I myself find much to agree with in these studies, and gratefully acknowledge how useful they have been as keys to understanding Vergil's imitation of Homer in the *Georgics*. On the other hand, we must be careful of how we define the philosophical element in the *Aeneid*. Again, we may look to the *Georgics* for guidance. Philosophy and philosophical poetry are centrally important themes in the *Georgics*; in addition to those factors that I have examined here, I would point to the work of Ross, who has so brilliantly elucidated the poem's thematic structure in terms of the four elements and the humoral theory of Greek philosophy and science. Ross traces this motif throughout the poem, and makes some particularly apt observations about its significance even in the most overtly narrative and heroic part of the poem, the "Aristaeus."[5] But Ross also traces the poem's preoccupation with history and, specifically, with Troy.[6] As the poem progresses in theme from inanimate to animate nature and to society, the historical dimension becomes more

2. Hardie 1986.5.
3. Hardie 1985.85–86.
4. Wlosok 1985.80.
5. Ross 1987.224–226, 231.
6. Ross 1987.254 s.v. "Troy."

and more pronounced; particularly suggestive is Ross's discovery of a diachronic subtext in the bee discourse of Book 4.[7] We should see this tendency as parallel to the growing importance of Homer and heroic epos in the poem. Vergil is a deeply and seriously philosophical poet—we are only beginning to realize how true this is—but one who did not express himself in conventional philosophical terms. If we can believe the ancient biographical tradition, he might eventually have done something of the kind.[8] But we can be more than content with what we have, a poetry that does not serve philosophy, as does that of Lucretius, nor one that makes its philosophical subject the paradoxical foil of its poetic qualities, like that of Aratus. There is no gap in Vergil between philosophy and his poetry; each element is coextensive with the other.

In the *Georgics* a poetics of philosophy is discovered within myth and history, and vice versa. The perpetual cycle of birth, death, and rebirth is situated in the practical venue of the farm, where, year by year, the farmer attempts to manage and benefit from this natural, inevitable process. Ultimately, like Orpheus, he will fail: each and every thing under his control, and he himself, will die. But, like Aristaeus, he plays a crucial role in the process that enables life itself to continue. Vergil acknowledges the cruelty and the inherent tragedy of the farmer's position, and makes him a representative of humanity at large. He does not allow the fact that life as a whole continues to outweigh any individual death. And yet, unlike the *Eclogues*, the *Georgics* is not a sentimental lament over the irrevocable loss of an ideal, or better, an idealized world. The georgic world is, by definition, imperfect; when it dies, our lament is conditioned by the knowledge that it is so. Whatever replaces it, too, will be imperfect; but our desire that it be reborn as something new is grounded in an understanding of the alternative, that we surrender, finally and irrevocably, to death. This rebirth will cost: as in Lucretius, so in the the *Georgics*, something must die before anything can be born. In the *Georgics*, what has died is, as in the *Eclogues*, Rome, a Rome that is repeatedly linked to its Trojan prototype. But the *Georgics*, unlike the *Eclogues*, looks anxiously ahead to the rebirth of society under its new leader, Caesar. That this rebirth will cost, has indeed already cost the lives of many individuals is

7. Ross 1987.191–200, 206–214.

8. *Vita Donati* 35 (Hardie 1966.14), very possibly no more than an elaborate inference from the interest that the *Georgics* poet expresses in philosophical themes at *G.* 2.475–482.

not left unsaid; but that sad fact is measured against the more general annihilation that threatens. In this way, the *Georgics* is perhaps Vergil's most hopeful poem.

Vergil's treatment of death and rebirth in the *Georgics* allows for allusion to Homer's treatment of the same themes. This interpretation of Homer, particularly as it concerns the birth of the cosmos itself, is closely bound up with the tradition of philosophical allegory; hence, Vergil's imitation of Homer takes on the appearance of that tradition. But in the *Aeneid*, his focus is more openly on history itself; consequently, the poem's Homeric program, which is also much more ambitious than that of the *Georgics*, is also more concerned with Homeric poetry per se than with the interpretive traditions that grew up around it. To be sure, these traditions inform Vergil's imitation, and an acquaintance with them can help us to understand specific ways in which the Homeric poems were transformed into the *Aeneid*. Juno's role as "Queen of the Air," Iopas' song at Dido's banquet, Anchises' eschatological discourse in the underworld, and the politico-cosmological significance of Aeneas' shield—all of these factors must be explained, at least in part, with reference to the allegorical tradition. Similarly, the enemies of Aeneas and Augustus are repeatedly associated with enemies of the universal order—Giants, Titans, Typhoeus et al.—a topos of political panegyric. Finally, Vergil exploits to the fullest measure all the resources of epic grandiloquence at his disposal as a means of suggesting the universal significance of his theme.

A natural conclusion would therefore be that the *Aeneid* develops in full the conception of Homer adumbrated in the *Georgics*, rivaling the *Iliad* and the *Odyssey* as philosophical allegories; and that the equation between the georgic world and the universe together with the parallelism between the georgic world and the Roman state in the *Georgics* anticipates the cosmic significance with which Roman history is endowed in the *Aeneid*, where Vergil assimilates Homeric epic to a panegyric of the Augustan regime. But this would be, in my opinion, an overstatement. There can be no question that Vergil shows himself familiar in the *Aeneid* with Homeric allegories on the one hand and with the related employment of cosmological themes such as gigantomachy in the service of political panegyric. The inference that he is therefore writing an allegorical epic as a panegyric of Rome or Augustus is however not warranted. It is easy to assume that the presence of any given commonplace, generic element, or token of cultural bias in Vergil forces us to conclude that he accepts the attitudes that accompany the motif elsewhere. But this is hardly ever the case. Knowing, for example, that the Romans generally regarded untamed nature not with modern admiration, but rather with something like abhorrence, and knowing further that they held the farmer's profession in the highest esteem, we are not justified in using this knowledge as a skeleton key to the *Georgics*. We do not applaud when the noble plowman uproots the ancient homes of birds (*G.* 2.209) in order to clear the land, any more than when he callously or robs a nest (*G.* 4.412–413); some scholars, however, like Page almost a century ago, have recommended that we do so.[9] Perhaps few would today. But G. Karl Galinsky has recently argued that the theme of Aeneas' *ira* is best explained with reference to ancient discussions of anger, and particularly of its place in the psychology of the general and the judge.[10] Thus, when Aeneas "rouses himself to anger" (*se suscitat ira*, *Aeneid* 12.108), and later, when he lets anger overcome the *clementia* aroused by Turnus' final speech, he is merely doing what Aristotle would say it is the duty of any man in this situation to do.[11] But why should the theories of Aristotle take precedence over the evidence of the *Aeneid* itself, in which anger is consistently represented as an ignoble, bestial emotion, a force that opposes Aeneas throughout his mission, and, finally, as a seductive temptation to which he at last, in spite of better instincts, succumbs in such a way as to suggest that he is no different from the enemy he slays?[12]

This last example relates very directly to the question of allegory and panegyric, in that the decisions one makes about both issues materially affect how one interprets the poem as a whole. One might well be tempted, because the *Aeneid* employs philosophical allegory in a way that is related to the conventions of political panegyric, to conclude that Vergil's main purpose in the epic is to lend grandeur to Augustus' regime.[13] But it is quite characteristic of Vergil to turn the conventions

9. E.g. Page 1898.262 *ad G.* 2.210.

10. Galinsky 1988.

11. Galinsky 1988.333: "In the final scene, Aeneas' behavior is almost a textbook illustration of Aristotle's view of anger."

12. The best analysis of the poem's end remains, in my view, that of Putnam 1965. 151–201. See also Johnson 1976.114–134.

13. Hardie (1986.1), while disavowing pretensions to a "global critical approach" to the poem, seems to look toward such a conclusion, noting that "those whose critical synthesis tends to dismiss the panegyrical content of the poem as so much 'surface ideology' would find that the 'surface' runs rather deep, if my analyses are correct."

of a given genre to new, even subversive purposes. Where Theocritus' *Idylls* construct an imaginary pastoral world out of myth and literary allusion, Vergil's *Eclogues* use the same techniques to depict a similar world in dissolution. Where Hesiod, Aratus, and Lucretius all, though for different reasons, wish to impart accurate information, Vergil in the *Georgics* concerns himself with his ostensible subject only in so far as it enables him to develop a larger, metaphorical and symbolic argument—and does not stop at warping or falsifying well-known facts in order to further that end. By the same token, in the *Aeneid* he uses the forms and procedures of Homeric poetry and of the panegyric tradition that derives from Homer to conduct a searching critique of the heroic ideal which that tradition celebrates. Most careful readers find that this ideal is presented at different times in a favorable light and otherwise; and the same applies to Augustus' regime, to which so much of Aeneas' experience corresponds.

Thus its use of Homeric allegory need not force us to regard the *Aeneid* as a straightforward panegyric of Augustus. Neither should we regard this exegetical tradition as the main formative influence on Vergil's imitation of Homer. In the *Aeneid*, his use of this tradition appears to be confined to the few passages listed above and to his treatment of divinity. These are important aspects of the poem; but they are not its core. The main point of the poem, and of its Homeric program, is the experience of Aeneas and how we view that experience in the light of the Homeric poems. In this respect, allegory plays only a subsidiary role.

Comparison with the *Georgics* both confirms the presence of allegory in the *Aeneid* and illustrates its lesser importance. In terms of Vergil's poetic development, it seems to me that allegory served him well by making possible what must have been an extremely difficult approach to Homer. Once Vergil had successfully engaged Homer on the so to speak favorable terms offered by didactic epos, he was ready to enter his last and greatest poetic agon. Allegory continues to play a role, but one no greater than those of many other factors that had, over the years, become important parts of the Homeric tradition. Allegory remained a part of the European epic tradition, and the *Aeneid* is no doubt in part responsible for this fact. Nevertheless, the abiding importance of the allegorical tradition in Vergil's work is, in my view, that it provided Vergil, who in his early career did not count Homer one of his literary forbears, with a means of coming to terms with Homer. It is not, I think, going too far to infer that, had Vergil not made the first step that we have traced in the *Georgics*, the *Aeneid* would never have been written.

Reception and Influence

For all Latin poets, the *Aeneid* immediately became and long remained the most important factor to be reckoned with in their literary heritage. The impact of the *Georgics* is more difficult to assess. The reactions of Horace and Propertius to Vergil's poetic development are expressed in highly conventional ways; but even these suggest the revolutionary nature of his achievement. In Ovid, we see indications that this fact had been forgotten—that the revolution had been so successful as to establish immediately a new orthodoxy.[14]

Horace

In his *Sermones* Horace mentions his friends Varius and Vergil as preeminent in two very different branches of the epic tradition: *forte epos acer* | *ut nemo Varius ducit, molle atque facetum* | *Vergilio adnuerunt gaudentes rure Camenae* (1.10.43–45). This statement was made when Vergil's career, like Horace's, was still young: it refers to the *Eclogues* alone, not to the *Georgics*, which was barely under way. But Horace returned to this motif in that portion of his masterpiece, the three-book collection of *Odes*, in which he defines his own poetic ambition.

One of the primary motives of *Carmina* 1.1–9—the so-called "Parade *Odes*"—is to provide the reader with a set of critical parameters for the collection that it introduces. The sequence accomplishes this end most notably in its metrical variety, but we must not ignore other strategies. Matthew Santirocco has shown convincingly that one of Horace's primary aims in these poems is to define his new endeavor in contrast to his prior accomplishments and also to the poetry of his contemporaries. The terms of these contrasts include subject matter, style, genre, and so forth. With respect to his own work, for example, Horace models the first poem of the collection on the beginning of *Sermo* 1.1; but here, instead of drawing upon the diatribe tradition to create the persona of a social critic, Horace invokes the *mempsimoiria* topos to draw quite

14. In its general tendencies my analysis parallels Newman's observations concerning the *vates* concept (1967.101–206 *et passim*).

explicit attention to the persona that he adopts in this new collection, that of the lyric poet. This persona is presented as an ambition that, if attained, confers on the poet a kind of immortality: Horace alludes to his eventual catasterism (*Carmen* 1.1.36) as a symbol of his success. Gone, then, is the pretence of the *Sermones* that Horace is not really writing poetry. In the *Odes*, he frankly admits the high ambition of his new poetic program. But not too high; and here he explicitly distances himself from the grander pursuits of his contemporaries. In *Carmen* 1.6 (*Scriberis Vario*) he refers again, as in *Sermo* 1.10, to Varius as the preeminent Roman poet of heroic epos. Horace politely declines to write a *Vipsaniad* in honor of Agrippa's military achievements; this is better left to Varius, who has proved himself a worthy follower of Homer (*Maeonii carminis alite* 2). The poet protests his Muse's unwillingness, despite the fact that she is described as *potens* (10), to sing about heroic—and specifically Homeric[15]—themes, describing the them as *grandia* while he himself is *tenues* (9). No denigration of Varius is intended,[16] still less of Agrippa, to whom this poem is a graceful compliment, although there is a suggestion that heroic subjects per se are distasteful.[17] But the main point is simply to set a limit to Horace's lofty lyric ambitions, to define by contrast with Varius' accomplishments what he himself hopes to achieve.

It is in much the same vein that Horace conceived his propempticon to Vergil (*Carmen* 1.3). The most convincing interpretations of this poem argue that the journey on which Vergil has embarked is as much metaphorical as real—a literary voyage that will carry him away from the safety of land and *ländliche Gedichte* onto the high seas of heroic epos.[18] The voyage is a newly begun poem, namely, the *Aeneid*. Once we recognize this fact, Horace's position becomes much clearer. His agitation at his friend's supposed departure is expressed in distressing terms that seem inappropriate to a poem of farewell. The first sailor is described in terms that suggest not merely bravery, but an actual lack of most human feelings (9–20). His achievement is characterized as a violation of divine

15. Both the *Iliad* (*gravem* | *Pelidae stomachum cedere nescii* 5–6) and the *Odyssey* (*cursus duplicis per mare Ulixei* 7) are mentioned, along with an epic on Pelops (8).

16. Nisbet–Hubbard 1970.84 *ad Carm.* 1.6.2 *alite* note that "elsewhere the ablative of agent [sc. without a preposition] refers to a quasi-instrument, i.e. something more inanimate than a poet"; but everything about this passage tells against such an interpretation.

17. *Stomachum* 6 is an odd way to translate the Homeric μῆνιν (*Iliad* 1.1). See Nisbet–Hubbard 1970.85 *ad loc.*

18. Santirocco 1986.27–30, with further references.

will, and made an emblem of human impiety (21–40). The associations of this venture are all heroic: the motif of the *protos heuretes* in this context points to the unnamed Jason. Also unnamed is Prometheus, who is called *audax Iapeti genus* (27): thus by antonomasia Horace alludes to the grand epic theme of Titanomachy. The twin assaults of Daedalus on the upper and Hecules on the lower worlds (34–36) round off the catalogue of *exempla;* and the poem then concludes with the rueful observation that human nature prevents Jupiter from ever laying down his thunderbolts (38–40).

These are rather surprising sentiments to associate with Vergil's ocean voyage. But if we view the voyage as a metaphor for poetic composition, it makes good sense. Horace addresses Vergil almost as an *alter ego*, calling him *animae dimidium meae* (8). This is an expression of great intimacy. It contains, in fact, more than a hint of eroticism. Its clearest parallel is an epigram of Callimachus that begins *ἥμισύ μευ ψυχῆς ἔτι τὸ πνέον, ἥμισυ δ' οὐκ οἶδ' | εἴτ' Ἔρος εἴτ' Ἀΐδης ἥρπασε, πλὴν ἀφανές.*[19] We may compare, for instance, the erotic language of Catullus' poem to Licinius Calvus (*Carmen* 50), another poem best read as a meditation on literature.[20] There the erotic element indicates Catullus' admiration for Calvus the poet, his shared admiration for essentially Callimachean principles. This parallel strongly corroborates the view that Horace, too, by addressing Vergil in erotic language that is actually borrowed right from Callimachus, is indicating their shared acceptance of these literary principles.

The allusion to Callimachus, however, sounds an ominous note in a poem on Vergil's ambition to become a poet of heroic epos. It is difficult not to recall the terms in which Vergil had addressed Pollio in *Eclogue* 3. Despite the Callimachean principles that the two poets share, Vergil has been lured into the deep water of Homeric epos. Just as in the *recusatio* to Agrippa, Horace makes it clear that his new, lofty ambitions are conceived in strict Callimachean terms. As Vergil himself had done in the case of Pollio's tragedies, Horace acknowledges Vergil's Homeric program as something altogether different from his own humbler pursuit.

19. Callimachus *Epigr.* 41 (Pfeiffer 1953.92). On the "half of my soul" motif see Nisbet–Hubbard 1970.48 *ad Carm.* 1.3.8.

20. See especially Buchheit 1976.

Propertius

Propertius, like Horace, uses Vergil as a foil to his own poetic interests; but he is much more explicit than Horace in drawing the contrast. It is very likely that at first Propertius too saw Vergil as a poet very much like himself. Certainly the affinities between the *Eclogues* and the *Monobiblos* are well known, and it is likely that both collections drew on substantially the same models.[21] In *Carmen* 2.34, when he describes the *Eclogues*, Propertius is able to cite specific evidence that their primary subject is love:

> tu canis umbrosi subter pineta Galaesi
>   Thyrsin et attritis Daphnin harundinibus,
> utque decem possint corrumpere mala puellas
>   missus et impressis haedus ab uberibus.
> felix qui vilis pomis mercaris amores!
>   huic licet ingratae Tityrus ipse canat.
> felix intactum Corydon qui temptat Alexin
>   agricolae domini carpere delicias!
> quamvis ille sua lassus requiescat avena,
>   laudatur facilis inter Hamadryadas.
>
> *Carmen* 2.34.67–76

To be sure, Propertius playfully distinguishes between the "fiscal realities" of the bucolic and elegiac worlds; but he presents the concerns of those who inhabit those worlds, and of the poets who fashion them, as essentially similar. In the lines that follow, the *Georgics* gets comparatively short shrift:

> tu canis Ascraei veteris praecepta poetae,
>   quo seges in campo, quo viret uva iugo.
>
> *Carmen* 2.34.77–78

There is no similarity of content between the *Georgics* and Propertius' elegies, and thus he does not dwell on Vergil's middle poem. His way of denoting it, however, is significant. The reference to the advice of the old poet of Ascra looks at once to the Vergil's Callimacheanism and to his successful poetic agon with the authentic Hesiod. Nevertheless, it is the former element that is uppermost in Propertius' mind. Twice elsewhere in the same poetry book Propertius uses Hesiod's *ethnikon* with

21. See Ross 1975.51–84.

programmatic significance. The two passages do not display a thoroughly consistent attitude. In the first—a famous interpretive crux—the poet asserts that his *carmina* are as yet unfamiliar with *Ascraeos fontis,* since Love bathes only in the stream of Permessus (*Carmen* 2.10.25–26). This passage seems to make the epithet a symbol of some higher genre than elegy. But in the second passage, Propertius does, in effect, call his poetry "Ascraean":

> non tot Achaemeniis armatur †etrusca† sagittis,
> spicula quot nostro pectore fixit Amor.
> hic me tam gracilis vetuit contemnere Musas,
> iussit et Ascraeum sic habitare nemus,
> non ut Pieriae quercus mea verba sequantur,
> aut possim Ismaria ducere valle feras,
> sed magis ut nostro stupefiat Cynthia versu:
> tunc ego sim Inachio notior arte Lino.
>
> *Carmen* 2.13.1–8

Here Propertius does inhabit the Ascraean grove, whereas before his verses did not know the waters of the Ascraean spring. His residence there, however, is conditional: his poetry is to work its charm on Cynthia, rather than on trees and beasts. Propertius thus claims the Hesiodic associations that he had forsworn in poem 10, but now denies any specifically Orphic pretensions. All of this is certainly familiar, and when Propertius states that success with Cynthia will make him a more famous poet even than Linus, it is difficult not to think of *Eclogue* 6, where Linus presents Gallus with the pipes previously used by both Hesiod and Orpheus. Propertius must be alluding to this passage, and perhaps to a lost passage of Gallus as well. Clearly he is distinguishing his own poetry from the higher poetry of nature adumbrated in *Eclogue* 6, where it is symbolized by Orpheus and the *Ascraeus senex;* and it is equally clear that he views the *Georgics*—the advice of the old poet of Ascra—as poetry of this higher mode. Yet even this has its affinity with Propertius' own work. In summing up Vergil's career to date, Propertius notes that he sings a song on his *docta testudine* such as Cynthius himself might produce (*Carmen* 2.34.79–80). *Doctrina* might be regarded as the hallmark of the Orphic poetry to which Vergil aspires and from which Propertius wishes to distance himself. On the other hand, Propertius too admires *doctrina:* in the same poem cited above, he refers to the pleasure of reading while lying in the lap of a *docta puella* (*Carmen* 2.13.11), to cite only one example. Further, the Cynthius to whom Propertius compares Vergil is not only the god who appears to Tityrus in *Eclogue* 6, but also the one who gives his name to Propertius' *puella.* We must, I think, conclude that Propertius sees himself and the Vergil of the *Eclogues* and the *Georgics* as poets of one tradition who have developed in different ways, Propertius remaining content with the humbler strains of elegy[22] while Vergil aspires to an Orphic mysticism in his poetry of nature.

What, then, of Vergil's masterpiece, the *Aeneid?* This is, after all, the poem that Propertius heralds in his *Carmen* 34 as "something greater than the *Iliad*"—a remark that has been interpreted in diametrically opposed ways: either Propertius is speaking of Vergil's ambitious undertaking in terms of the highest possible praise, or he is depreciating it as something fundamentally incompatible with his own stubbornly maintained Callimachean principles. It is not equivocal to maintain that there are elements of truth in both views. I think that the latter is closer to the truth, but I do not think that Propertius' evaluation of the *Aeneid* is hostile: like Horace's *Sic te diva potens Cypri*, the poem simply takes Vergil's colossal Homeric agon as a clear and convenient point of contrast with the elegiac poet's own agenda.

The contrast is spelled out with painstaking clarity in the first two-thirds of the poem, before Vergil is even mentioned. Lynceus, a friend of the poet, has made a play for Propertius' girlfriend (9–10). This is no wonder, since Amor has power to corrupt relations of friendship, kinship, and of even peaceful nations: both Iliadic and Argonautic stories are adduced as evidence (5–8). The point of these *exempla* is immediately made clear: Lynceus is a devotee of serious literature who has no use for elegy. Thus his sudden susceptibility to Cynthia's charms gives Propertius reason to gloat:

> quid tua Socraticis tibi nunc sapientia libris
>   proderit aut rerum dicere posse vias?
> aut quid Erecthei tibi prosunt carmina lecta?
>   nil iuvat in magno vester amore senex.
> tu satius memorem Musis imitere Philitan
>   et non inflati somnia Callimachi.
>
> *Carmen* 2.34.27–32

22. Propertius expresses his pride in this "humbler" tradition by concluding this poem with a recitation of the names of the great poets who created it (85–94).

The field of reference here, as often in Propertius, is deliberately left rather vague. We need not wonder what Socratic texts are meant, or what "the poems of Erectheus" might specifically denote: Lynceus likes serious literature, philosophy and the classics. The next line, however, is more puzzling: who is Lynceus' *senex?* Socrates, possibly. Erectheus, hardly. To infer from what follows, should we perhaps think of the *Ascraeus senex* in this context? We next learn that Lynceus would do better to forget him and to take the elegists Philitas and Callimachus as models. Callimachean stylistic principles are recommended (*non inflati* 32), and reference is made to Callimachus' most famous and important programmatic passage (*somnia* 32). Because of Hesiod's importance in this regard, it is unlikely that the *Ascraeus senex* is meant. Perhaps, then, Lynceus' *senex* is some other figure symbolically opposed to Hesiod, one that sums up the essence of the poetry that Lynceus admires, the antitype of Propertius' own work. No doubt the vagueness is deliberate; but we can infer all that we need to know from what we are told in the following lines.

Propertius' comments only make sense if we assume that he regards Lynceus' putative behavior as suggesting some sort of poetic composition. He has already told Lynceus to imitate Philitas and Callimachus, and it is difficult to see how this could be done, except in a literary sense. He then lists the sort of themes on which Lynceus has presumably written in the past, themes that will do him no good in his current predicament: the course of the rivers Achelous and Maeander, the tales of Arion and Archemorus, of Amphiareus and Capaneus (33–40). All of this Propertius sums up in the injunction *desine et Aeschyleo componere verba coturno, | desine, et ad mollis membra resolve choros* (41–42). Lynceus is, then, a poet in the high and serious tradition of Attic tragedy who must learn the softer strains of elegy if he would succeed at love. The same thing happened even to Antimachus and Homer (45). Elegiac *puellae* don't much care about celestial mechanics (51–52), eschatological speculation (53), or the philosophical aspects of meteorology (54). This brings us to the point, namely, Propertius himself, the perfect lover, reigning at a banquet amidst all sorts of girls, without an ancestral triumph to his name, by virtue of his *ingenium* (55–58).

It is here that Vergil is finally introduced. Propertius, wounded by *Amor*, loves to languish among the garlands of last night's party,[23] while

23. In light of what follows, should we think here too of *Eclogue* 6, where Silenus is introduced in the very state that Propertius here claims as his own?

Vergil enjoys singing of Actium, Caesar, Aeneas, and the foundation of Rome (59–64). Vergil, then, as the poet of the *Aeneid*, is revealed as Propertius' antitype. It is, I think, in terms of this revelation that the entire poem should be read. The wide-ranging examples of themes and models ascribed to Lynceus and declared useless to a poet of love in fact characterize Vergil's heroic ambitions. The fact that Propertius mentions the *Aeneid* first (61–66) and in terms of sharp contrast to his own poetry (55–60), but then turns to Vergil's earlier, more Callimachean achievements (67–80), supports the interpretation that he, like Horace, perceived the *Aeneid* as a departure from Vergil's earlier interests. Both Horace and Propertius probably wrote before they had a clear idea of what the *Aeneid* would be like; how could they know that the final product would be at once thoroughly Homeric and consummately Callimachean? But their reactions closely parallel Vergil's earlier treatment of Pollio, and corroborate the modern view that Vergil's own career underwent a kind of revolution. It was the overwhelming importance of Vergil's masterpiece that made Vergil's personal revolution perhaps the most important single development in Roman literary history since the time of Ennius.

Ovid

The last great Augustan poet, Ovid, is really of a different generation from Vergil, Horace, and Propertius. His concerns are different from theirs, and he can take for granted their hard-won solutions to weighty poetic problems. One sees this especially in his literary-historical observations, which are knowing and shrewd, but not pioneering. Where Vergil and his contemporaries worked deliberately to construct for themselves a usable literary past, Ovid, like a pattern-book architect, was free to borrow and combine even the most diverse elements from the rich trove that his predecessors had laboriously recovered from the past.

I will confine my illustrations to two points. The first concerns the topic with which this study began, literary antonomasia. In the final poem of *Amores* 1, Ovid predicts the eternal fame that his poetry will win him. In so doing, he runs through a list of the immortals whom he hopes to join:[24]

24. For an interesting counterpoint to this catalogue see the proemium to Manilius *Astr.* 2 (1–52).

vivet Maeonides, Tenedos dum stabit et Ide,
dum rapidas Simois in mare volvet aquas;
vivet et Ascraeus, dum mustis uva tumebit,
dum cadet incurva falce resecta Ceres;
Battiades semper toto cantabitur orbe:
quamvis ingenio non valet, arte valet;
nulla Sophocleo veniet iactura cothurno;
cum sole et luna semper Aratus erit;
dum fallax servus, durus pater, improba lena
vivent et meretrix blanda, Menandros erit;
Ennius arte carens animosique Accius oris
casurum nullo tempore nomen habent;
Varronem primamque ratem quae nesciet aetas
aureaque Aesonio terga petita duci?
carmina sublimis tunc sunt peritura Lucreti,
exitio terras cum dabit una dies;
Tityrus et fruges Aeneiaque arma legentur
Roma triumphati dum caput orbis erit;
donec erunt ignes arcusque Cupidinis arma
discentur numeri, culte Tibulle, tui;
Gallus et Hesperiis et Gallus notus Eois,
et sua cum Gallo nota Lycoris erit.

Amores 1.15.9–30

This catalogue is rather miscellaneous. It is, simply, a list of those poets whom Ovid wishes to single out for their excellence, either because he particularly admired them, or because he expected his reader to do so, or for some other reason. If we compare it to earlier, more purposive catalogues—like the list of elegists with which Propertius ends Book 2—it becomes obvious that Ovid's poem is not ideologically charged. To be sure, elsewhere in the *Amores* Ovid declares his preference for the slender genres, in contrast to heroic epos (cf., e.g., *Amores* 2.1) and, especially, to tragedy (cf. *Amores* 3.1). Yet here he takes no pains to distinguish himself from Homer, Sophocles, Ennius, Accius, Varro, Lucretius, or Vergil, whose fame, it is suggested, rests mainly on the *Aeneid.* In other lists where such poets appear, the point is that even poets who dealt with the grandest themes were subject to the power of love. Propertius, for instance, in the poem discussed above, mentions Varro of Atax as having turned to elegy after establishing himself as an epic poet (*Carmen* 2.34.85–86)—a succession of topics that makes him a virtual antitype of Vergil in that poem. Hermesianax makes this the central motif of his poetic catalogue, which I discussed in Chapter 2.[25] In these and other similar poems, the implied or overt argument is that the slender genre of love poetry is not really inferior to the heroic genres because even powerful warriors and the poets who celebrate them succumb to the power of love. But in Ovid's catalogue, this is simply not the case. Ovid is proud to claim his place beside such poets as Homer, Hesiod, Callimachus, and the rest. The familiar contrast between heroic and unheroic genres is simply not present.

In my discussion of Hermesianax' catalogue, I was concerned to illustrate the Hellenistic habit of using antonomasia as an emblem of the poet's independence, of his refusal to be dominated by the great figures of the past, particularly Homer. This element, too, is not discernable in Ovid's catalogue. The first three poets mentioned—Homer, Hesiod, and Callimachus—are named through antonomasia: *Maeonides* (9), *Ascraeus* (11), and *Battiades* (13). Here, too, Ovid is not using the figure in a tendentious way, such as we find in Vergil, Horace, and Propertius. Instead, it has become for him a merely decorative element, as it remained for later European poets. This fact can hardly be more clearly illustrated than by the proem of *Ars Amatoria* 2:

> dicite "io Paean" et "io" bis dicite "Paean":
>   decidit in casses praeda petita meos.
> laetus amans donat viridi mea carmina palma
>   praelata *Ascraeo Maeonioque seni.*
>
> *Ars Amatoria* 2.1–4

Such a declaration would make no sense in the literary culture of the previous generation. We need only think of *Eclogue* 6, where Gallus is so carefully placed in the tradition of Linus, Orpheus, and the *Ascraeus senex*, a tradition from which Homer is pointedly excluded. Here, however, Ovid blithely claims to be the heir of both Hesiod and Homer, doing so, moreover, on the basis of his achievement in *Ars* 1. It would be easy to follow other scholars who have doubted Ovid's sensitivity to the subtleties of other poets and to ascribe his apparently promiscuous use of antonomasia here to his general obtuseness. But such despicably facile interpretations are now being replaced by a more generous and accurate estimate of Ovid's abilities.[26] Here, too, it seems to me that

25. See pp. 42–43.
26. I think in particular of Knox 1986 and Hinds 1987.

Ovid cannot be unaware of the difference between the *Ascraeus* and the *Maeonius senex*. His ability to claim both Hesiod and Homer as his forerunners depends directly on Vergil's achievement in permanently combining these two traditions.

Ovid's understanding of Vergil's achievement informs more than such passing statements as these. It becomes the basis of his masterpiece, the *Metamorphoses*, as well.

> in nova fert animus mutatas dicere formas
> corpora. di coeptis (nam vos mutastis et illa)
> adspirate meis primaque ab origine mundi
> ad mea perpetuum deducite tempora carmen.
>
> *Metamorphoses* 1.1–4

The paradox of these opening lines is by now familiar. Ovid uses the terminology of the Roman Callimacheans to make what would be to those poets a self-contradictory statement. He proposes to make a *carmen perpetuum* that is also a *deductum carmen*. This contradiction in terms accurately describes the poem that follows, which in subject matter resembles *Eclogue* 6 and in scope surpasses the *Aeneid*. In fact, Ovid subjects these two Vergilian models to converse treatments. Where the scheme of Silenus' song—cosmogony, stories of disastrous love and metamorphosis, and barely a nod to heroic epos—is played out over fifteen books, the plot of the *Aeneid* is condensed into a few hundred lines. This is only one of the many ways in which Ovid remakes his models as he fashions his greatest work—the last great poem of its age. Ovid stands squarely on Vergil's shoulders as he celebrates the possibility of moving freely between what had once been treated as discrete traditions, the tradition of Hesiod and that of Homer, and his treatment of Vergilian material in the *Metamorphoses* declares his debt to Vergil's pioneering reconciliation of them. But Ovid's achievement is unique, standing as the logical conclusion of the development that we have traced throughout Vergil's career and in the reaction of his contemporaries.

In later times, of course, an accurate understanding of these achievements grew inaccessible as Hesiod, Homer, Callimachus, Vergil, and Ovid came to be regarded, indifferently, as classics. Recovery of the sophistication with which the ancient poets reacted to their literary past thus became the specialized interest of professional philologists. But the careful study of these traditions speaks to the concerns of all who interest themselves in the literature of the West. It is perhaps a truism that Vergil's career occupies a central position in the history of that literature, and history, in retrospect, tends to take on an aspect of inevitability. But such truisms and tendencies should not be allowed to blind us to the extraordinary and improbable creative achievement by which Vergil, in effect, defined what we naturally think of as the main stream of the western poetic tradition. This is his greatest and most lasting accomplishment; and it is largely the remarkable allusive program of the *Georgics* that made it possible.

Bibliography

Ackermann 1978. Erich Ackermann. *Lukrez und der Mythos.* Palingenesia 13. Wiesbaden 1978.

Adler 1928–1935. *Lexicographi Graeci recogniti et apparatu critico instructi.* Vol. 1. Suidae lexicon. Ed. Ada Adler. 4 Parts. Leipzig 1928–1935.

Allen–Halliday–Sykes 1936. *The Homeric Hymns.* Ed. T. W. Allen, W. R. Halliday, and E. E. Sykes. 2d ed. Oxford 1936.

Altevogt 1952. Heinrich Altevogt. *Labor improbus. Eine Vergilstudie.* Orbis Antiquus 8. Münster 1952.

Anderson 1933. W. B. Anderson. "Gallus and the Fourth *Georgic.*" *Classical Quarterly* 27.1933.36–45 and 73.

Austin 1955. *P. Vergili Maronis Aeneidos liber quartus.* Edited with a commentary by R. G. Austin. Oxford 1955.

Austin 1971. *P. Vergili Maronis Aeneidos liber primus.* With a commentary by R. G. Austin. Oxford 1971.

Austin 1977. *P. Vergili Maronis Aeneidos liber sextus.* With a commentary by R. G. Austin. Oxford 1977.

Axelson 1945. Bertil Axelson. *Unpoetische Wörter. Ein Beitrag zur Kenntnis der lateinischen Dichtersprache.* Skrifter Utgivna av Vetenskaps-Societeten i Lund 29. Lund 1945.

Bailey 1931. Cyril Bailey. "Virgil and Lucretius." *Proceedings of the Cambridge Philological Association* 28.1931.21–39.

Bailey 1947. *Titi Lucreti Cari De Rerum Natura libri sex.* Edited with prolegomena, critical apparatus, translation, and commentary by Cyril Bailey. 3 volumes. Oxford 1947.

Barchiesi 1984. Alessandro Barchiesi. *La traccia del modello. Effetti omerici nella narrazione virgiliana.* Biblioteca di *Materiali e discussioni per l'annalisi dei testi classici* 1. Pisa 1984.

Bardon 1952. Henry Bardon. *La Littérature latine inconnue.* Vol. 1. Paris 1952.

Bayet 1930. Jean Bayet. "Les Premiers 'Géorgiques' de Virgile." *Revue philologique* 56.1930.128–150 and 227–247.

Becker 1937. Otfrid Becker. *Das Bild des Weges und verwandte Vorstellungen im frühgriechischen Denken. Hermes* Einzelschriften 4. Berlin 1936.

Beede 1936. Grace Lucile Beede. *Vergil and Aratus. A Study in the Art of Translation.* Diss. Chicago 1936.

Benediktson 1977. D. T. Benediktson. "Vocabulary Analysis and the Generic Classification of Literature." *Phoenix* 31.1977.341–348.

Berg 1974. William Berg. *Early Virgil.* London 1974.

Bernardakis 1896. *Plutarchi Chaeronensis Moralia*. Recognovit Gregorius N. Bernardakis. Vol. 7. Plutarchi fragmenta vera et spuria accessionibus locupletata continens. Leipzig 1896.

Betensky 1979. Aya Betensky. "The Farmer's Battles." *Virgil's Ascraean Song. Ramus Essays on the Georgics*. Ed. A. J. Boyle. Melbourne 1979. Pp. 108–119.

Beutler 1940. R. Beutler. "Zur Komposition von Vergils Georgica 1.43–159." *Hermes* 75.1940.410–421.

Bing 1988. Peter Bing. *The Well-Read Muse. Present and Past in Callimachus and the Hellenistic Poets*. Hypomnemata Heft 90. Göttingen 1988.

Bing 1990. Peter Bing. "A Pun on Aratus' Name in Verse 2 of the *Phaenomena*." *Harvard Studies in Classical Philology* 93.1990.281–285.

Boeck 1958. Urs Boeck. "Über den Fragmenten des Lukrez." *Hermes* 86.1958. 243–246.

Bömer 1976. Franz Bömer. *P. Ovidius Naso. Metamorphosen, Buch VI–VII*. Heidelberg 1976.

Bona Quaglia 1973. Luciana Bona Quaglia. *Gli "Erga" di Esiodo*. Università di Torino. Faccoltà di lettere e filosofia. Filologia classica e glottologia. Vol. 6. Turin 1973.

Bonnafé 1987. Annie Bonnafé. *Poésie, nature et sacré. 2. L'Âge archaïque*. Collection de la Maison de l'Orient méditerrain 17. Série littéraire & philosophique 4. Lyons 1987.

Bosworth 1972. A. B. Bosworth. "Asinius Pollio and Augustus." *Historia* 21. 1972.441–473.

Bowersock 1971. G. W. Bowersock. "A Date in the *Eighth Eclogue*." *Harvard Studies in Classical Philology* 75.1971.73–80.

Briggs 1980. Ward W. Briggs, Jr. *Narrative and Simile from the Georgics in the Aeneid. Mnemosyne* Supplement 58. Leiden 1980.

Bright 1978. David F. Bright. *Haec Mihi Fingebam. Tibullus in His World*. Cincinnati Classical Studies n.s. 3. Leiden 1978.

Brown 1963. Edwin L. Brown. *Numeri Vergiliani. Studies in "Eclogues" and "Georgics."* Collection *Latomus*. Vol. 63. Brussels–Berchem 1963.

Brown 1982. Robert D. Brown. "Lucretius and Callimachus." *Illinois Classical Studies* 7.1982.77–97.

Brummer 1912. *Vitae Vergilianae. Vitae Vergilianae*. Recensuit Iacobus Brummer. Leipzig 1912.

Buchheit 1971. Vinzenz Buchheit. "Epikurs Triumph des Geistes (Lucr. I, 62–79)." *Hermes* 99.1971.303–323.

Buchheit 1972. Vinzenz Buchheit. *Der Anspruch des Dichters in Vergils Georgika. Dichtertum und Heilsweg*. Impulse der Forschung 8. Darmstadt 1972.

Buchheit 1976. Vinzenz Buchheit. "Catull c. 50 als Programm und Bekenntnis." *Rheinisches Museum für Philologie* 119.1976.162–180.

Büchner 1959. Karl Büchner. *P. Vergilius Maro. Dichter der Römer.* Stuttgart 1959 [= *RE* 8A 1021–1486].

Buffière 1956. Félix Buffière. *Les Mythes d'Homère et la pensée grecque.* Association Guillaume Budé. Collection d'études anciennes. Paris 1956.

Bulloch 1985. *Callimachus. The Fifth Hymn.* Edited with introduction and commentary by A. W. Bulloch. Cambridge Classical Texts and Commentaries 26. Cambridge 1985.

Burck 1929. Erich Burck. "Die Komposition von Vergils Georgica." *Hermes* 64. 1929.279–321.

Burkert 1979. Walter Burkert. "Kynaithos, Polycrates, and the Homeric Hymn to Apollo." *Arktouros. Hellenic Studies Presented to Bernard M. W. Knox on the Occasion of His 65th Birthday.* Ed. Glen W. Bowersock, Walter Burkert, and Michael C. J. Putnam. Berlin–New York 1979.

Byl 1980. Simon Byl. *Recherches sur les grands traitées biologiques d'Aristote: sources écrites et préjugés.* Académie royale de Belgique. Mémoires de la Classe des lettres. Collection in 8°. 2e série. T. 64. Fascicule 3. Brussels 1980.

Cairns 1969. Francis Cairns. "Catullus 1." *Mnemosyne* 22.1969.153–158.

Cairns 1972. Francis Cairns. *Generic Composition in Greek and Roman Poetry.* Edinburgh 1972.

Cairns 1989. Francis Cairns. *Virgil's Augustan Epic.* Cambridge 1989.

Campbell 1967. David A. Campbell. *Greek Lyric Poetry. A Selection of Early Greek Lyric, Elegiac and Iambic Poetry.* London–New York 1967.

Cartault 1897. A. Cartault. *Étude sur les Bucoliques de Virgile.* Paris 1897.

Cerda 1647. *P. Virgilii Maronis Bucolica et Georgica argumentis, explicationibus, notis illustrata.* Auctore Ioanne Ludovico de la Cerda. Cologne 1647.

Chryssafis 1981. G. Chryssafis. *A Textual and Stylistic Commentary on Theocritus' Idyll XXV.* London Studies in Classical Philology 1. Amsterdam 1981.

Clausen 1964. Wendell V. Clausen. "Callimachus and Latin Poetry." *Greek, Roman, and Byzantine Studies* 5.1964.181–196.

Clausen 1972. Wendell Clausen. "On the Date of the *First Eclogue.*" *Harvard Studies in Classical Philology* 76.1972.201–205.

Clausen 1976. Wendell Clausen. "*Cynthius.*" *American Journal of Philology* 97.1976.245–247 and 362.

Clausen 1982. W. V. Clausen. "The New Direction in Poetry." *The Cambridge History of Classical Literature.* Vol. 2. Latin Literature. Ed. E. J. Kenney and W. V. Clausen. Cambridge 1982. Pp. 178–206.

Clausen 1987. Wendell Clausen. *Virgil's* Aeneid *and the Tradition of Hellenistic Poetry.* Sather Classical Lectures. Vol. 51. Berkeley–Los Angeles–London 1987.

Clauss 1988. James J. Clauss. "Vergil and the Euphrates Revisited." *American Journal of Philology* 109.1988.309–320.

Clay 1977. Diskin Clay. "Lucretius *contra Empedoclen.* A Textual Note." *Classical Journal* 73.1977.27–29.

Clay 1983. Diskin Clay. *Lucretius and Epicurus.* Ithaca 1983.

Cody 1976. John V. Cody. *Horace and Callimachean Aesthetics.* Collection *Latomus.* Vol. 147. Brussels 1976.

Cole 1979. Judith Mae Cole. "Vergil and His Sources. *Georgics* 1.311–464." Diss. Brown 1979.

Coleman 1977. Robert Coleman. *Vergil. Eclogues.* Cambridge Greek and Latin Classics. Cambridge 1977.

Conington–Nettleship 1898. *The Works of Virgil* with a commentary by John Conington and Henry Nettleship. Vol. 1. Eclogues and Georgics. 5th ed. Rev. F. Haverfield. London 1898. Rpt. Hildesheim 1963.

Conte 1986. Gian Biagio Conte. *The Rhetoric of Imitation. Genre and Poetic Memory in Virgil and Other Latin Poets.* Translated from the Italian. Edited and with a foreword by Charles Segal. Ithaca 1986.

Cook 1895. A. B. Cook. "The Bee in Greek Mythology." *Journal of Hellenic Studies* 15.1895.1–24.

Cook 1914. A. B. Cook. *Zeus. A Study in Ancient Religion.* Vol. 1. The God of the Bright Sky. Cambridge 1914.

Courcelle 1955. P. Courcelle. "Histoire du cliché virgilien des cent bouches (Georg. II, 42–44 = Aen. VI, 625–627)." *Revue des études latines* 33.1955. 231–240.

Cumont 1933. Franz Cumont. "Les Présages lunaires et les sélénodromia." *L'Antiquité classique* 2.1933.259–270.

Curtius 1973. Ernst Robert Curtius. *European Literature in the Latin Middle Ages.* Tr. Willard R. Trask. Bollingen Series 36. Princeton 1973.

Dahlmann 1954. Hellfried Dahlmann. *Der Bienenstaat in Vergils Georgica.* Abhandlungen der Geistes- und Sozialwissenschaftlichen Classe. Akademie der Wissenschaft und der Literatur in Mainz 10. Mainz 1954. Pp. 547–562.

Davies–Kathirithamby 1986. Malcolm Davies and Jeyaraney Kathirithamby. *Greek Insects.* New York–Oxford 1986.

DeLacy 1948. Phillip DeLacy. "Stoic Views of Poetry." *American Journal of Philology* 69.1948.241–271.

Delatte 1915. Armand Delatte. *Étude sur la littérature pythagoricienne*. Bibliotheque de l'École des hautes études 217. Paris 1915.

Delatte 1924. *Catalogus Codicum Astrologorum*. Tomus X. Codices Athenienses. Descripsit Armandus Delatte. Brussels 1924.

Desport 1952. Marie Desport. *L'Incantation virgilienne. Essai sur les mythes du poète enchanteur et leur influence dans l'oeuvre de Virgile*. Bordeaux 1952.

Deutsch 1939. Rosamund E. Deutsch. *The Pattern of Sound in Lucretius*. Diss. Bryn Mawr 1939.

Di Benedetto 1956. Vincenzo Di Benedetto. "Omerismi e struttura metrica negli idilli dorici di Teocrito." *Annali della Scuola normale superiore di Pisa*. Classe di lettere e filosofia. 2d ser. 25.1966.48–60.

Dickie 1983. Matthew Dickie. "Phaeacian Athletes." *Papers of the Liverpool Latin Seminar*. Fourth Volume 1983. Ed. Francis Cairns. ARCA Classical and Medieval Texts, Papers and Monographs, 11. Liverpool 1983. Pp. 237–276.

Diels–Kranz 1951. *Die Fragmente der Vorsokratiker*. Griechisch und Deutsch von Hermann Diels. Sechste Auflage. Hrsg. Walther Kranz. Erster Band. Berlin 1951.

Dindorf 1855. *Scholia Graeca in Homeri Odysseam ex codicibus aucta et emendata*. Ed. Gulielmus Dindorfius. 2 vol. Oxford 1855.

Drachmann 1903–1927. *Scholia vetera in Pindari carmina*. Recensuit A. B. Drachmann. 3 vol. Leipzig 1903–1927.

Edelstein 1967. Ludwig Edelstein. *The Idea of Progress in Classical Antiquity*. Baltimore 1967.

Erbse 1960. Hartmut Erbse. *Beitrage zur Überlieferung der Iliasscholien*. Zetemata 24. Munich 1960.

Erbse 1969–1988. *Scholia Graeca in Homeri Iliadem (scholia vetera)*. Recensuit Hartmut Erbse. 7 vol. Berlin 1969–1988.

Erdmann 1912. Otto Erdmann. *Beiträge zur Nachahmungskunst Vergils in den Georgika*. Teil 1. Wissenschaftliche Beigabe zum Jahresbericht des Königlichen Domgymnasiums zu Halberstadt 336. Halberstadt 1912.

Farrell 1983. Joseph Farrell. "Thematic Allusion to Lucretius in Vergil's *Georgics*." Diss. Chapel Hill 1983.

Farrington 1963. Benjamin Farrington. "Polemical Allusions to the De Rerum Natura of Lucretius in the Works of Vergil." *Geras. Studies Presented to George Thomson*. Ed. L. Varcl and R. F. Willetts. Acta Universitatis Carolinae Philosophica et Historica 1. Graecolatina Pragensia 2. Prague 1963.87–94.

Feeney 1988. Denis Feeney. "The Paradoxical Country." *The Times Literary Supplement* no. 4,439. April 29–May 5 1988. P. 476 [Review of Ross 1987].

Fordyce 1961. *Catullus.* A commentary by C. J. Fordyce. Oxford 1961.

Fraenkel 1957. Eduard Fraenkel. *Horace.* Oxford 1957.

Fränkel 1921. Hermann Fränkel. *Die homerischen Gleichnisse.* Göttingen 1921. 2., unveränderte Auflage mit einem Nachwort und einem Literaturverzeichnis. Hrsg. Ernst Heitsch. Göttingen 1977.

Frazer 1913. James George Frazer. *The Golden Bough. A Study in Magic and Religion.* Part 1. *The Magic Art and the Evolution of Kings.* Vol. 2. 3d ed. London 1911. Rpt. 1922.

Frentz 1967. Willi Frentz. *Mythologisches in Vergils Georgica.* Beiträge zur klassischen Philologie 21. Meisenheim am Glan 1967.

Friedländer 1941. Paul Friedländer. "Pattern of Sound and Atomistic Theory in Lucretius." *American Journal of Philology* 62.1941.16–34 [= *Studien zur antiken Literatur und Kunst.* Berlin 1969. Pp. 337–353].

Frisk 1960. Hjalmar Frisk. *Griechisches etymologisches Wörterbuch.* Band 1. Indogermanische Bibliothek. Heidelberg 1960.

Galinsky 1988. Karl Galinsky. "The Anger of Aeneas." *American Journal of Philology* 109.1988.321–348.

Garson 1972. R. W. Garson. "Homeric Echoes in Apollonius Rhodius' *Argonautica.*" *Classical Philology* 77.1972.1–9.

Giancotti 1976. Francesco Giancotti. "Aerea vox. Un frammento attribuito da Servio a Lucrezio e consimili espressioni di altri poeti in Macrobio, Servio, e altri." *Grammatici latini dell' età imperiale. Miscellanea filologica.* Università di Genova. Pubblicazioni dell' Istituto di filologia classica e medievale 45. Genoa 1976. Pp. 41–95.

Giangrande 1967. Giuseppe Giangrande. "'Arte Allusiva' and Alexandrian Poetry." *Classical Quarterly* n.s. 17.1967.85–97.

Giangrande 1970. Giuseppe Giangrande. "Hellenistic Poetry and Homer." *L'Antiquité classique* 39.1970.46–77.

Gomme 1956. A. W. Gomme. *A Historical Commentary on Thucydides. TheTen Years' War.* Vol. 2. Books 2–3. Oxford 1956.

Gow 1952. *Theocritus.* Edited with a translation and commentary by A. S. F. Gow. Cambridge 1950. Rev. 1952.

Gow–Page 1965. *The Greek Anthology. Hellenistic Epigrams.* Ed. A. S. F. Gow and D. L. Page. Cambridge 1965.

Gow–Scholfield 1953. *Nicander. The Poems and Poetical Fragments.* Edited with a translation and notes by A. S. F. Gow and A. F. Scholfield. Cambridge 1953.

Greene 1982. Thomas M. Greene. *The Light in Troy. Imitation and Discovery in Renaissance Poetry.* New Haven 1982.

Griffin 1981. Jasper Griffin. "Haec super arvorum cultu." *Classical Review* n.s. 31.1981.23–27.

Griffin 1986a. Jasper Griffin. *Virgil.* Oxford 1986.

Griffin 1986b. Jasper Griffin. "Virgil." *The Oxford History of the Classical World.* Ed. John Boardman, Jasper Griffin, and Oswyn Murray. Oxford–New York 1986.

Grimm 1965. Jürgen Grimm. *Die literarische Darstellung der Pest in der Antike und in Romania.* Freiburger schriften zur romanischen Philologie 6. Munich 1965.

Hagen 1867. *Scholia Bernensia ad Vergili Bucolica atque Georgica.* Edidit emendavit praefatus est Hermannus Hagen. Jahrbücher für classische Philologie. 4. Supplement. Leipzig 1867. Rpt. Hildesheim 1967.

Hagen 1902. *Servii grammatici qui feruntur in Vergilii carmina commentarii.* Recensuerunt Georgius Thilo et Hermannus Hagen. Vol. 3. Fasc. 2. *Appendix Serviana ceteros praeter Servium et scholia Bernensia Vergilii commentatores continens.* Recensuit Hermannus Hagen. Leipzig 1902.

Halperin 1983. David M. Halperin. *Before Pastoral. Theocritus and theTradition of Ancient Bucolic Poetry.* New Haven 1983.

Hamilton 1989. Richard Hamilton. *The Architecture of Hesiodic Poetry. AJP* Monographs in Classical Philology 3. Baltimore–London 1989.

Hardie 1966. *Vitae Vergilianae antiquae. Vita Donati, Vita Servii, Vita Probiana, Vita Focae, S. Hieronymi excerpta.* Edidit Colinus Hardie. Editio altera. Oxford 1966. Rpt. 1967.

Hardie 1984. P. R. Hardie. "The Sacrifice of Iphigeneia. An Example of 'Distribution' of a Lucretian Theme in Virgil." *Classical Quarterly* n.s. 34.1984.406–412.

Hardie 1985. P. R. Hardie. "Cosmological Patterns in the *Aeneid.*" *Papers of the Liverpool Latin Seminar.* Fifth volume 1985. Ed. Francis Cairns. ARCA Classical and Medieval Texts, Papers and Monographs, 19. Liverpool 1986. Pp. 85–98.

Hardie 1986. Philip R. Hardie. *Virgil's* Aeneid. Cosmos *and* Imperium. Oxford 1986.

Harrison 1979. E. L. Harrison. "The Noric Plague in Vergil's Third Georgics." *Papers of the Liverpool Latin Seminar.* Second volume 1979. Vergil and Roman Elegy, Medieval Latin Poetry and Prose, Greek Lyric and Drama. Ed. Francis Cairns. ARCA Classical and Medieval Texts, Papers and Monographs 3. Liverpool 1979. Pp. 1–65.

Heinze 1915. Richard Heine. *Virgils epische Technik.* 3d ed. Leipzig–Berlin 1915.

Helm 1913. *Die griechischen christlichen Schriftsteller der ersten drei Jahrhunderte.* Vol. 7. Eusebius. 7. Band. Die Chronika des Hieronymus. Hrsg. Rudolf Helm. 1. Teil. Text. Leipzig 1913.

Herter 1929. Hans Herter. "Kallimachos und Homer. Ein Beitrag zur Interpretation des Hymnos auf Artemis." *Xenia Bonnensia.* Festschrift zum 75jahrigen Bestehen des Philologischen Vereins und Bonner Kreises. Bonn 1929. Pp. 50–105 [=*Kleine Schriften.* Hrsg. Ernst Vogt. Munich 1975. Pp. 370–416].

Higgins 1987. Dick Higgins. *Pattern Poetry. Guide to an Unknown Literature.* Albany 1987.

Hiller 1872. *Eratosthenis carminum reliquiae.* Disposuit et explicavit Eduardus Hiller. Leipzig 1872.

Hinds 1987. Stephen Hinds. *The Metamorphosis of Persephone. Ovid and the Self-conscious Muse.* Cambridge Classical Studies. Cambridge 1987.

Holder 1894. *Pomponi Porfyrionis commentum in Horatium Flaccum.* Recensuit Alfred Holder. Innsbruck 1894.

Hommel 1958. Hildebrecht Hommel. "Das hellenische Ideal vom einfachen Leben." *Studium Generale* 11.1958.742–751.

Hopkinson 1984. *Callimachus. Hymn to Demeter.* Edited with an introduction and commentary by N. Hopkinson. Cambridge Classical Texts and Commentaries 27. Cambridge 1984.

Innes 1979. D. C. Innes. "Gigantomachy and Natural Philosophy." *Classical Quarterly* n.s. 29.1979.165–171.

Jacobson 1984. Howard Jacobson. "Aristaeus, Orpheus, and the *laudes Galli.*" *American Journal of Philology* 105.1984.271–300.

Jacques 1960. J.-M. Jacques. "Sur un acrostiche d'Aratos (Phén., 783–787)." *Revue des études anciennes* 62.1960.48–61.

Jahn 1897–1899. Paul Jahn. *Die Art der Abhängigkeit Vergils von Theokrit und anderen Dichtern.* 3 parts. Programmen des köllnischen Gymnasiums zu Berlin. 1897, 1898, 1899.

Jahn 1903a. Paul Jahn. "Die Quellen und Muster des ersten Buchs der Georgica Vergils (bis Vers 350) und ihre Bearbeitung durch den Dichter." *Rheinisches Museum für Philologie* 58.1903.391–426.

Jahn 1903b. Paul Jahn. "Eine Prosaquelle Vergils und ihre Umsetzung in Poesie durch den Dichter." *Hermes* 38.1903.244–264, 480.

Jahn 1904. Paul Jahn. "Aus Vergils Dichterwerkstätte (Georgica IV 1–280)." *Philologus* 63.1904.66–93.

Jahn 1905a. Paul Jahn. *Aus Vergils Dichterwerkstätte (Georg. IV 281–558).* Programmen des köllnischen Gymnasiums zu Berlin. Berlin 1905.

Jahn 1905b. Paul Jahn. "Aus Vergils Dichterwerkstätte (Georgica III 49–170)." *Rheinisches Museum für Philologie* 60.1905.361–387.

Jahn 1915. *Vergils Gedichte.* Erklärt von Th. Ladewig, C. Schaper und P. Deuticke. 1. Bukolika und Georgika. 9. Auflage bearbeitet von Paul Jahn. Berlin 1915.

Janko 1982. Richard Janko. *Homer, Hesiod and the Hymns. Diachronic Development in Epic Diction.* Cambridge Classical Studies. Cambridge 1982.

Jermyn 1951. L. A. S. Jermyn. "Weather-Signs in Vergil." *Greece & Rome* 20. 1951.26–37, 49–59.

Jocelyn 1967. *The Tragedies of Ennius.* The fragments, edited with an introduction and commentary by H. D. Jocelyn. Cambridge Classical Texts and Commentaries 10. Cambridge 1967.

Johnson 1976. W. R. Johnson. *Darkness Visible. A Study of Vergil's* Aeneid. Berkeley–Los Angeles–London 1976.

Johnston 1980. Patricia A. Johnston. *Vergil's Agricultural Golden Age. A Study of the Georgics. Mnemosyne* Supplement 60. Leiden 1980.

Kaiser 1964. Erich Kaiser. "Odyssee-Szenen als Topoi." *Museum Helveticum* 21. 1963.109–224.

Kambylis 1965. Athanasios Kambylis. *Die Dichterweihe und ihre Symbolik. Untersuchungen zu Hesiod, Kallimachos, Properz, und Ennius.* Bibliothek der klassischen Altertumswissenschaften. Neue Folge 2. Heidelberg 1965.

Kassegger 1967. Helga Kassegger. "Das homerische Formengut bei Apollonios Rhodios (Argonautica, 1. und 2. Gesang)." Diss. Vienna 1967.

Kassel–Austin 1983. *Poetae comici Graeci.* Ed. R. Kassel and C. Austin. Vol. 4. Aristophon–Crobylus. Berlin–New York 1983.

Kenney 1971. *Lucretius. De Rerum Natura Book III.* Ed. E. J. Kenney. Cambridge Greek and Latin Classics. Cambridge 1971.

Keil 1857. *Grammatici Latini* ex recensione Henrici Keilii. Vol. 1. Flavi Socipatri Charisii *Artis Grammaticae* libri V. Diomedis *Artis Grammaticae* libri III. Ex Charisii *Arte Grammatica* excerpta. Leipzig 1857.

Kenney 1970. E. J. Kenney. "Doctus Lucretius." *Mnemosyne* 23.1970.366–392.

Klepl 1940. Herta Klepl. *Lukrez und Virgil in ihren Lehrgedichten.* Dresden 1940. Rpt. 1967.

Klingner 1931. Friedrich Klingner. "Über das Lob des Landlebens in Virgils Georgica." *Hermes* 66.1931.159–189.

Klingner 1967. Friedrich Klingner. *Virgil. Bucolica, Georgica, Aeneis.* Zurich–Stuttgart 1967.

Knauer 1964a. Georg Nicolaus Knauer. *Die Aeneis und Homer. Studien zur poetischen Technik Vergils mit Listen der Homeritate in der Aeneis.* Hypomnemata 7. Göttingen 1964.

Knauer 1964b. Georg Nicolaus Knauer. "Vergil's *Aeneid* and Homer." *Greek, Roman, and Byzantine Studies* 5.1964.61–84 [Rpt. Knauer 1981.870–890].

Knauer 1981. Georg Nicolaus Knauer. "Vergil and Homer." *Aufstieg und Niedergang der römischen Welt. Geschichte und Kultur Roms im Spiegel der neueren Forschung.* II. Prinzipat. Hrsg. Hildegard Temporini und Wolfgang

Haase. 31. Sprache und Literatur. (Literatur der augusteischen Zeit; einzelne Autoren, Forts. [Vergil, Horaz, Ovid]). Berlin 1981. Part 1. "The 'Aeneid' and Homer." Pp. 870–890 [= Knauer 1964b]. Part 2. "The 'Georgics' and Homer." Pp. 890–918.

Knight 1932. W. F. J. Knight. *Vergil's Troy. Essays on the Second Book of the Aeneid.* Oxford 1932.

Knight 1966. W. F. J. Knight. *Roman Vergil.* 3d ed. Harmondsworth 1966.

Knox 1986. Peter E. Knox. *Ovid's "Metamorphoses" and the Traditions of Augustan Poetry.* Cambridge Philological Society supplementary vol. 11. Cambridge 1986.

Kollmann 1971. E. D. Kollman. "Lucretius' Criticism of the Early Greek Philosophers." *Studii clasice* 13.1971.79–83.

Koster 1970. Severin Koster. *Antike Epostheorien.* Palingenesia 5. Wiesbaden 1970.

Krenkel 1970. *Lucilius. Satiren.* Lateinisch und Deutsch von Werner Krenkel. Leiden 1970.

Krevans 1983. N. Krevans. "Geography and the Literary Tradition in Theocritus 7." *Transactions of the American Philological Association* 113.1983.201–220.

Kroll 1924. Wilhelm Kroll. *Studien zum Verständnis der römischen Literatur.* Stuttgart 1924.

Kuiper 1896. Koenraad Kuiper. *Studia Callimachea.* Vol. 1. *De hymnorum I–IV dictione epica.* Leiden 1896.

Kumpf 1974. Michael M. Kumpf. "The Homeric Hapax Legomena and their Literary Use by Later Authors, Especially Euripides and Apollonius Rhodius." Diss. Ohio State 1974.

Lachmann 1850. *T. Lucreti Cari De Rerum Natura libri sex.* Carolus Lachmannus recensuit et emendavit. *Caroli Lachmanni in T. Lucreti Cari De Reum Natura libros commentarius.* 2 vol. Berlin 1850. Rpt. 1853–1855.

Lamberton 1986. Robert Lamberton. *Homer the Theologian. Neoplatonist Allegorical Reading and the Growth of the Epic Tradition.* The Transformation of the Classical Heritage 9. Berkeley–Los Angeles–London 1986.

La Penna 1962. Antonio La Penna. "Esiodo nella cultura e nella poesia di Virgilio." *Hésiode et son influence.* Fondation Hardt. Entretiens sur l'antiquité classique 7. Geneva 1962. Pp. 225–247.

Latte 1924–1925. Kurt Latte. "Glossographika." *Philologus* 80.1924–1925.136–175.

Lausberg 1983. Marion Lausberg. "Iliadisches im ersten Buch der Aeneis." *Gymnasium* 90.1983.204–239.

Leach 1974. Eleanor Winsor Leach. *Vergil's* Eclogues. *Landscapes of Experience.* Ithaca and London 1974.

Leach 1981. Eleanor Winsor Leach. "*Georgics* 2 and the Poem." *Arethusa* 14.1981.35–48.

Le Grelle 1949. Guy Le Grelle. "Le Premier livre des Géorgiques, poème pythagoricien." *Les Études classiques* 17.1949.139–235.

Lehnert 1896. Hermann Georg Lehnert. *De scholiis ad Homerum rhetoricis.* Diss. Leipzig 1896.

Lenaghan 1967. Lydia Lenaghan. "Lucretius 1.921–950." *Transactions of the American Philological Association* 98.1967.221–251.

Liebeschuetz 1965. W. Liebeschuetz. "Beast and Man in the Third Book of Virgil's *Georgics.*" *Greece & Rome* 12.1965.64–77.

Liebeschuetz 1967–1968. W. Liebeschuetz. "The Cycle of Growth and Decay in Lucretius and Virgil." *Proceedings of the Virgil Society* 7.1967–1968.30–40.

Linnaeus 1735. Carolus Linnaeus. *Systema naturae sive Regna tria naturae systematice proposita per classes, ordines, genera, & species.* Leiden 1735.

Livrea 1973. *Apollonii Rhodii liber quartus.* Introduzione, testo critico, traduzione e commento a cura di Enrico Livrea. Biblioteca di studi superiori 60. Florence 1973.

Lloyd-Jones and Parsons 1983. *Supplementum Hellenisticum.* Ed. Hugh Lloyd-Jones and Peter Parsons. Texte und Kommentare 11. Berlin–New York 1983.

Long 1991. A. A. Long. "Stoic Readings of Homer." *Homer's Ancient Readers. Hermeneutic Strategies of the Earliest Readers of the* Iliad *and* Odyssey. Ed. Robert Lamberton and John Keaney. Princeton 1991 (forthcoming).

Ludwig 1963. Walther Ludwig. "Die Phainomena Arats als hellenistische Dichtung." *Hermes* 91.1963.425–448.

Lyne 1978a. *Ciris. A Poem Attributed to Vergil.* Edited with an introduction and commentary by R. O. A. M. Lyne. Cambridge Classical Texts and Commentaries 20. Cambridge 1978.

Lyne 1978b. R. O. A. M. Lyne. "The Neoteric Poets." *Classical Quarterly* n.s. 28.1978.167–187.

Lyne 1987. R. O. A. M. Lyne. *Further Voices in Vergil's* Aeneid. Oxford 1987.

Maass 1898. *Commentariorum in Aratum reliquiae.* Collegit recensuit prolegomenis indicibusque instruxit Ernst Maass. Berlin 1898.

Mankin 1988. David Mankin. "The Addressee of Virgil's Eighth Eclogue. A Reconsideration." *Hermes* 116.1988.63–76.

Martin 1956. *Arati Phaenomena.* Introduction, texte critique, commentaire et introduction...par Jean Martin. Florence 1956.

Martin 1963. *T. Lucreti Cari De Rerum Natura libri sex.* Quintum recensuit Joseph Martin. Leipzig 1963.

Martini–Bassi 1901. *Catalogus codicum astrologorum.* Tomus 3. Codices Mediolanenses. Descripserunt Aemygdius Martini et Dominicus Bassi. Brussels 1901.

Marx 1904. *C. Lucilii carminum reliquiae.* Recensuit enarravit Fridericus Marx. Volumen prius. Prolegomena testimonia fasti Luciliani carminum reliquiae Indices. Leipzig 1904.

Marx 1905. *C. Lucilii carminum reliquiae.* Recensuit enarravit Fridericus Marx. Volumen posterius. Commentarius. Leipzig 1905.

Mayer 1983. Roland Mayer. "Missing Persons in the Eclogues." *Bulletin of the Institute for Classical Studies* 30.1983.17–30.

Merkelbach–West 1967. *Fragmenta Hesiodea.* Ed. R. Merkelbach and M. L. West. Oxford 1967.

Merrill 1905. W. A. Merrill. "On the Influence of Lucretius on Horace." *University of California Publications in Classical Philology* 1.4.1905.111–129.

Merrill 1907. *T. Lucreti Cari De Rerum Natura libri sex.* Ed. William Augustus Merrill. Morris and Morgan's Latin Series. New York 1907. Rpt. 1935.

Merrill 1909. W. A. Merrill. "Cicero's Knowledge of Lucretius' Poem." *University of California Publications in Classical Philology* 2.2.1909.35–42..

Merrill 1917. *Lucreti De Rerum Natura libri sex.* Recognovit Guilielmus Augustus Merrill. University of California Publications in Classical Philology 4. Berkeley 1917.

Merrill 1918a. W. A. Merrill. "Parallels and Coincidences in Lucretius and Vergil." *University of California Publications in Classical Philology* 3.3.1918.135–247.

Merrill 1918b. W. A. Merrill. "Parallelisms and Coincidences in Lucretius and Ennius." *University of California Publications in Classical Philology* 3.4.1918.249–264.

Merrill 1921. W. A. Merrill. "Lucretius and Cicero's Verse." *University of California Publications in Classical Philology* 5.9.1921.143–154.

Merrill 1924. W. A. Merrill. "The Metrical Technique of Lucretius and Cicero." *University of California Publications in Classical Philology* 7.10.1924.293–306.

Merrill 1929. W. A. Merrill. "Lucretian and Virgilian Rhythm." *University of California Publications in Classical Philology* 9.10.1929.373–404.

Mette 1936. H. J. Mette. *Sphairopoiia. Untersuchungen zur Kosmologie des Krates von Pergamon.* Munich 1936.

Meyer 1855. J. B. Meyer. *Aristoteles Thierkunde.* Berlin 1855.

Michels 1944. Agnes Kirsopp Michels. "Lucretius and the Sixth Book of the Aeneid." *American Journal of Philology* 65.1944.135–148.

Miles 1980. Gary Miles. *Virgil's* Georgics. *A New Interpretation.* Berkeley–Los Angeles–London 1980.

Miller 1986. Andrew M. Miller. *From Delos to Delphi. A Literary Study of the Homeric Hymn to Apollo. Mnemosyne* Supplement 93. Leiden 1986.

Minadeo 1969. Richard Minadeo. *The Lyre of Science. Form and Meaning in Lucretius' De Rerum Natura*. Detroit 1969.

Minyard 1978. John Douglas Minyard. *Mode and Value in the De Rerum Natura. A Study in Lucretius' Metrical Language. Hermes* Einzelschriften 39. Wiesbaden 1978.

Morel–Büchner 1982. *Fragmenta poetarum Latinorum epicorum et lyricorum praeter Ennium et Lucilium.* Post W. Morel novis curis adhibitis edidit Carolus Buechner. Leipzig 1982.

Moskalew 1982. Walter Moskalew. *Formular Language and Poetic Design in the* Aeneid. *Mnemosyne* Supplement 73. Leiden 1982.

Mueller 1872. *C. Lucilii Saturarum reliquiae*. Emendavit et adnotavit Lucianus Mueller. Accedunt Acci et Suei carminum reliquiae. Leipzig 1872.

Müller 1848. *Fragmenta historicorum Graecorum*. Collegit, disposuit, notis et prolegomenis illustravit, indicibus instruxit Carolus Müllerus. Vol. 2. Paris 1848.

Munro 1886. *T. Lucreti Cari De Rerum Natura libri sex*. With notes and a translation by H. A. J. Munro. Cambridge 1886.

Murrin 1980. Michael Murrin. *The Allegorical Epic. Essays in its Rise and Decline*. Chicago 1980.

Mylius 1946. Kläre Mylius. *Die wiederholten Verse bei Vergil.* Diss. Freiburg in Breslau 1946.

Nauck 1964. *Tragicorum Graecorum fragmenta*. Recensuit Augustus Nauck. Supplementum continens nova fragmenta Euripidea et adespota apud scriptores veteres reperta adiecit Bruno Snell. 4th ed. Hildesheim 1964.

Neitzel 1975. Heinz Neitzel. *Homer-Rezeption bei Hesiod. Interpretation ausgewählter Passagen*. Anhandlungen zur Kunst-, Musik-, und Literaturwissenschaft 189. Bonn 1975.

Nelis 1988. Damien Nelis. "The *Aeneid* and the *Argonautica* of Apollonius Rhodius." Diss. Belfast 1988.

Nethercut 1973. William R. Nethercut. "Virgil's De Rerum Natura." *Ramus* 2.1973.41–52.

Newman 1967. J. K. Newman. *Augustus and the New Poetry*. Collection *Latomus*. Vol. 88. Brussels 1967.

Newman 1986. John Kevin Newman. *The Classical Epic Tradition*. Madison 1986.

Newton 1953. Francis Lannaeu Newton. "Studies in Verbal Repetition in Virgil." Diss. Chapel Hill 1953.

Nisbet 1987. R. G. M. Nisbet. "Pyrrha Among the Roses. Real Life and Poetic Imagination in Augustan Rome." *Journal of Roman Studies* 77.1987.184–190.

Nisbet–Hubbard 1970. *A Commentary on Horace. Odes. Book 1*. By R. G. M. Nisbet and Margaret Hubbard. Oxford 1970.

Norden 1934. Eduard Norden. "Orpheus und Eurydice. Ein nachträgliches Gedenkblatt für Vergil." *Sitzungsberichte der preussischen Akademie der Wissenschaften* 12.1934.626–683.

Norden 1957. Eduard Norden. *P. Vergilius Maro. Aeneis Buch VI.* Vierte Auflage. Darmstadt 1957.

Ohlert 1912. K. Ohlert. *Rätsel und Rätselspiele der alten Griechen.* 2d ed. Berlin 1912.

Otis 1964. Brooks Otis. *Virgil. A Study in Civilized Poetry.* Oxford 1964.

Otis 1972. Brooks Otis. "A New Study of the *Georgics.*" *Phoenix* 26.1972.40–62.

Otis 1976. Brooks Otis. "Virgilian Narrative in the Light of its Predecessors and Successors." *Studies in Philology* 73.1976.1–28.

Ott 1972. Ulrich Ott. "Theokrits 'Thalysien' und ihre literarische Vorbilder." *Rheinisches Museum für Philologie* 115.1972.134–149.

Page 1898. T. E. Page. *P. Vergilii Maronis Bucolica et Georgica.* London–New York 1898.

Paratore 1938. Ettore Paratore. *Introduzione alle Georgiche di Virgilio.* Palermo 1938.

Parry 1972. Adam Parry. "The Idea of Art in Virgil's Georgics." *Arethusa* 5.1972.35–52.

Parsons 1974. *The Oxyrhynchus Papyri.* Vol. 42. Edited with a translation and notes by P. J. Parsons. Graeco-Roman Memoirs, No. 58. London 1974.

Parsons 1977. P. J. Parsons. "Callimachus. Victoria Berenices." *Zeitschrift für Papyrologie und Epigraphik* 25.1977.1–50.

Pascucci 1959. G. Pascucci. "Ennio, Ann., 561–6 vs 2 e un tipico procedimento di ΑΥΞΗΣΙΣ nella poesia latina." *Studi italiani di filologia classica* n.s. 31.1959.79–99.

Pasquali 1942. Giorgio Pasquali. "Arte allusiva." *Italia che scrive* 5.1942.185–187 [= *Pagine stravagane quarte e supreme.* Venice 1951. Pp. 11–20].

Perkell 1981. Christine Godfrey Perkell. "On the Corycian Gardener of Vergil's Fourth *Georgic.*" *Transactions of the American Philological Association* 111.1981.167–177.

Perkell 1989. Christine G. Perkell. *The Poet's Truth. A Study of the Poet in Virgil's* Georgics. Berkeley–Los Angeles–London 1989.

Perrotta 1924. G. Perrotta. "Virgilio e Arato." *Atene e Roma* 5.1924.3–19.

Pertusi 1955. *Scholia vetera in Hesiodi Opera et Dies.* Recensuit Augustinus Pertusi. Pubblicazioni dell' Università catolica del S. Cuore. Nuova serie. Vol. 53. Milan [n.d.].

Pfeiffer 1949. *Callimachus.* Ed. Rudolf Pfeiffer. Vol. 1. Fragmenta. Oxford 1949.

Pfeiffer 1953. *Callimachus.* Ed. Rudolf Pfeiffer. Vol. 2. Hymni et epigrammata. Oxford 1953.

Pfeiffer 1955. Rudolf Pfeiffer. "The Future of Studies in the Field of Hellenistic Poetry." *Journal of Hellenic Studies* 75.1955.69–73.

Pfeiffer 1968. Rudolf Pfeiffer. *History of Classical Scholarship from the Beginnings to the End of the Hellenistic Age.* Oxford 1968.

Posch 1969. Sebastian Posch. *Beobachtungen zur Theokritnachwirkung bei Vergil.* Commentationes Aenipontanae 19. Innsbruck–Munich 1969.

Pöschl 1962. Viktor Pöschl. *The Art of Vergil. Image and Symbol in the Aeneid.* Tr. Gerda Seligson. Ann Arbor 1962.

Powell 1925. *Collectanea Alexandrina. Reliquiae minores poetarum Graecorum aetatis Ptolemaicae 323–146 A.C. epicorum, elegiacorum, lyricorum, ethicorum.* Cum epimetris et indice nominum edidit Iohannes U. Powell. Oxford 1925.

Pucci 1977. Pietro Pucci. *Hesiod and the Language of Poetry.* Baltimore 1977.

Puelma Piwonka 1949. Mario Puelma Piwonka. *Lucilius und Kallimachos.* Frankfurt 1949.

Putnam 1965. Michael C. J. Putnam. *The Poetry of the* Aeneid. *Four Studies in Imaginative Unity and Design.* Cambridge, Massachusetts, 1965.

Putnam 1970. Michael C. J. Putnam. *Virgil's Pastoral Art. Studies in the* Eclogues. Princeton 1970.

Putnam 1975. Michael C. J. Putnam. "Italian Virgil and the Idea of Rome." *Janus. Essays in Ancient and Modern Studies.* Ed. Louis L. Orlin. Ann Arbor 1975. Pp. 171–199.

Putnam 1979. Michael C. J. Putnam. *Virgil's Poem of the Earth. Studies in the* Georgics. Princeton 1979.

Raabe 1974. Hermann Raabe. *Plurima mortis imago. Vergleichende Interpretationen zur Bildersprache Vergils.* Zetemata 59. Munich 1974.

Radt 1977. *Tragicorum Graecorum fragmenta.* Vol. 4. Sophocles. Ed. Stefan Radt. Göttingen 1977.

Radt 1985. *Tragicorum Graecorum fragmenta.* Vol. 2. Aeschylus. Ed. Stefan Radt. Göttingen 1985.

Rand 1946. *Servianorum in Vergilii carmina commentariorum editionis Harvardianae vol. II.* Ed. E. K. Rand et al. Lancaster 1946.

Regel 1907. G. E. F. Regel. *De Vergilio poetarum imitatore testimonia.* Diss. Göttingen 1907.

Reiff 1959. Arno Reiff. *Interpretatio, imitatio, aemulatio. Begriff und Vorstellung literarische Abhängigkeit bei der Römer.* Diss. Cologne 1959.

Reinhardt 1910. Karl Reinhardt. *De Graecorum theologia capita duo.* Diss. Berlin 1910.

Reinsch-Werner 1976. Hannelore Reinsch-Werner. *Callimachus Hesiodicus.* Berlin 1976.

Richter 1957. *Vergil. Georgica*. Hrsg. und erklärt von Will Richter. Das Wort der Antike 5. Munich 1957.

Roberts 1968. Louis Roberts. *A Concordance of Lucretius*. ΑΓΩΝ Supplement 2. Berkeley 1968. Rpt. with corrections New York 1977.

Rosen 1988. Ralph M. Rosen. *Old Comedy and the Iambographic Tradition*. American Classical Studies 19. Atlanta 1988.

Rosen 1990. Ralph M. Rosen. "Poetry and Sailing in Hesiod's *Works and Days*." *Classical Antiquity* 9.1990.99–113.

Ross 1966. David O. Ross, Jr. "Catullus' Poetic Vocabulary and the Roman Poetic Traditions." Diss. Harvard 1966.

Ross 1969. David O. Ross, Jr. *Style and Tradition in Catullus*. Loeb Classical Monographs. Cambridge, Massachusetts, 1969.

Ross 1975. David O. Ross, Jr. *Backgrounds to Augustan Poetry. Gallus, Elegy, and Rome*. Cambridge 1975.

Ross 1987. David O. Ross, Jr. *Virgil's Elements. Physics and Poetry in the* Georgics. Princeton 1987.

Russell 1979. D. A. Russell. "*De imitatione.*" *Creative Imitation and Latin Literature*. Ed. David West and Tony Woodman. Cambridge 1979. Pp. 1–16 and 201–202.

Rusten 1989. *Thucydides. Book 2*. Ed. J. S. Rusten. Cambridge Greek and Latin Classics. Cambridge 1989.

Ryberg 1958. Inez Scott Ryberg. "Vergil's Golden Age." *Transactions of the American Philological Association* 89.1958.112–131.

Sandbach 1967. *Plutarchi Moralia*. Vol. 7. Recensuit et emendavit F. H. Sandbach. Leipzig 1967.

Santirocco 1982. Matthew S. Santirocco. "Poet and Patron in Ancient Rome." *Book Forum* 6.1982.56–62.

Santirocco 1984. Matthew Santirocco. "The Maecenas Odes." *Transactions of the American Philological Association* 114.1984.241–253.

Santirocco 1986. Matthew S. Santirocco. *Unity and Design in Horace's* Odes. Chapel Hill–London 1986.

Scaliger 1561. Julius Caesar Scaliger. *Poetices libri septem*. Faksimile–Neudruck der Ausgabe von Lyon 1561 mit einer Einleitung von August Buck. Stuttgart–Bad Cannstatt 1964.

Scaliger 1606. *Thesaurus Temporum Eusebii Pamphili...Chronicorum canonum historiae libri duo....* Opera et studio Iosephi Iusti Scaligeri..., etc. Leiden 1606.

Schanz–Hosius 1927–1935. Martin Schanz. *Geschichte der römischen Literatur bis zum Gesetzgebung des Kaisers Justinian*. 1. *Die römische Literatur in der Zeit der Republik*. 2. *Die römische Literatur in der Zeit der Monarchie bis auf Hadrian*. Vierte neubearbeiteten Auflage von Carl Hosius. Handbuch der

Altertumswissenschaft. Begründet von Iwan von Müller. Hrsg. von Walter Otto. 8. Abteilung. 1. und 2. Teil. Munich 1927–1935. Rpt. 1967.

Scheinberg 1979. Susan Scheinberg. "The Bee Maidens of the Homeric *Hymn to Hermes.*" *Harvard Studies in Classical Philology* 83.1979.1–28.

Schlunk 1974. Robin R. Schlunk. *The Homeric Scholia and the* Aeneid. *A Study of the Influence of Ancient Literary Criticism on Vergil.* Ann Arbor 1974.

Schmidt 1930. Magdalena Schmidt. *Die Komposition von Vergils Georgica.* Paderborn 1930.

Schmidt 1972. Ernst A. Schmidt. *Poetische Reflexion. Vergils Bukolik.* Munich 1972.

Schneider 1856. *Nicandrea. Theriaca et Alexipharmaca* recensuit et emendavit, fragmenta collegit, commentationes addidit Otto Schneider. Accedunt scholia in *Theriaca* ex recensione Henrici Keil, scholia in *Alexipharmaca* ex recensione Bussmeleri et R. Bentlei emendationes partim ineditae. Leipzig 1856.

Scodel–Thomas 1984. Ruth S. Scodel and Richard F. Thomas. "Virgil and the Euphrates." *American Journal of Philology* 105.1984.339.

Schrijvers 1970. P. H. Schrijvers. *Horror ac divina voluptas. Études sur la poétique et la poésie de Lucrèce.* Amsterdam 1970.

Schwabl 1972. Hans Schwabl. "Zur Mimesis bei Arat." *Antidosis. Festschrift für Walther Kraus zum 70. Geburtstag.* Hrsg. Rudolf Hanslik, Albin Lesky, Hans Schwabl. Vienna–Cologne–Graz 1972.

Segal 1967. Charles Segal. "Virgil's *caelatum opus.* An Interpretation of the Third *Eclogue.*" *American Journal of Philology* 88.1967.279–308.

Segal 1977. Charles Segal. "Pastoral Realism and the Golden Age. Correspondence and Contrast between Virgil's Third and Fourth *Eclogues.*" *Philologus* 121.1977.158–163.

Segal 1981. Charles Segal. *Poetry and Myth in Ancient Pastoral. Essays on Theocritus and Virgil.* Princeton 1981.

Sellar 1897. W. Y. Sellar. *The Roman Poets of the Augustan Age. Virgil.* 3d ed. Oxford 1897. Rpt. 1908.

Skutsch 1901. Franz Skutsch. *Aus Vergils Frühzeit.* Leipzig 1901.

Skutsch 1985. *The* Annals *of Quintus Ennius.* Edited with introduction and commentary by Otto Skutsch. Oxford 1985.

Smith 1982. *Scholia vetera in Aeschylum quae extant omnia.* Pars 1. Fasc. 2. Scholia in *Septem adversus Thebas* continens. Ed. Ole Langwitz Smith. Leipzig 1982.

Snell–Maehler 1975. *Pindari carmina cum fragmentis.* Pars 2. Fragmenta, indices. Post Brunonem Snell edidit Henricus Maehler. Leipzig 1975.

Snyder 1980. Jane McIntosh Snyder. *Puns and Poetry in Lucretius' De Rerum Natura.* Amsterdam 1980.

Solmsen 1963. Friedrich Solmsen. "The 'Days' of the *Works and Days.*" *Transactions of the American Philological Association* 94.1963.293–320.

Solmsen 1966. Friedrich Solmsen. "Aratus on the Maiden and the Golden Age." *Hermes* 94.1966.124–128.

Solmsen 1970. *Hesiodi Theogonia, Opera et Dies, Scutum.* Ed. Friedrich Solmsen. *Fragmenta selecta.* Ed. R. Merkelbach and M. L. West. Oxford 1970.

Soubiran 1972. *Cicéron. Aratea, fragments poétiques.* Texte établi et traduit par Jean Soubiran. Association Guillaume Budé. Collection des universités de France. Paris 1972.

Sparrow 1931. John Sparrow. *Half-lines and Repetitions in Virgil.* Oxford 1931.

Spurr 1986. M. S. Spurr. "Agriculture and the *Georgics.*" *Greece & Rome* 33.1986.164–187.

Starobinski 1979. Jean Starobinski. *Words upon Words. The Anagrams of Ferdinand de Saussure.* Tr. Olivia Emmet. New Haven 1979.

Steiner 1986. Deborah Steiner. *The Crown of Song. Metaphor in Pindar.* New York 1986.

Steinmetz 1964. Peter Steinmetz. "Gattungen und Epochen der griechischen Literatur in der Sicht Quintilians." *Hermes* 92.1964.454–466.

Stewart 1959. Zeph Stewart. "The Song of Silenus." *Harvard Studies in Classical Philology* 64.1959.179–205.

Summers 1989. David Summers. "'Form,' Nineteenth-Century Metaphysics, and the Problem of Art Historical Description." *Critical Inquiry* 15.1989.372–406.

Tatum 1984. W. Jeffrey Tatum. "The Presocratics in Book One of Lucretius' *De Rerum Natura.*" *Transactions of the American Philological Association* 114.1984.177–189.

Tavenner 1918. Eugene Tavenner. "The Roman Farmer and the Moon." *Transactions of American Philological Association* 49.1918.67–82.

Taylor 1955. Margaret E. Taylor. "Primitivism in Virgil." *American Journal of Philology* 76.1955.261–178.

Thill 1979. Andrée Thill. *Alter ab illo. Recherches sur l'imitation dans la poésie personnelle a l'époque augustéenne.* Association Guillaume Budé. Collection d'études anciennes. Paris 1979.

Thilo 1884. *Servii grammatici qui feruntur in Vergilii carmina.* Recensuerunt Georgius Thilo et Hermannus Hagen. Vol. 2. *Aeneidos* librorum VI–XII commentarii. Recensuit Georgius Thilo. Leipzig 1884.

Thilo 1887. *Servii grammatici qui feruntur in Vergilii carmina.* Recensuerunt Georgius Thilo et Hermannus Hagen. Vol. 3. Fasc. 1. Servii grammatici qui feruntur in Vergilii *Bucolica* et *Georgica* commentarii. Recensuit Georgius Thilo. Leipzig 1887.

Thomas 1982a. Richard F. Thomas. "Catullus and the Polemics of Poetic Reference (64.1–18)." *American Journal of Philology* 103.1982.144–164.

Thomas 1982b. Richard F. Thomas. *Lands and Peoples in Roman Poetry. The Ethnographical Tradition.* Cambridge Philological Society supplementary volume no. 7. Cambridge 1982.

Thomas 1985. Richard F. Thomas. "From *Recusatio* to Commitment. The Evolution of the Vergilian Programme." *Papers of the Liverpool Latin Seminar.* Fifth volume 1985. Ed. Francis Cairns. ARCA Classical and Medieval Texts, Papers and Monographs, 19. Liverpool 1986. Pp. 61–73.

Thomas 1986. Richard F. Thomas. "Virgil's *Georgics* and the Art of Reference." *Harvard Studies in Classical Philology* 90.1986.171–198.

Thomas 1987. Richard F. Thomas. "Prose into Poetry. Tradition and Meaning in Virgil's *Georgics.*" *Harvard Studies in Classical Philology* 91.198. 229–260.

Thomas 1988. Richard F. Thomas. *Virgil. Georgics.* 2 vol. Cambridge Greek and Latin Classics. Cambridge 1988.

Ursinus 1568. *Virgilius.* Collatione scriptorum Graecorum illustratus opera et industria Fulvii Ursini. Antwerp 1568.

Vahlen 1903. *Ennianae poesis reliquiae.* Iteratis curis recensuit Iohannes Vahlen. Leipzig 1903.

van der Valk 1971–1987. *Eustathii...commentarii ad Homeri Iliadem pertinentes.* Curavit Marchinus van der Valk. 4 vol Leiden 1971–1987.

Van Sickle 1975. John Van Sickle. "Epic and Bucolic (Theocritus, *Id.* viii/Virgil, *Ecl.* 1)." *Quaderni urbinati di cultura classica* 19.1975.3–30.

Van Sickle 1976. John Van Sickle. "Theocritus and the Development of the Conception of Bucolic Genre." *Ramus* 5.1976.18–44.

Van Sickle 1978. John Van Sickle. *The Design of Virgil's* Bucolics. Filologia e critica 24. Istituto di filologia classica. Università di Urbino. Rome 1978.

Van Sickle 1981. John Van Sickle. "*Commentaria in Maronem commenticia.* A Case History of 'Bucolics' Misread." *Arethusa* 14.1981.17–34.

Vischer 1965. R. Vischer. *Das einfache Leben. Wort- und motivgeschichtliche Untersuchungen zu einem Wertbegriff der antiken Literatur.* Studienhefte zur Altertumswissenschaft 11. Göttingen 1965.

Vogt 1967. Ernst Vogt. "Das Akrostichon in der griechischen Literatur." *Antike und Abendland* 13.1967.80–95.

von Jan 1893. F. de Ian. *De Callimacho Homeri interprete.* Diss. Strasbourg 1893.

Walter 1933. Otto Walter. *Die Entstehung der Halbverse in der Aeneis.* Diss. Giessen 1933.

Waszink et al. 1962. J. H. Waszink et al. Discussion of La Penna 1962. *Hésiode et son influence.* Fondation Hardt. Entretiens sur l'antiquité classique 7. Geneva 1962. Pp. 253–270.

Waszink 1974. J. H. Waszink. *Biene und Honig als Symbol des Dichters und der Dichtung in den griechisch-römischen Antike.* Geisteswissenschaftliche Vorträge der Rheinisch-Westfälische Akademie der Wissenschaften G 196. Opladen 1974.

Weinreich 1916–1919. Otto Weinreich. "Religiöse Stimmen der Völker." *Archiv für Religionswissenschaft* 19.1916–1919.158–173.

Wendel 1935. *Scholia in Apollonium Rhodium vetera.* Recensuit Carolus Wendel. Berlin 1935. Rpt. 1974.

Wender 1979. Dorothea Wender. "From Hesiod to Homer by Way of Rome." *Virgil's Ascraean Song. Ramus Essays on the Georgics.* Ed. A. J. Boyle. Berwick 1979. Pp. 59–64.

West 1966. *Hesiod. Theogony.* Edited with prolegomena and commentary by M. L. West. Oxford 1966.

West 1974. M. L. West. *Studies in Greek Elegy and Iambus.* Untersuchungen zur antike Literatur und Geschichte 14. Berlin–New York 1974.

West 1972. *Iambi et elegi Graeci ante Alexandrum cantati.* Edidit M. L. West. Vol. 2. Oxford 1972.

West 1978. *Hesiod. Works and Days.* Edited with an introduction and commentary by M. L. West. Oxford 1978.

West 1979. David West. "Two Plagues. Virgil, *Georgics* 3.478–566 and Lucretius 6.1090–1286." *Creative Imitation and Latin Literature.* Ed. David West and Tony Woodman. Cambridge 1979. Pp.71–88 and 221–222.

White 1967. K. D. White. *Agricultural Implements of the Roman World.* Cambridge 1967.

Wigodsky 1972. Michael Wigodsky. *Vergil and Early Latin Poetry. Hermes* Einzelschriften 24. Wiesbaden 1972.

Wilamowitz 1906. Ulrich von Wilamowitz-Moellendorf. *Die Textgeschichte der griechischen Bukoliker.* Philologische Untersuchungen 18. Berlin 1906.

Wilamowitz 1916. Ulrich von Wilamowitz-Moellendorf. *Die Ilias und Homer.* Berlin 1916. Rpt. 1966.

Wilamowitz 1924. Ulrich von Wilamowitz-Moellendorf. *Hellenistische Dichtung in der Zeit des Kallimachos.* 2 vol. Berlin 1924.

Wilhelm 1982. R. M. Wilhelm. "The Plough-Chariot. Symbol of Order in the *Georgics.*" *Classical Journal* 77.1982.213–230.

Wilkinson 1950. L. P. Wilkinson. "The Intention of Virgil's *Georgics.*" *Greece & Rome* 19.1950.19–28.

Wilkinson 1963. L. P. Wilkinson. "Virgil's Theodicy. *Classical Quarterly* n.s. 13.1963.75–84.

Wilkinson 1969. L. P. Wilkinson. *The Georgics of Virgil. A Critical Survey.* Cambridge 1969.

Williams 1978. *Callimachus. Hymn to Apollo.* A commentary by Frederick Williams. Oxford 1978.

Wills 1987. Jeffrey Wills. "Scyphus—A Homeric Hapax in Virgil." *American Journal of Philology* 108.1987.455–457.

Wimmel 1960. Walter Wimmel. *Kallimachos in Rom Die Nachfolge seines apologetischen Dichtens in der Auggusteerzeit. Hermes* Einzelschriften 16. Wiesbaden 1960.

Wimsatt–Beardsley 1946. W. K. Wimsatt and Monroe C. Beardsley. "The Intentional Fallacy." *Sewanee Review* 54.1946.466–488 [Rpt. in W. K. Wimsatt, Jr. *The Verbal Icon. Studies in the Meaning of Poetry.* Lexington, Kentucky, 1954. Pp. 2–18].

Wissowa 1917. Georg Wissowa. "Das Prooemium von Vergils Georgica." *Hermes* 52.1917.92–104.

Wlosok 1985. Antonie Wlosok. "*Gemina doctrina.* On Allegorical Interpretation." *Papers of the Liverpool Latin Seminar.* Fifth volume 1985. Ed. Francis Cairns. ARCA Classical and Medieval Texts, Papers and Monographs, 19. Liverpool 1986. Pp. 75–84.

Zetzel 1977. J. E. G. Zetzel. "Lucilius, Lucretius, and Persius 1.1." *Classical Philology* 72.1977.40–42.

Zetzel 1982. James E. G. Zetzel. "The Poetics of Patronage in the Late First Century B.C." *Literary and Artistic Patronage in Ancient Rome.* Ed. Barbara Gold. Austin 1982. Pp. 87–102

Zetzel 1983a. J. E. G. Zetzel. "Catullus, Ennius, and the Poetics of Allusion." *Illinois Classical Studies* 8.1983.251–266.

Zetzel 1983b. James E. G. Zetzel. "Re-creating the Canon. Augustan Poetry and the Alexandrian Past." *Critical Inquiry* 10.1983.83–105.

Zetzel 1984. James E. G. Zetzel. "Servius and Triumviral History in the *Eclogues.*" *Classical Philology* 79.1984.139–142.

Index of Passages Cited

General Index

This book was printed from camera-ready sheets
prepared by the author on an Apple Macintosh SE/30
and produced on a Linotronic 300 imagesetter.
The text was created in Microsoft Word 4.0,
and the page makeup executed in QuarkXPress 2.12.
The Roman typeface is Janson,
a late seventeenth-century design erroneously named
for the Leipzig-based Dutch typographer Anton Janson,
but now attributed to Nicholas Kis,
a Hungarian printer who worked in Amsterdam.
This book uses a PostScript version of the font by Adobe Systems.
The Greek typeface is Bosporos by Allotype Typographics,
a PostScript version of Porson,
a font based on the calligraphy of the great British Hellenist
Richard Porson.